Y0-BSD-897

ALLE · ZEIT · WACH
1842

Springer Series in Synergetics

Synergetics, an interdisciplinary field of research, is concerned with the cooperation of individual parts of a system that produces macroscopic spatial, temporal or functional structures. It deals with deterministic as well as stochastic processes.

Volume 1 **Synergetics** An Introduction 2nd Edition
By H. Haken

Volume 2 **Synergetics** A Workshop
Editor: H. Haken

Volume 3 **Synergetics** Far from Equilibrium
Editors: A. Pacault and C. Vidal

Volume 4 **Structural Stability in Physics**
Editors: W. Güttinger and H. Eikemeier

Volume 5 **Pattern Formation** by Dynamic Systems and **Pattern Recognition**
Editor: H. Haken

Volume 6 Dynamics of **Synergetic Systems**
Editor: H. Haken

Volume 7 Problems of **Biological Physics**
By L. A. Blumenfeld

Volume 8 **Stochastic Nonlinear Systems**
in Physics, Chemistry, and Biology
Editors: L. Arnold and R. Lefever

Stochastic Nonlinear Systems

in Physics, Chemistry, and Biology

Proceedings of the Workshop
Bielefeld, Fed. Rep. of Germany, October 5–11, 1980

Editors:
L. Arnold and R. Lefever

With 48 Figures

Springer-Verlag Berlin Heidelberg New York 1981

Professor Dr. *Ludwig Arnold*

Forschungsschwerpunkt Dynamische Systeme, Universität Bremen,
D-2800 Bremen 33, Fed. Rep. of Germany

Professor Dr. *René Lefever*

Service de Chemie Physique II,
C.P. 231 Campus Plaine, Université Libre de Bruxelles,
B-1050 Bruxelles, Belgium

ISBN 3-540-10713-4 Springer-Verlag Berlin Heidelberg New York
ISBN 0-387-10713-4 Springer-Verlag New York Heidelberg Berlin

Offset printing: Beltz Offsetdruck, Hemsbach. Bookbinding: J. Schäffer OHG, Grünstadt
2153/3130-543210

Preface

This book contains the invited papers of the interdisciplinary workshop on "Stochastic Nonlinear Systems in Physics, Chemistry and Biology" held at the Center for Interdisciplinary Research (ZIF), University of Bielefeld, West Germany, October 5-11, 1980.

The workshop brought some 25 physicists, chemists, and biologists - who deal with stochastic phenomena - and about an equal number of mathematicians - who are experts in the theory of stochastic processes - together.

The Scientific Commitee consisted of L. Arnold (Bremen), A. Dress (Bielefeld), W. Horsthemke (Brussels), T. Kurtz (Madison), R. Lefever (Brussels), G. Nicolis (Brussels), and V. Wihstutz (Bremen).

The main topics of the workshop were the transition from deterministic to stochastic behavior, external noise and noise induced transitions, internal fluctuations, phase transitions, and irreversible thermodynamics, and on the mathematical side, approximation of stochastic processes, qualitative theory of stochastic systems, and space-time processes.

The workshop was sponsored by ZIF, Bielefeld, and by the Universities of Bremen and Brussels. We would like to thank the staff of ZIF and H. Crauel and M. Ehrhardt (Bremen) for the perfect organization and their assistance. In addition, our thanks go to Professor H. Haken for having these Proceedings included in the Series in Synergetics.

Bremen and Brussels
December 1980

L. Arnold and R. Lefever

Contents

Part I

Introduction: From Deterministic to Stochastic Behavior

On the Foundations of Kinetic Theory

B. Misra and I. Prigogine
Chimie-Physique II, C.P. 231, Université Libre de Bruxelles,
B-1050 Bruxelles, Belgium

Abstract

We discuss the problem of deriving an *exact* Markovian master equation from dynamics *without* resorting to approximation schemes such as the weak coupling limit, Boltzmann-Grad limit, etc. Mathematically, it is the problem of the existence of a suitable positivity preserving operator Λ such that the unitary group U_t induced from dynamics satisfies the intertwining relation

$$\Lambda U_t = W_t^* \Lambda \quad , \quad t \geqq 0$$

with the contraction semigroup W_t of a strongly irreversible stochastic Markov process. Two cases are of special interest: i) $\Lambda = P$ is a projection operator, ii) Λ has a densely defined inverse. Our recent work, which we summarize here, shows that the class of (classical) dynamical systems for which a suitable projection operator satisfying the above intertwining relation exists is identical with the class of K flows or K systems. As a corollary of our consideration it follows that the function $\int \hat{\rho}_t \ln \hat{\rho}_t d\mu$ with $\hat{\rho}_t$ denoting the coarse-grained distribution with respect to a K partition obtained from $\rho_t \equiv U_t \rho$ is a Boltzmann-type H function for K flows. This is not in contradiction with the time-reversal (velocity-inversion) symmetry of dynamical evolution as the suitably constructed projection operator or the Λ transformation are dynamics dependent and break the time reversal.

1. Introduction

The study of the possible connections that may exist between deterministic dynamics and probabilistic processes is of obvious importance for the foundation of nonequilibrium statistical mechanics. As it is well known, stochastic Markov processes provide suitable models to represent irreversible evolution admitting a Liapounov functional or H function. The important question, thus, is how the passage from deterministic dynamics to probabilistic Markov processes is to be achieved.

Our work described below shows that in the presence of suitable instability of motion, described by the condition of K flow, the dynamical evolution indeed becomes *similar* in a well-defined sense to the stochastic evolution of a Markov process.

Let us recall that the procedure for obtaining a Markovian master equation from dynamics usually starts with an initial "contraction of description" or coarse graining brought about by a projection operator P. The operation of "coarse graining" alone, however, generally leads to the so-called generalized master equation which is non-Markovian in character [1]. To obtain a Markovian evolution equation one needs to consider a special asymptotic limit (e.g., the weak coupling limit etc.). Thus, even when this program succeeds, the resulting master equation is only an approximation. To lay a more satisfactory foundation of nonequilibrium statistical mechanics it seems desirable to investigate the possibility of establishing *exact* Markovian master equations whose validity does not depend on sepcial approximation schemes. This paper summarizes our recent work in this direction [2-6]. We shall discuss the problem for classical dynamical systems. Our main result is that the class of dynamical systems for which an *exact* Markovian master equation follows from a suitable projection operation alone is not empty, but is precisely the class of so-called K flow. The condition of K flow is thus seen to play the same role in the foundation of nonequilibrium statistical mechanics as that of ergodicity in the foundation of equilibrium statistical mechanics.

Let us, however, mention that just as the method of replacing time average by ensemble average can be justified for a special class of functions on phase space even if the system as a whole is not ergodic, one may be able to derive a exact Markovian master equation for special subclass of initial distributions even if the system is not a K flow. But it is only for K flows that one can derive an exact master equation through a projection operator for all initial distribution functions in $L^2_\mu \cap L^1_\mu$.

2. Formulation of the Problem

Consider an abstract dynamical system $(\Gamma, \mathcal{B}, \mu, T_t)$. Here Γ denotes the phase space of T_t the system equipped with a σ algebra of measurable subsets, T_t a group of measurable transformations mapping Γ onto itself and preserving the measure μ. For example, Γ could be the energy surface of a classical dynamical system, T_t the group of dynamical evolution and μ the invariant measure whose existence is assured by Liouville's theorem. For convenience we shall assume the measure μ to be normalized: $\mu(\Gamma) = 1$. As is well known, the evolution $\rho \to \rho_t$ of density functions under the given deterministic dynamics is described by the unitary group U_t induced by T_t

$$\rho_t(\omega) = (V_t \rho)(\omega) = (T_{-t}\omega) \quad .$$

The generator L of the unitary group U_t is called the Liouvillian operator of the system: $U_t = \exp(-itL)$. It is given by Poisson bracket with the Hamiltonian H:

$$L\rho = i[H,\rho]_{P.B.}$$

for Hamiltonian evolution.

On the other hand, stochastic Markov processes on the state space Γ, preserving μ, are associated with contraction semigroups of L^2_μ 8 . In fact, let $P(t,\omega,\Delta)$ denote the probability of transition from the point $\omega \in \Gamma$ to the region Δ in time t. Then the operators W_t defined by

$$(W_t f)(\omega) = \int f(\omega')P(t,\omega,d\omega')$$

form a contraction semigroup for $t \geq 0$. Moreover, W_t has the following properties:

i) W_t preserves positivity (i.e., $f \geq 0$ implies $W_t f \geq 0$ for $t \geq 0$),

ii) W_t .1. = 1.

The evolution of the distribution functions $\hat{\rho}$ under the Markov process is described now by the ajoint semigroup W^*_t which also preserves positivity since W_t does: $\hat{\rho}_0 \to \hat{\rho}_t = W^*_t\hat{\rho}_0$. Since the measure μ is an invariant measure for the process (or equivalently the *microcanonical distribution function* 1 is the equilibrium state of the process) we also have

iii) $W^*_t \cdot 1 = 1$.

Every Markov process on Γ with stationary measure μ is thus associated with a contraction semigroup satisfying the conditions i-iii). Conversely, every contraction semigroup W_t on Γ satisfying the above conditions comes from a stochastic Markov process, the transition probabilities $P(t,\omega,\Delta)$ being given by

$$P(t,\omega,\Delta) = (W_t \varphi_\Delta)(\omega) \quad .$$

Here φ_Δ denotes the characteristic (or indicator) function of the set Δ.

In the following we are interested in a special class of Markov processes whose semigroups W_t satisfy [in addition to conditions i-iii)] the condition:

iv) $\| W^*_t\rho - 1\|^2$ decreases strictly monotonically to 0 as $t \to +\infty$; for all states $\rho \neq 1$ (i.e., for all nonnegative distribution functions with $\int\rho \, d\mu = 1$). This condition expresses the requirement that any initial distribution ρ tends strictly monotonically in time to the equilibrium distribution 1. For such processes the functional

$$\int_\Gamma \hat{\rho}_t \log \hat{\rho}_t d\mu \quad , \quad \hat{\rho}_t = W^*_t\hat{\rho}_0$$

and indeed any other convex functional of $\hat{\rho}_t$ is an H function. Such Markov processes thus provide the best possible model of irreversible evolution obeying the law of monotonic increase of entropy. Semigroups satisfying the conditions i-iv) will be called *strongly irreversible Markov semigroups*.

The problem before us is to determine the class of dynamical systems for which one can construct a bounded operator Λ having the following properties

i) Λ preserve positivity,

ii) $\Lambda 1 = 1$

iii) $\int_\Gamma \Lambda\rho \, d\mu = \int_\Gamma \rho \, d\mu$

iv) The dynamical group $U_t = \exp(-itL)$ satisfies the intertwining relation: $\Lambda U_t = W_t^* \Lambda$ (for $t \geq 0$) with a strongly irreversible Markov semigroup W_t.

We shall consider two cases:

First, Λ has a densely defined inverse Λ^{-1}. In this case Λ may be interpreted as defining a "change of representation" of dynamics $\rho_t = U_t\rho_0 \to \Lambda\rho_t = \hat{\rho}_t$. Condition iv) then means that the evolution of transformed states obeys the master equation of the Markov process W_t. For dynamical systems admitting an invertible operator Λ satisfying conditions i-iv), the dynamical group U_t is similar to a strongly irreversible Markov semigroup $\Lambda U_t \Lambda^{-1} = W_t^*$ for $t \geq 0$.

Note that the demanded invertibility of Λ assures that the passage $\rho \to \Lambda\rho = \hat{\rho}$ involves no "loss of information". Dynamical systems admitting such a Λ may, hence, be said to be *intrinsically random*.

The other case we consider is: Λ is a projection operator P. Such projection operators [i.e., projections P satisfying conditions i-iii), with Λ replaced by P] correspond to operations of "coarse graining". The existence of such a projection P satisfying the intertwinning relation iv) thus implies an *exact* Markovian master equation for the system which does not depend on special approximation schemes, but results solely from the projection operator P.

As described below, it is a remarkable fact that there is a rather general class of dynamical systems for which an exact master equation holds in this sense.

3. Dynamical Systems Admitting Exact Markovian Master Equations

A K flow is, by definition [9], an abstract dynamical system $(\Gamma,\mathcal{B},\mu,T_t)$ for which there exists a distinguished (measurable) partition ξ_0 of the phase space into disjoint cells having the following properties

i) $\xi_t = T_t\xi_0 \geq \xi_s \quad \text{if} \quad t \geq s$.

Here $T_t\xi_0 = \xi_t$ is the partition into which the original partition ξ_0 is transformed in time under the dynamical evolution. The notation $\xi_t \geq \xi_s$ signifies that the partition ξ_t is "finer" than ξ_s (i.e., every cell of ξ_t is entirely contained in one of the cells of ξ_s).

ii) The (least fine) partition $\bigvee_{t=-\infty}^{\infty}$ which is finer than each ξ_t, $-\infty < t < +\infty$ is the partition of the phase space into distinct phase points.

iii) The (finest) partition $\bigcap_{t=-\infty}^{\infty}\xi_t$ which is less fine than every ξ_t $-\infty < t < +\infty$ is a trivial partition consisting of a cell of measure 1.

A partition ξ_0 with the above-stated properties is called a K partition. Many systems of physical interest have been recently found to be K flows, for instance, the motion of hard spheres within a finite box [10], the geodesic flow on space of constant negative curvature [11], the Lorentz gas, and infinite ideal gas and hard rod systems, etc. Generally, a K partition consists of an uncountable number of cells, each of null μ measure. The notion of coarse graining with respect to a K partition cannot, therefore, be defined directly as the operation of taking averages over the cells of the partition. The appropriate extension of the usual concept of coarse graining is provided by the projection operator P_0 of $L^2\mu$ onto the subspace $L^2[a(\xi_0),\mu]$. Here $a(\xi_0)$ denotes the σ subalgebra of $\mathcal{B}$ consisting of only those measurable subsets of Γ that are unions of cells in ξ_0 and $L^2[a(\xi_0),\mu]$ is the subspace of all $f \in L^2\mu$ that are measurable with respect to $a(\xi_0)$. In fact, it is clear that for any nonnegative density function ρ the function $P_0\rho$ has the following characteristic properties of coarse graining distribution with respect to the partition

i) $P_0\rho = \hat{\rho} \geqq 0$,

ii) $\int_\Gamma \hat{\rho}\, d\mu = \int_\Gamma \rho\, d\mu$,

iii) $\hat{\rho}$ being measurable with respect to $a(\xi_0)$, can only assume constant values on individual cells of ξ_0,

iv) $\int_\Delta \hat{\rho}\, d\mu = \int_\Delta \rho\, d\mu$
for any measurable Δ_0 that is a union of cells in ξ_0.

The (self-adjoint) projection operation P_0 of coarse graining with respect to a partition ξ_0 obviously preserves positivity and maps the constant function (microcanonical ensemble) onto itself. It is interesting that the converse is also true. For standard measure spaces $(\Gamma,\mathcal{B},\mu)$ with $\mu(\Gamma) = 1$ every self-adjoint projection operation P of $L^2\mu$ that preserves positivity and maps the constant function onto itself is the projection operator onto $L^2[a(\xi),\mu]$ for some measurable partition ξ of Γ. The operations of coarse graining may thus be identified with (self-adjoint) projection operators P satisfying

i) $f \geqq 0 \Rightarrow Pf \geqq 0$ and

ii) $P1 = 1$.

The following theorem tells that the condition of K flow is both necessary and sufficient for the existence of an operation of coarse graining that converts the dynamical evolution into that of a strongly irreversible stochastic Markov process.

4. Theorem 1

Suppose a dynamical system with induced unitary group U_t of dynamical evolution admits a positivity preserving projection P mapping the unit function 1 to itself such that

i) $PU_t\rho = PU_t\rho'$ for all $t \Rightarrow \rho = \rho'$, and

ii) $PU_t = W_t^*P$ for $t \geq 0$,

where W_t is a *strongly irreversible Markov semigroup* (see Sect.2 for definition). Then the dynamical system is necessarily a K flow. Conversely, ever K flow admits a positivity preserving projection {namely, the projection onto $L^2[a(\xi_0),\mu]$ with ξ_0 denoting a K partition} satisfying the conditions i) and ii) above.

Remark: Condition i) means that the coarse graining under consideration is sufficiently fine so that a knowledge of the coarse grained distribution $PU_t\rho$ during the entire history, both past and future, of the system's evolution is equivalent to a knowledge of the original distribution.

We shall not stop here for a proof of the theorem which may be found elsewhere, but let us mention an important corollary of this result.

5. Corrollary 2

For K flows the (negative) entropy functional

$$\Omega(\rho_t) = \int \hat{\rho}_t \log \hat{\rho}_t \, d\mu \quad , \quad \hat{\rho}_t = P_0U_t\rho$$

with P_0 denoting the coarse graining projection operation with respect to a K partition, is a monotonically decreasing function (H function) of t which attains the fine-grained value $\int \rho \ln \rho \, d\mu$ at $t \to -\infty$ and the equilibrium value (0 due to our normalization of μ) at $t \to +\infty$.

This result follows immediately from theorem 1 because $\hat{\rho}_t = P_0U_t\rho = W_t^*P_0\rho = W_t^*\rho_0$ where W_t is the semigroup of a *strongly irreversible Markov process*. It is well-known that for such processes the functional Ω, and indeed any convex functional of $\hat{\rho}_t$, is an H function. Let us mention that this result has recently been obtained by Penrose and Goldstein by an independent argument [12].

Finally, let us mention that for K flow one cannot only construct a coarse graining operation that leads from the deterministic and reversible evolution U_t to that of a strongly irreversible stochastic process, but one can also construct an *invertible* and positivity preserving transformation Λ such that $\Lambda U_t \Lambda^{-1}$ is a strongly irreversible Markov semigroup for $t \geq 0$. In other terms, the unitary group U_t induced from every K flow is nonunitarily equivalent, through a positivity preserving similarity Λ, to the contraction semigroup of a stochastic Markov process.

A detailed discussion of this construction is described in our previous publications [4,5]. Let us only mention that the operator Λ establishing nonunitary equivalence between U_t and a stochastic Markov process W_t^* may be constructed as a *suitable* function of operator time T [3]

$$\Lambda = h(T) + P_{-\infty}$$

$$T = \int_{-\infty}^{+\infty} \lambda dF_\lambda \quad , \quad F_\lambda = U_\lambda P_0 U_\lambda^* - P_{-\infty} \quad .$$

Here P_0 denotes the projection of coarse graining with respect to the K partition ξ_0 and $P_{-\infty}$, the projection to the equilibrium ensemble. With Λ constructed as above, the functional

$$\int_\Gamma \tilde{\rho}_t \log \tilde{\rho}_t \, d\mu \quad , \quad \tilde{\rho}_t \equiv \Lambda U_t \rho$$

is again an H function.

6. Concluding Remarks

Let us emphasize that the possibility of obtaining an exact master equation from dynamics as discussed here comes from the existence of suitable *dynamics-dependent* operator Λ or P_0 that explicitly break the time-reversal (or velocity-inversion) symmetry. The necessity of considering such dynamics-dependent transformations for the passage from dynamics to irreversible evolution have been discussed in a previous publication [2]. This should be distinguished from the usual procedure of obtaining a master equation via a generalized master equation, where one starts with a projection that does not depend on dynamics and does not break the time-reversal symmetry. In fact, as is well known, it is the approximation schemes (such as weak coupling limit) that breaks the time-reversal symmetry in the conventional approach.

To see explicitly the noninvariance of the projection P_0 under velocity inversion let us consider the operation of velocity inversion which, by definition, has the following properties

i) V preserves positivity,

ii) $V^2 = I$,

iii) $VU_t = U_{-t}V$.

Now, P_0 being the projection with respect to the K partition ξ_0, it is clear that $U_t P_0 U_t^* = P_t$ is the projection with respect to $T_t \xi_0$. The defining properties of ξ_0 (see Sect.2) translate into the following properties of P_t

i) $P_t \geqq P_s$ if $t \geqq s$,

ii) $\lim_{t \to \infty} P_t = I$, and

iii) $\lim_{t \to -\infty} P_t = P_{-\infty}$

the projection onto the equilibrium ensemble. Suppose now that P_0 is invariant under V, $VP_0V = P_0$. This would then imply that

$$P_t = U_tP_0U_t^* = U_tVP_0VU_t^* = VU_{-t}P_0U_{-t}^*V = VP_{-t}V \quad .$$

Since $P_t \geqq P_0$ for $t \geqq 0$ it follows that

$$P_{-t} = VP_tV \geqq VP_0V = P_0 \quad \text{for} \quad t \geqq 0 \quad .$$

Thus we have

$$P_t = U_tP_0U_t^* \geqq P_0 \quad \text{for } all \text{ real t}$$

which is possible only if $P_0 = U_tP_0^{-1}U_t^*$ or P_0 commutes with U_t. This contradicts, however, the properties that $P_t \to I$, $t \to +\infty$ and $P_t \to P_{-\infty}$, $t \to -\infty$.

A variant of this argument shows also that no projection operator P (whether coming from a K partition or not) which commutes with velocity inversion V can yield an exact master equation, i.e., satisfy the relation $PU_t = W_t^*P$ (for $t \geqq 0$) with a strongly irreversible Markov semigroup W_t.

Similar considerations show that the symmetry of velocity inversion is also broken by invertible transformation Λ. It fact if $V\Lambda V = \Lambda$ it would follow that together with $\Lambda U_t\Lambda^{-1}$ (for $t \geqq 0$) the operation $V\Lambda U_t\Lambda^{-1}V = \Lambda U_{-t}\Lambda^{-1}$ also preserves positivity. However, $\Lambda U_{-t}\Lambda^{-1}$ is the inverse of $\Lambda U_t\Lambda^{-1}$ and they both can preserve positivity if and only if $\Lambda U_t\Lambda^{-1}$ is unitary [7]. This, however, contradicts the established fact [4,5] that, with Λ constructed as described in a previous section, $\Lambda U_t\Lambda^{-1}$ is a strongly irreversible Markov semigroup and hence nonunitary.

Finally, let us briefly show that the classical objection (such as those of Zermelo and Loschmidt) against the derivation of Boltzmann's H theorem does not apply to the H functions $\Omega(\rho_t)$ considered by us

$$\Omega(\rho_t) = \int_\Gamma \hat{\rho}_t \log \hat{\rho}_t \, d\mu$$

with

$$\hat{\rho}_t = P_0\rho_t \quad \text{or} \quad \Lambda\rho_t \quad , \quad \hat{\rho}_t = U_t\rho \quad .$$

Zermelo's objection, which is based on Poincare's recurrence theorem, obviously does not apply as there can be no recurrence theorem for regular density functions ρ. Let us emphasize in this connection that both P_0 and Λ cannot be defined to act on phase points or singular distributions, but must necessarily be defined as only acting on regular density functions.

Loschmidt's objection which is based on symmetry of dynamical evolution under velocity inversion is deeper. Indedd, computer calculation of Boltzmann's H quantity

$$H_B = \int f(v) \log f(v) dv \quad ,$$

where f(v) is the one-particle velocity distribution, has been made for a system of hard disks. If one starts with an initial distribution based on lattice sites and with isotropic velocity distribution, then one finds that H_B does indeed decrease with time t. However, if all velocities are inverted at time t_0, it is found that H_B behaves antithermodynamically (i.e., increases rather than decreases) in the interval $[t_0, 2t_0]$. This clearly demonstrates that Boltzmann's H theorem cannot be valid for all initial density functions. In particular, it cannot be valid both before and after velocity inversion.

In contrast the H function $\Omega(\rho_t)$ is a monotonically decreasing function of t for *all initial* density ρ as long as the system remains isolated. Velocity inversion at time t_0, may cause a discontinuous jump in $\Omega(\rho_t)$ due to entropy (or information) flow from the external apparatus implementing the velocity inversion; but $\Omega(\rho_t)$ again continues to decrease as soon as the operation of velocity inversion is completed and the system is again isolated from the outside. At no stage is there any antithermodynamic behavior. In any cycle in which the system returns to its initial state ρ_0 the net entropy flow dS_e from the outside to the system is negative in agreement with the thermodynamic principle that the entropy production $dS_i = -dS_e$ inside the system is positive. Let us illustrate this explicitly in the case that the initial state is symmetric under velocity inversion $V\rho = \rho$, and one takes $\hat{\rho}_t = P_0 U_t \rho$ in defining $\Omega(\rho_t)$.

Now, using the notation $\Omega(t) \equiv \Omega(\rho_t)$

$$\Omega(0) = \int (P_0\rho) \ln (P_0\rho) d\mu$$

$$\begin{aligned}
\Omega(t_{0-}) &= \int_\Gamma (P_0 U_{t_0} \rho) \ln (P_0 U_{t_0} \rho) d\mu \\
&= \int_\Gamma (U_t^* P_0 U_t \rho) \ln (U_{t_0}^* P_0 U_{t_0} \rho) d\mu \\
&= \int (P_{-t_0} \rho) \ln (P_{-t_0} \rho) d\mu \leq \Omega(0) \quad , \quad t_0 \geq 0 \ .
\end{aligned}$$

(Here the first equality follows because the expression $\int \rho \ln \rho \, d\mu$ is invariant under dynamical evolution U_t, and the inequality follows because P_{-t_0} is a coarse graining operation with respect to a coarser partition than that of P_0).

If velocities are inversed at t_0 the H function $\Omega(t_{0+})$ immediately after this operation is given by

$$\begin{aligned}
\Omega(t_{0+}) &= \int_\Gamma (P_0 V U_{t_0} \rho) \ln (P_0 V U_{t_0} \rho) d\mu \\
&= \int_\Gamma (P_0 U_{t_0}^* V\rho) \ln (P_0 U_t^* V\rho) d\mu \\
& \int_\Gamma (P_{t_0} V\rho) \ln (P_{t_0} V\rho) d\mu = \int (P_{t_0} \rho) \ln (P_{t_0} \rho) d\mu \ .
\end{aligned}$$

Thus $\Omega(t_{0+}) \geq \Omega(0) \geq \Omega(t_{0-})$.

After velocity inversion at t_0 the quantity $\Omega(t_{0+})$ again decreases. At time $2t_0$ the state is $U_{t_0}VU_{t_0}\rho = V\rho = \rho$ and $\Omega(2t_0) = \Omega(0)$ and the cycle is completed. The diagram representing the change of Ω in this cycle (with $V\rho = \rho$) is thus of the form

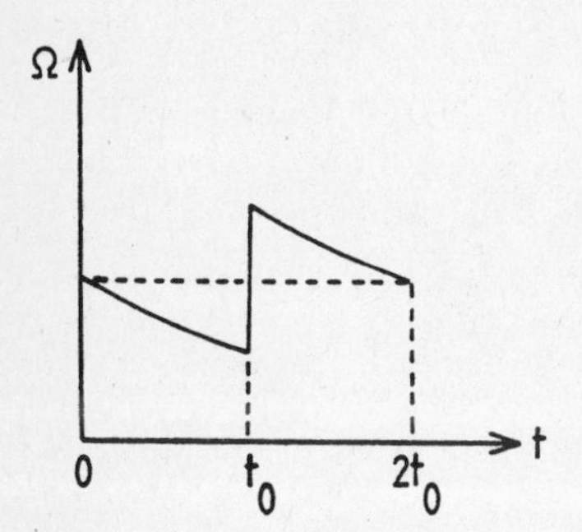

Fig.1

The net entropy flow dS_e from outside during the cycle is $-[\Omega(t_{0+}) - \Omega(t_{0-})]$ which is obviously negative in conformity with the second law. Similar considerations apply to an arbitrary initial distribution ρ and also for the expression Ω defined with $\rho_t = \Lambda U_t\rho$, Λ invertible. For further discussions of this question the reader may see [2].

Kinetic theory in the form we have presented here thus avoids all the difficulties associated with the classical "paradoxes". It is amusing to remember in this connection that Poincaré wrote that he could not recommend that anybody read Boltzmann's work as he could not recommend the study of proofs in which the conclusions clash with the premises. We believe that the results discussed here take a step towards validating Boltzmann's program. However, we have till now not shown the role of "collisions" in our theory or the explicit link between the stochastic processes we obtain and transport processes such as diffusion, etc. For this we need to study specific models (e.g., the Lorentz gas) in the framework of the approach discussed here. We hope to come back to these questions soon.

Reference

1. I. Prigogine: *Nonequilibrium Statistical Mechanics* (Wiley, New York 1962)
2. I. Prigogine, C. George, F. Henin, L. Rosenfeld: Chemica Scripta *4*, 5-32 (1973)
3. B. Misra: Proc. Natl. Acad. Sci. U.S.A. *75*, 1627 (1978)
4. B. Misra, I. Progogine, M. Courbage: Physica *98*A, 1 (1979)
5. S. Goldstein, B. Misra, M. Courbage: to appear in J. Stat. Phys.
6. B. Misra, I. Prigogine: to appear
7. K. Goodrich, K. Gustafson, B. Misra: Physica *102*A, 379 (1980)
8. E.G. Dynkin: *Markov Processes*, Grundlehren der mathematischen Wissenschaften, Vols.121-122 (Springer, Berlin, Heidelberg, New York 1965)
9. V.I. Arnold, A. Avez: *Ergodic Problems in Classical Mechanics* (Benjamin, New York 1968)
10. Y. Sinai: Sov. Math. Dokl. *4*, 1818 (1963)
11. D. Anosov: Proc. Steklov Inst. (1967)
12. S. Goldstein, O. Penrose: Univ. of Rutgers, preprint

Transition Phenomena in Nonlinear Systems

H. Haken

Institut für Theoretische Physik der Universität Stuttgart,
D-7000 Stuttgart 80, Fed. Rep. of Germany

1. Introduction

Since this workshop brings together mathematicians, on the one hand, and research workers in the field of natural sciences, on the other, I will try to present the basic ideas of my talk in a form which is not too technical. Furthermore I can omit details because they are laid down in a number of recent articles and a book of mine [1],[2]. I shall first outline what I mean by transition phenomena. As will transpire from the examples, the general feature common to the transition phenomena I treat here are dramatic changes on macroscopic scales.

2. Examples from Physics, Chemistry, and Biology

The best studied system showing transition phenomena is probably the laser, a by now well-known device [3]. When a laser is pumped weakly, the emitted lightfield consists of noise. Beyond a first critical pump strength a sinusoidal wave is emitted, which beyond a second threshold is replaced by ultrashort pulses. From the dynamic systems point of view, first a stable focus is realized which is replaced by a limit cycle and eventually by a torus. Under different conditions the periodic motion is replaced by deterministic chaos. We have studied in particular the impact of noise on transitions between the different states. In an old paper of mine [4] I have shown that the statistical properties of laser light differ drastically below and above a critical pump parameter. This paper was followed by an avalanche of theoretical and experimental studies of the noise properties close to laser threshold. For reviews of this work see e.g. [3], [5]. More recently we found that noise plays an important role at the second threshold when ultrashort pulses occur [6]. Fluid dynamics provides us with a variety of patterns, for instance the Taylor instability. A similar sequence of ordered states is observed. In the sense of dynamic systems theory nodes, foci, limit cycles, tori and again chaos occur. Precisely the same sequence is observed in gas discharge plasmas. In the field of chemistry similar types of ordered states may occur like oscillations or spatial patterns [8]. Experimentally, some of these ordered states occur during a time evolution so that one might speak of "order through transients". Biology provides us with a wealth of transition phenomena from homogeneous states to patterns.

3. Basic Equations

When we wish to treat these phenomena (part of which have been even first predicted theoretically and then checked experimentally) we usually resort to suitable equations. I will here briefly describe what we do in physics. In classical physics the fundamental equations are provided by such fields as mechanics, electrodynamics and relativity.

These equations are deterministic and time reversal invariant. Fluctuations or, in other words, stochasticity is not intrinsic to these fundamental equations. Rather stochasticity arises because we do not follow up all microscopic events but rather try to describe the systems macroscopically. The microscopic events are taken care of by statistical approaches. In that way fluctuations mirror a lack of knowledge. This lack of knowledge is introduced, so to speak, at our will in order not to be overloaded with unessential details. In view of the more recent results on chaotic motion, etc., it might well be that we have to abandon such interpretation, because any measurement we make cannot be done with absolute precision. But any inaccuracy of measurements will lead us via exponentially growing distances between trajectories to the incapability of predicting the future development of the system in the sense of "deterministic" behavior.

In quantum physics fluctuations are there in principle, for instance the spontaneous emission of a photon by an atom can be described only in terms of a statistical theory. In addition we may do away with superfluous degrees of freedom and then we introduce a lack of knowledge at our will. I should like to add a comment on the question of irreversibility and damping. There is, of course, the fundamental problem to understand why the originally time reversal invariant equations of classical or quantum physics can be replaced by equations which describe the actual processes in physics where damping is observed. One possibility is here to treat specific models in which irreversibility can be deduced rigorously (compare in particular the contribution by Prigogine to this volume). My own attitude to this problem is a more pragmatic one which can best be illustrated by the coupling of an oscillator to many others. If the coupling is bilinear the problem can be solved exactly and the motion is quasiperiodic. However, when we follow up the motion of the harmonic oscillator under consideration for a finite time, we shall first observe a damping due to dephasing processes. The time during which we observe the usual damping can be made sufficiently large provided the number of reservoir oscillators is big enough. Therefore for practical purposes it suffices to describe the oscillator or any other system by equations which contain the damping and which are good enough approximations to the behavior of a system described by the original equations. It is noteworthy that there exist in the quantum domain procedures which allow us to establish quantum mechanically consistent equations which contain damping. It is interesting that quantum mechanical consistency is brought about by appropriate fluctuating forces.

In general it may be said that the original equations from mechanics, electrodynamics or relativity contain far too many variables so that one of the main ideas of macroscopic physics is to devise elimination procedures for these variables. This is done in various ways. I just mention the laser case in which all individual steps can be done explicitly starting from first principles. In other cases, such as in fluids, it is useful to start from a more macroscopic description, for instance on the basis of conservation laws.

In chemistry the situation is, of course, still more difficult than in physics and there is a long way from individual processes such as collisions between molecules, leading to binding by quantum mechanical processes etc., to macroscopic behavior. Therefore phenomenological equations such as those based on birth and death processes and diffusion are most useful. This statement holds still more true for biology. A good deal of mathematical work with respect to morphogenesis is nowadays done on reaction diffusion equations.

4. Types of Equations

For mathematicians it might be interesting to see what kind of equations are presently used in theoretical physics to model nonlinear stochastic processes. Those are

1) Langevin-type equations for a set of variables $\underset{\sim}{q}$.

$$\dot{\underset{\sim}{q}} = \underset{\sim}{N}(\underset{\sim}{q}) + \underset{\sim}{F}(t) \tag{1}$$

in which $\underset{\sim}{N}$ is a nonlinear function and $\underset{\sim}{F}$ represents stochastic forces which in most cases are assumed to be δ-correlated in time and Gaussian distributed.

2) If $\underset{\sim}{F}$ depends on the variables itself we use the Îto or Stratonovich stochastic equations of the form

$$d\underset{\sim}{q} = \underset{\sim}{N}(\underset{\sim}{q})dt + d\underset{\sim}{F}(t,\underset{\sim}{q}). \tag{2}$$

3) Another kind of approach which in well defined cases is equivalent to (1) or (2) is the Fokker-Planck equation for a distribution function f

$$\dot{f}(\underset{\sim}{q}) = L(\underset{\sim}{q},\nabla_{\underset{\sim}{q}})f(\underset{\sim}{q}), \tag{3}$$

L is a linear operator on f. While the nonlinearity of the problem in the case (1) stems from $\underset{\sim}{N}$, the nonlinearity is now inherent in the nonlinear dependence of L on $\underset{\sim}{q}$.

4) A fourth possibility rests on the use of path integrals which may be derived as solutions of eq.(3). It can be shown that there are whole classes of such path integral solutions to (3) which are equivalent with each other. Though path integrals are very appealing, their practical use has been rather limited so far because aside from very few exceptions only integrals with quadratic or bilinear terms can be evaluated explicitly.

5) To treat Markov processes we may resort to the Chapman-Kolmogorov equation or to the master equation of the distribution function P

$$\dot{P}(\underset{\sim}{m}) = \sum_{\underset{\sim}{n}} w_{\underset{\sim}{m},\underset{\sim}{n}} P(\underset{\sim}{n}) - P(\underset{\sim}{m}) \sum_{\underset{\sim}{n}} w_{\underset{\sim}{n},\underset{\sim}{m}}. \tag{4}$$

6) Finally a class of equations which becomes more and more fashionable not only in mathematics but also in physics are evolution equations with discrete time intervals e.g. the logistic equation

$$x_{n+1} = ax_n(1-x_n). \tag{5}$$

5. Transformation of Evolution Equations

In my own work I have found the following procedure must useful. One starts with evolution equations of the type (1) - (3), which may be partial differential equations because $\underset{\sim}{q}$ may depend on space and $\underset{\sim}{N}$ may contain operators acting on space coördinates:

$$d\underset{\sim}{q} = \underset{\sim}{N}(\underset{\sim}{q},\nabla,\alpha)dt + d\underset{\sim}{F}. \tag{6}$$

I have described the procedure for treating these equations close to transition points at various occasions [1], [2] and I will just spend a few words on that. We first neglect $d\underset{\sim}{F}$ and perform a linear analysis. By means of boundary conditions and general theorems on the form of the solution of the linearization, namely $\underset{\sim}{u} = e^{\lambda t}\underset{\sim}{v}(t)$ where $\underset{\sim}{v}$ obeys certain growth limitations, the spectrum $\{\tilde{\lambda}\}$ of the linearization is determined. By change of control parameters the real parts of one or several eigenvalues λ can become positive. We have developed a general scheme to construct the new configuration $\underset{\sim}{q}$ close to instability points of control parameter space. As result we obtain a new set of equations for certain phase angles Φ_k and expansion amplitudes ξ_j

$$d\xi_j = \lambda_j \xi_j dt + \tilde{N}_j(\underset{\sim}{\xi},\phi)dt + d\underset{\sim}{F}_j \tag{7}$$

$$d\Phi_k = i\omega_k dt + \tilde{N}'_k(\underset{\sim}{\xi},\phi)dt + d\underset{\sim}{F}'_k . \tag{8}$$

6. Slaving Principle

The slaving principle which I am going to discuss now is central to the theory of transition phenomena of the kind I have described in the beginning. It allows for an enormous reduction of the degrees of freedom. Without going into mathematical details it can be stated as follows. Select first those variables ξ_j for which the real part of $\lambda_j \geqslant 0$, if the spectrum around 0 is discrete and if all other eigenvalues are bounded away from the imaginary axis. If the spectrum around 0 is continuous we select those ξ_j for which

$$\left| \mathrm{Re}\ \lambda_j \right| < \delta \tag{9}$$

holds. Similarly if relatively strong fluctuations are present or the damped modes are not sufficiently far bounded from the imaginary axis we select a suitable set of ξ_u's with real part of λ around 0 in a finite range as order parameters. The slaving principle can be derived mathematically and rigorously states the following. We may express all other variables called ξ_s by the set of order parameters ξ_j, ϕ_k uniquely and explicitly. In particular we may construct the dependence of ξ_s by ξ_u, ϕ and dF by a rapidly converging continued fraction of matrices. I have derived several forms of the slaving principle (cf. [1], [9]) which also applies to Ito or Stratonovich-type equations [1o]. The slaving principle contains a number of special cases known in the literature. It is related to the center manifold theorem [11] (or similar theorems) but it is more general in several respects. It includes stochastic processes. It contains finite band width effects in continuous media and it comprises a region around the center manifold. More precisely speaking it is in particular applicable to a situation in which the attracting order parameter set is replaced by a repeller in whose neighbourhood new attracting sets are established. In contrast to the usual center manifold theorem which is an existence theorem our approach is constructive. If the eigenvalues of the order parameters are real, as another special case, elimination techniques based on the slow manifold concept or different time scales (including scaling techniques and adiabatic elimination) are included in the slaving principle. Finally when steady states, $\dot{\xi} = 0$, are treated and oscillations and fluctuations are not present, contact can be made with conventional bifurcation theory based on the implicit function theorem. The slaving principle allows us to eliminate all slaved variables and to obtain closed equations for the order parameter set ξ_u, Φ alone. The resulting equations are stochastic integro-differential equations. Because noise induced transitions (or, in other words, multiplicative noise) are of great current interest (cf. the contributions by Horsthemke, Lefever, Schenzle and others to this volume), I should like to mention that the slaving principle can be used also in these cases. It again allows us to reduce a multivariable problem to a problem of a single variable (as treated by these authors), provided the spectrum of the damped modes is sufficiently bounded away from the imaginary axis.

7. Nonequilibrium Phase Transitions

In many cases close to the transition point order parameters show critical slowing down and they are still small. Exploiting symmetry in addition, the integro-differential equations can be considerably simplified. An explicit presentation would go far beyond the scope of the present article, however.

A large class of the resulting order parameter equations can be considered as a generalization [1] of the Ginzburg-Landau equations which were originally introduced to deal with equilibrium phase transitions. The laser example showed that the same equations occur in nonequilibrium phase transitions [12] .The same type of equation is found in many other instances, e.g. chemical reactions (cf. the contributions by Malek-Mansour, Nicolis, Walgraef and others to this volume, and [1]). If no fluctuations are present, a Kolmogorov-Arnold-Moser condition for the frequencies is fulfilled, and some technical conditions apply, we find quasi-periodic motion on the bifurcating tori. A variety of other phenomena may be found if these conditions are not fulfilled. For instance, we treated explicit cases of phase locking. As a result of stochastic forces phase diffusion on the torus takes place. More recently, I have studied still more complicated cases in which quasi-periodic motion on a torus bifurcates into motion on new tori ([9], also my contributions to [2] and work to be published).

A further class of problems we have treated concerns eq.(5), but subjected to noise [13]. Our calculations show that chaotic solutions are robust against not too strong fluctuating forces. As an interesting effect we have found that chaos can become "visible" due to fluctuating forces, e.g., in the period three case of Li and Yorke [14].

8. Equilibrium and Nonequilibrium Phase Transitions: Analogies and Differences

It has been shown [12] , [1], that there are pronounced analogies between the class of transitions I have been describing above and equilibrium phase transitions. These analogies include critical slowing down, critical fluctuations, symmetry breaking, in many cases existence of a potential function, etc. Equilibrium phase transitions are grouped nowadays into universality classes. Within such a class the phase transition is independent of mechanism and of geometry. The latter

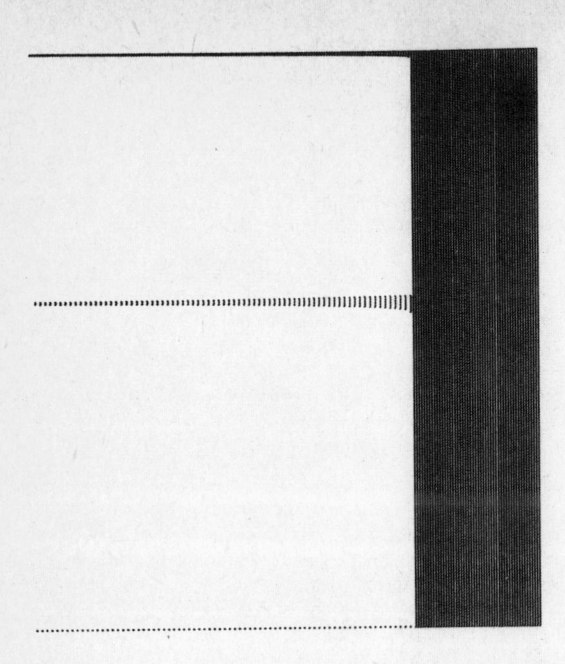

This figure (from G.Mayer-Kress and H. Haken, to be published) shows, how chaotic behavior may become dominant through the addition of small fluctuations. According to Li and Yorke [14] there is an uncountable set of points in the unit-interval which behaves chaotically under iterations of the logistic function if there is an orbit of period three. The behavior of the system as it is seen in computer experiments is, however, periodic since the "chaotic set" is not attracting. This can be seen in the figure where the support of the invariant measure (the set of points which is asymptotically approached by all trajectories) is drawn as ordinate versus the standard deviation of the fluctuating forces (abscissa). For fluctuations smaller than $5.6 \cdot 10^{-4}$ all trajectories are attracted by the period 3 orbit. Above that critical value the support consists of the whole invariant interval. Further numerical studies by G.Mayer-Kress have shown that the periodic behavior is interrupted by "turbulent bursts" during which the system behaves chaotically, i.e. here we have a transition to chaos via "intermittency" as it is also observed in jet streams [15] and large-aspect-ratio Benard cells [16].

---

one is a trivial statement because the limit: volume $\longrightarrow \infty$ is treated. Universality classes depend on dimensions.

In the present paper we have considered systems far away from thermal equilibrium and their transitions which are often called nonequilibrium phase transitions [1],[8]. They are much richer than equilibrium phase transitions because they include formation of limit cycles, motion on tori and chaos. But still here we can establish universality classes. This rests on the fact that we may obtain order parameter equations of exactly the same type for quite different systems. As we have shown (see e.g. [1],[12]) order parameter equations may be of the Ginzburg-Landau type and one may construct

Ginzburg-Landau functionals as stationary solutions of the Fokker-Planck equation. These concepts are now widely used and carried further including the application of renormalization group ideas (compare the paper of Walgraef to this volume). These new universality classes are in a well defined way independent of microscopic mechanisms but depend on geometry and size. In my opinion this is the most striking difference between equilibrium and nonequilibrium phase transitions. Of course, one may study also special cases which are artificial from a physical point of view but interesting from a theoretical point of view. Here, one lets the volume tend to infinity so that a continuous spectrum of the order parameters is caused. Such a model has been treated, for instance, in the continuous mode laser [12]. On the other hand, one should bear in mind that by the limit $V \longrightarrow \infty$ a number of many important facts can get lost, for instance, that patterns are selected by geometry, or the phase diffusion of the laser which is an important physical effect.

9. Concluding Remarks

With my above remarks I have tried to sketch a new field of research. I am fully aware that my remarks are far too sketchy but the material as it is known at present would fill more than one book. I do hope, however, that the reader is stimulated to do his own research in a still very young and promising field in which the interplay between nonlinearity and stochasticity is vital.

Appendix

I use this occasion(and the remaining space) to supplement the Appendix of my contribution to Vol. 6 of the Springer Series in Synergetics [2]. In that appendix I considered a set of first order linear differential equations with quasi-periodic coefficients and presented theorems on their solutions, which have forms quoted above in section 5. The supplement of the assumptions (A 8) and (A 9) of the former appendix should read:

It is assumed that each pair of solutions of the fundamental set keeps a finite minimal angle when $t \rightarrow \pm \infty$. Or in other words, we assume that the solutions remain linearly independent when $t \rightarrow \pm \infty$.

References

1. H. Haken: *Synergetics. An Introduction*, 2nd ed., Springer Series in Synergetics, Vol. I (Springer, Berlin, Heidelberg, New York 1978)
2. Springer Series in Synergetics (Springer, Berlin, Heidelberg, New York)
   Vol. 2: *Synergetics. A Workshop*, ed. by H. Haken (1977)
   Vol. 3: *Synergetics, Far from Equilibrium*, ed. by A. Pacault, C. Vidal (1979)
   Vol. 4: *Structural Stability in Physics*, ed. by W. Güttinger, H. Eikemeier (1979)
   Vol. 5: *Pattern Formation by Dynamic Systems and Pattern Recognition*, ed. by H. Haken (1979)
   Vol. 6: *Dynamics of Synergetic Systems*, ed. by H. Haken (1980)
3. H. Haken: "Laser Theory", in *Light and Matter Ic*, ed. by L. Genzel, Encyclopedia of Physics, Vol. XXV/2c (Springer, Berlin, Heidelberg, New York 1970)
4. H. Haken: Z. Phys. 181, 96 (1964)

5. M. Sargent III, M.O. Scully, W.E. Lamb, Jr.: *Laser Physics* (Addison-Wesley, Reading, Mass. 1974)
6. H. Haken, M. Nuoffer: To be published
7. For recent reviews consult the contributions of Busse, Gollub, Koschmieder, Swinney and others to Ref. 2
8. G. Nicolis, I. Prigogine: *Self-Organization in Nonequilibrium Systems* (Wiley-Interscience, New York 1977)
9. H. Haken: Z. Phys. B29, 61 (1978); B30, 423 (1978)
10. H. Haken: To be published
11. A. Kelley: "Appendix C" in *Transversal Mappings and Flows*, ed. by R. Abraham, J. Robbin (Benjamin, New York 1967) and J. Differ. Equ. 3, 546 (1967)
12. R. Graham, H. Haken: Z. Phys. 237, 31 (1970)
13. G. Mayer-Kress, H. Haken: To be published
14. T.Y. Li, J.A. Yorke: Am. Math. Mon. 82, 985 (1975)
15. e.g., Yih Chia-Shun: *Fluid Mechanics* (McGraw-Hill, New York 1969)
16. e.g., G. Ahlers, R.W. Walden: Phys. Rev. Lett. 44, 445 (1980)

Part II

Approximation of Stochastic Processes

Approximation of Discontinuous Processes by Continuous Processes

Thomas G. Kurtz

Department of Mathematics, University of Wisconsin-Madison
Madison, WI 53706, USA

1. Introduction

There is a close relationship between methods of characterizing stochastic processes and methods for justifying approximations. In this paper we will consider two methods of characterizing processes, (1) as solutions of stochastic equations involving multidimensional random time transformations, and (2) (for Markov processes) by specifying their generators, and we will consider examples of how these methods lead naturally to approximation theorems.

First, however, let us consider some of the basic ideas in the modern view of stochastic processes that are needed to understand the nature of the approximations we will be discussing. We do not give all the formal definitions (these can be bound in such books as BREIMAN [3] and BILLINGSLEY [1, 2]), but we hope to give an intuitive feel for what is involved.

A probability space $(\Omega, \mathfrak{F}, P)$ is a mathematical abstraction of a lottery. Ω, called the sample space, is the set of possible outcomes of the lottery (the tickets); $\mathfrak{F}$, called the collection of events, is a collection of statements about the outcome of the lottery which we identify with a collection of subsets of Ω, (a σ-algebra to be precise) that is a statement is identified with the set of elements (tickets) $\omega \in \Omega$ for which the statement is true; P, the probability measure, is a measure defined on $\mathfrak{F}$ (considered as a σ-algebra of subsets of Ω) which "measures" the tickets in each $E \in \mathfrak{F}$, that is $P(E)$ measures the chance that E will occur. (An event E occurs if the "ticket" $\omega_0 \in \Omega$ that is actually drawn is in E.)

A random variable Y is a quantity (say with values in $\mathbb{R}^d$) that depends on the outcome of the lottery, that is Y is a function defined on Ω which we require to be measurable (i.e. $\{\omega : Y(\omega) \in \Gamma\} \in \mathfrak{F}$ for all $\Gamma \in \mathcal{B}(\mathbb{R}^d)$ where $\mathcal{B}(\mathbb{R}^d)$ denotes the Borel sets in $\mathbb{R}^d$). A stochastic process is a function $X : [0,\infty) \times \Omega \to \mathbb{R}^d$, such that for each $t \geq 0$, $X(t,\cdot)$ is a random variable. We interpret the variable t as time. In other words for each "ticket" $\omega \in \Omega$ we have a function $X(\cdot,\omega) : [0,\infty) \to \mathbb{R}^d$. We will always assume that these functions are right continuous and have limits from the left, i.e. $\lim_{s \searrow t} X(s,\omega) = X(t,\omega)$ and $\lim_{s \nearrow t} X(s,\omega)$ exists for all $t \geq 0$.

Suppose an observer watches X during an interval of time $[0,t]$. At the end of this time interval the observer will be able to make certain statements based on his observations. Again we identify each statement with the set of outcomes for which it is true and assume that the collection of subsets form a σ-algebra $\mathfrak{F}_t \subset \mathfrak{F}$. Of course the observer is watching X so we must have $\{\omega : X(s,\omega) \in \Gamma\} \in \mathfrak{F}_t$ for $0 \leq s \leq t$ and Borel Γ. We will think of $\mathfrak{F}_t$ as representing the information available to the observer at time t. The expectation $E[Y]$ of a random variable Y is just the integral (if it exists)

$$E[Y] = \int_{\Omega} Y(\omega)\, P(d\omega) , \tag{1.1}$$

and the <u>conditional</u> <u>expectation</u> of Y given $\mathcal{F}_t$ (denoted $E[Y|\mathcal{F}_t]$) is the best approximation of Y the observer can make using the information available at time t. If $E[|Y|^2] < \infty$, then $Z = E[Y|\mathcal{F}_t]$ is "best" in the sense that it is the unique (up to changes on sets of probability zero) $\mathcal{F}_t$-measurable random variable that minimizes $E[|Y-Z|^2]$. Note that a random variable is $\mathcal{F}_t$-measurable if its value is known to the observer at time t.

The <u>distribution</u> of Y is the measure on $\mathbb{R}^d$ defined by

$$\mu_Y(\Gamma) = P\{\omega : Y(\omega) \in \Gamma\} , \qquad \Gamma \in \mathcal{B}(\mathbb{R}^d) . \tag{1.2}$$

We note that

$$E[Y] = \int_{\mathbb{R}^d} y\mu_Y(dy) \tag{1.3}$$

and more generally

$$E[\varphi(Y)] = \int_{\mathbb{R}^d} \varphi(y)\, \mu_Y(dy) . \tag{1.4}$$

Let $D_{\mathbb{R}^d}[0,\infty)$ denote the set of all functions $x : [0,\infty) \to \mathbb{R}^d$ such that x is right continuous and has left limits. Then, under our assumptions, the process X can be thought of as a function $X : \Omega \to D_{\mathbb{R}^d}[0,\infty)$; we say that X has sample paths in $D_{\mathbb{R}^d}[0,\infty)$. A metric d (the Skorohod metric) can be defined so that $D_{\mathbb{R}^d}[0,\infty)$ becomes a complete, separable metric space (see BILLINGSLEY [1] or KURTZ [8]). The <u>distribution</u> of X is the measure defined on the Borel sets $\mathcal{B}(D_{\mathbb{R}^d})$ by

$$\mu_X(\Gamma) = P\{\omega : X(\cdot,\omega) \in \Gamma\} , \qquad \Gamma \in \mathcal{B}(\mathbb{R}^d) . \tag{1.5}$$

Two processes are called <u>equivalent</u> if they have the same distribution.

We are interested in <u>weak</u> <u>convergence</u> for sequences of stochastic processes $\{X_n\}$. Weak convergence is really a statement about the distributions $\{\mu_{X_n}\}$ of the processes (see BILLINGSLEY [1] for a detailed discussion), in fact the X_n need not be defined on the same sample space. However, the meaning of weak convergence can, perhaps, most easily be understood as follows: The distributions $\{\mu_{X_n}\}$ of a sequence of processes $\{X_n\}$ with sample paths in $D_{\mathbb{R}^d}[0,\infty)$ converge weakly to μ_X (we write $\mu_{X_n} \Longrightarrow \mu_X$ or $X_n \Longrightarrow X$) if there exist processes $\tilde{X}_n$, $\tilde{X}$ all defined on the same sample space, such that $\tilde{X}_n$ is equivalent to X_n and $\tilde{X}$ is equivalent to X and

$$\lim_{n\to\infty} d(\tilde{X}_n, \tilde{X}) = 0 \qquad \text{a.s.} \tag{1.6}$$

We will not go into the exact meaning of convergence in the Skorohod metric d (again see BILLINGSLEY [1]). However, in all the examples we will consider the limiting process has continuous sample paths, i.e. $t \to X(t,\omega)$ is continuous for each ω, and in this case (but not in general) (1.6) implies

$$\lim_{n\to\infty} \sup_{t \le T} |\tilde{X}_n(t) - \tilde{X}(t)| = 0 \qquad \text{a.s.} \tag{1.7}$$

for all $T > 0$. (We will frequently write $X(t)$ instead of $X(t,\omega)$.)

In general (1.6) implies

$$\lim_{n\to\infty} \int_0^T h(\tilde{X}_n(s))\,ds = \int_0^T h(\tilde{X}(s))\,ds \tag{1.8}$$

for all continuous h and

$$\lim_{n\to\infty} \sup_{t\le T} |\tilde{X}_n(t)| = \sup_{t\le T} |\tilde{X}(t)| \tag{1.9}$$

for all T such that $\tilde{X}(T) = \tilde{X}(T-) \equiv \lim_{t\nearrow T} \tilde{X}(t)$, to give some examples of useful consequences of weak convergence. Note that (1.8) implies

$$\lim_{n\to\infty} P\{\int_0^T h(X_n(s))\,ds \le x\} = P\{\int_0^T h(X(s))\,ds \le x\} \tag{1.10}$$

for all x that are points of continuity of the right side, i.e. all x such that $P\{\int_0^T h(X(s))\,ds = x\} = 0$, so that statements like (1.8) and (1.9) concerning $\tilde{X}_n$ and $\tilde{X}$ can be translated into statements about the distributions of the original processes X_n and X.

To motivate the discussion of how we will specify processes consider the differential equation

$$\dot{X} = F(X), \qquad X \in \mathbb{R}^d \tag{1.11}$$

where $F: \mathbb{R}^d \to \mathbb{R}^d$. Of course (1.12) implies that for small h

$$X(t+h) - X(t) = F(X(t))h + o(h) \tag{1.12}$$

so that (1.12) describes how X changes over a small time interval. We want to specify stochastic processes in the same way, that is by describing how they change over small time intervals. Of course the change $X(t+h) - X(t)$ for a process is a random variable and instead of a description like (1.12) we will describe the distribution of $X(t+h) - X(t)$ given the information $\mathcal{F}_t$. For example, in Section 3 we will consider Markov processes X_n with values in the d-dimensional lattice $E_n = \{n^{-1}k: k \in \mathbb{Z}^d\}$. We will specify these processes by giving functions $\beta_\ell(x) \ge 0$, $\ell \in \mathbb{Z}^d$, $x \in \mathbb{R}^d$ and requiring

$$P\{X_n(t+h) - X_n(t) = n^{-1}\ell \mid \mathcal{F}_t\} = n\beta_\ell(X_n(t))h + o(h), \tag{1.13}$$

that is we give the conditional probabilities of the different possible jumps. For more complicated Markov processes we can specify a generator A, that is an operator defined on some collection of functions $\mathcal{D}(A)$ and require

$$E[f(X(t+h)) - f(X(t))|\mathcal{F}_t] = Af(X(t))h + o(h). \tag{1.14}$$

For the processes in (1.13), A should have the form

$$A_n f(x) = \sum_\ell n\beta_\ell(x)\,(f(x+n^{-1}\ell) - f(x)) \tag{1.15}$$

and for one dimensional Brownian motion

$$Af(x) = af''(x) + bf'(x) \tag{1.16}$$

for some $a > 0$ and b. (We call X a <u>standard Brownian motion</u> if $a = 1/2$ and $b = 0$.)

As with (1.11), (1.13) and (1.14) present questions of existence and uniqueness. For processes satisfying (1.13) (continuous parameter Markov chains) see CHUNG [4]. For diffusion processes, that is processes for which A is an elliptic differential operator, see STROOCK and VARADHAN [10]. For a general discussion of Markov processes and generators see DYNKIN [5]. An existence theorem would state that given a probability measure ν on $\mathbb{R}^d$, there exists a stochastic process X with sample paths in $D_{\mathbb{R}^d}[0,\infty)$ such that (1.14) (or (1.13)) holds and $P\{X(0) \in \Gamma\} = \nu(\Gamma)$, and a uniqueness theorem would state that all such processes have the same distribution on $D_{\mathbb{R}^d}[0,\infty)$.

2. Convergence of Renormalized Poisson Processes

One of the simplest convergence results of the type we want to consider is for Poisson processes. We obtain a Poisson process $Y(t)$ on $\mathbb{Z}$ with parameter λ by taking $n = 1$, $\beta_1(x) = \lambda$ and $\beta_\ell(x) = 0$ for $\ell \neq 1$ in (1.13). The corresponding generator is

$$Af(k) = \lambda(f(k+1) - f(k)) . \tag{2.1}$$

If Y is a Poisson process with parameter λ, then

$$W_n(t) = n^{-1/2}(Y(nt) - \lambda nt) \tag{2.2}$$

is a Markov process with generator

$$A_n f(x) = n\lambda(f(x+n^{-1/2}) - f(x)) - n^{1/2}\lambda f'(x) \tag{2.3}$$

say for f continuously differentiable with compact support. If f is twice continuously differentiable with compact support then

$$\lim_{n\to\infty} A_n f(x) = \frac{1}{2}\lambda f''(x) , \tag{2.4}$$

and convergence is uniform in x. We can use this "convergence of generators" to obtain the second part of the following theorem (see for example Theorem 4.2 of KURTZ [8]). The first part is just the law of large numbers for independent random variables.

2.1 Theorem Let Y be a Poisson process with parameter λ and let W_n be given by (2.2). Then

$$\lim_{n\to\infty} \sup_{t\le T} n^{-1}(Y(nt) - \lambda nt) = 0 \qquad \text{a.s.} \tag{2.5}$$

for all $T > 0$ and $W_n \Longrightarrow W$, where W is a Brownian motion with generator $Af = \frac{1}{2}\lambda f''$.

3. Density Dependent Jump Markov Processes

Let $\beta_\ell(k) \ge 0$, $\ell, k \in \mathbb{Z}^d$, $\sum_\ell \beta_\ell(k) < \infty$, and let Y_ℓ, $\ell \in \mathbb{Z}^d$ be independent Poisson processes with parameter 1. Then the equation

$$X(t) = X(0) + \sum_\ell \ell Y_\ell\Big(\int_0^t \beta_\ell(X(s))\,ds\Big) \tag{3.1}$$

has a unique solution for $t < \tau_\infty$, where

$$\tau_\infty = \sup\{t : \sum_\ell Y_\ell\Big(\int_0^t \beta_\ell(X(s))\,ds\Big) < \infty\} . \tag{3.2}$$

We claim that X satisfies (1.13) with $n = 1$ for $t < \tau_\infty$. To see that this claim is reasonable, we have, at least intuitively,

$$\begin{aligned} P\{X(t+h) - X(t) = \ell \mid \mathcal{F}_t\} &= P\{Y_\ell(\int_0^{t+h} \beta_\ell(X(s))\,ds) - Y_\ell(\int_0^t \beta_\ell(X(s))\,ds) = 1 \mid \mathcal{F}_t\} + o(h) \\ &= E[\int_t^{t+h} \beta_\ell(X(s))\,ds \mid \mathcal{F}_t] + o(h) \\ &= \beta_\ell(X(t))\,h + o(h) \quad . \end{aligned} \tag{3.3}$$

We introduce three examples to motivate the class of processes we will consider.

Chemical Reaction

Consider a simple, reversible chemical reaction $A + B \rightleftarrows C$ taking place in a container with volumn n. Let $X_n(t) = (X_n^A(t), X_n^B(t), X_n^C(t))$ be the numbers of molecules per unit volume for each of the reactants. Two types of transitions are possible. For $k \in \mathbb{Z}^3$ and $\ell = (-1,-1,1)$, we have the transition $n^{-1}k \to n^{-1}(k+\ell)$ if the reaction takes place in the forward direction and for $\ell = (1,1,-1)$ we have the transition $n^{-1}k \to n^{-1}(k+\ell)$ if the reaction occurs in the reverse direction. Noting that $nX_n^C(t)$ is the number of molecules of C, a reasonable choice for transition intensities gives

$$P\{X_n(t+h) - X_n(t) = n^{-1}(1,1,-1) \mid \mathcal{F}_t\} = n\mu X_n^C(t)h + o(h) \tag{3.4}$$

and

$$P\{X_n(t+h) - X_n(t) = n^{-1}(-1,-1,1) \mid \mathcal{F}_t\} = n\lambda X_n^A(t) X_n^B(t)\,h + o(h) \quad , \tag{3.5}$$

that is, with reference to (1.13), $\beta_{(1,1,-1)}(x_1, x_2, x_3) = \mu x_3$ and $\beta_{(-1,-1,1)}(x_1, x_2, x_3) = \lambda x_1 x_2$.

Logistic Growth

We consider a model for a single species living in a region with area n. We assume that the death rate is an increasing function of the population density (i.e. the number per unit area). In particular, letting $X_n(t)$ denote the population density, we assume

$$P\{X_n(t+h) - X_n(t) = n^{-1} \mid \mathcal{F}_t\} = \lambda n X_n(t)\,h + o(h) \tag{3.6}$$

and

$$P\{X_n(t+h) - X_n(t) = -n^{-1} \mid \mathcal{F}_t\} = nX_n(t)\,(\mu + \gamma X_n(t)) + o(h) \quad . \tag{3.7}$$

Note $nX_n(t)$ is just the population size, so λ is the "per individual" birth rate and $(\mu + \gamma x)$ is the death rate when the density is x. We have $\beta_1(x) = \lambda x$ and $\beta_{-1}(x) = x(\mu + \gamma x)$.

Prey-Predator Interaction

We consider a prey species and a predator species living in a region with area n, and let $X_n(t) = (X_n^1(t), X_n^2(t))$ where $X_n^1(t)$ is the number of prey per unit area and $X_n^2(t)$ is the number of predators per unit area. Four transitions are possible, corresponding to the birth of a prey, the death of a prey, the birth of a predator and the death of a predator. We assume

$$\begin{aligned} \beta_{(1,0)}(x_1, x_2) &= a_1 x_1 \\ \beta_{(-1,0)}(x_1, x_2) &= a_2 x_1 + b x_1 x_2 \\ \beta_{(0,1)}(x_1, x_2) &= c_1 x_2 + d x_1 x_2 \\ \beta_{(0,-1)}(x_1, x_2) &= c_2 x_2 \ . \end{aligned} \tag{3.8}$$

In all the examples we can obtain X_n as the solution of an equation of the form

$$X_n(t) = X_n(0) + \sum_{\ell} n^{-1} \ell Y_\ell (n \int_0^t \beta_\ell (X_n(s))\, ds) \ . \tag{3.9}$$

With reference to (2.5), we see that (3.9) has a limiting form

$$X(t) = X(0) + \sum_{\ell} \ell \int_0^t \beta_\ell (X(s))\, ds = X(0) + \int_0^t F(X(s))\, ds \tag{3.10}$$

where $F(x) = \sum_{\ell} \ell \beta_\ell(x)$. Of course (3.10) is equivalent to the differential equation $\dot{X} = F(X)$.

Stating the "obvious" result precisely, we have

<u>3.1 Theorem</u> Suppose for all compact $K \subset \mathbb{R}^d$

$$\sum_{\ell} |\ell| \sup_{x \in K} \beta_\ell(x) < \infty \tag{3.11}$$

and there exists $M_K > 0$ such that

$$|F(x) - F(y)| \leq M_K |x - y| \ , \qquad x, y \in K \ . \tag{3.12}$$

Suppose $\lim_{n\to\infty} X_n(0) = X(0)$, and X satisfies (3.10) for all $t \geq 0$ (in particular $\sup_{s \leq t} |X(s)| < \infty$). Then

$$\lim_{n\to\infty} \sup_{s\leq t} |X_n(s) - X(s)| = 0 \qquad \text{a.s.} \quad \text{for all} \quad t > 0 \ . \tag{3.13}$$

Now let $\tilde{Y}_\ell(u) = Y_\ell(u) - u$ and $W_\ell^{(n)}(u) = n^{-1/2} \tilde{Y}_\ell(nu)$. By Theorem 2.1 $W_\ell^{(n)} \Rightarrow W_\ell$ where W_ℓ is a Brownian motion. Subtracting (3.10) from (3.9) and defining $V_n(t) = n^{-1/2}(X_n(t) - X(t))$ we have

$$\begin{aligned} V_n(t) &= V_n(0) + \sum_{\ell} n^{-1/2} \ell \tilde{Y}_\ell (n \int_0^t \beta_\ell (X_n(s))\, ds) \\ &\quad + \int_0^t n^{1/2} (F(X_n(s)) - F(X(s)))\, ds \\ &= V_n(0) + \sum_{\ell} \ell W_\ell^{(n)} (\int_0^t \beta_\ell (X_n(s))\, ds) \\ &\quad + \int_0^t n^{1/2} (F(X(s) + n^{-1/2} V_n(s)) - F(X(s)))\, ds \ . \end{aligned} \tag{3.14}$$

Again we have a "limiting form"

$$V(t) = V(0) + \sum_{\ell} \ell W_{\ell}\left(\int_0^t \beta_{\ell}(X(s))\,ds\right) + \int_0^t \partial F(X(s))\, V(s)\, ds \tag{3.15}$$

and a theorem.

3.2 Theorem   Suppose for all compact $K \subset \mathbb{R}^d$

$$\sum_{\ell} |\ell|^2 \sup_{x \in K} \beta_{\ell}(x) < \infty , \tag{3.16}$$

$\partial F(x)$ is continuous and X satisfies (3.10) for all $t \geq 0$. If $\lim_{n \to \infty} V_n(0) = V(0)$, then $V_n \Rightarrow V$.

We note that (3.15) is equivalent to the Ito equation

$$V(t) = V(0) + \sum_{\ell} \ell \int_0^t \sqrt{\beta_{\ell}(X(s))}\, d\tilde{W}_{\ell}(s) + \int_0^t \partial F(X(s)) V(s)\, ds \tag{3.17}$$

and setting $\Gamma(t) = \sum_{\ell} \ell W_{\ell}\left(\int_0^t \beta_{\ell}(X(s))\,ds\right)$

$$V(t) = \Phi(t,0)(V(0) + \Gamma(t)) + \int_0^t \Phi(t,s)\, \partial F(X(s))(\Gamma(s) - \Gamma(t))\, ds \tag{3.18}$$

where Φ satisfies

$$\frac{\partial}{\partial t} \Phi(t,s) = \partial F(X(t))\, \Phi(t,s) , \qquad \Phi(s,s) = I . \tag{3.19}$$

Since V is a linear transformation of a Gaussian process it is Gaussian.

A number of related results can be obtained. Suppose $d = 1$, $F(x_0) = 0$ and $F'(x_0) = 0$. If $X(0) = x_0$, then (3.15) becomes

$$V(t) = V(0) + \sum_{\ell} \ell W_{\ell}(\beta_{\ell}(x_0) t) , \tag{3.20}$$

that is V is a sum of independent Brownian motions and hence is a Brownian motion $(E[V(t)] = V(0)$, $\mathrm{Var}(V(t)) = t \sum \ell^2 \beta_{\ell}(x_0))$. Suppose $\gamma > 1$ and $\lim_{x \to x_0} (x - x_0)^{-\gamma}(F(x) - F(x_0)) = c$. Set $\alpha = (\gamma + 1)^{-1}$ and

$$\begin{aligned} U_n(t) &\equiv n^{\alpha}(X_n(n^{1-2\alpha} t) - x_0) \\ &= U_n(0) + \sum_{\ell} n^{\alpha - 1} \ell\, \tilde{Y}_{\ell}\left(n^{2(1-\alpha)} \int_0^t \beta_{\ell}(X_n(n^{1-2\alpha} s))\, ds\right) \\ &\quad + \int_0^t n^{1-\alpha} F(X_n(n^{1-2\alpha} s))\, ds . \end{aligned} \tag{3.21}$$

Note $1 - \alpha = \alpha\gamma$ and $n^{\alpha-1} \tilde{Y}_{\ell}(n^{2(1-\alpha)} \cdot) \Rightarrow W_{\ell}$, so the limiting form is

$$U(t) = U(0) + \sum_{\ell} \ell W_{\ell}(\beta_{\ell}(x_0) t) + \int_0^t c\, U(s)^{\gamma}\, ds . \tag{3.22}$$

Keeping $d = 1$, let $\beta_1(x) = 1 + b\varepsilon x^2 + \varepsilon^2 ab - (x \wedge 0)^3$ and $\beta_{-1}(x) = 1 + a\varepsilon x^2 + (x \vee 0)^3$. Note

$$F(x) = -x(x + \varepsilon a)(x - \varepsilon b) . \tag{3.23}$$

If we allow ε to vary with n, in particular taking $\varepsilon = n^{-1/4}$, we have

$$U_n(t) \equiv n^{1/4} X_n(n^{1/2} t) = U_n(0) + n^{-3/4} \tilde{Y}_1(n^{3/2} \int_0^t \beta_1^{(n)}(X_n(n^{1/2} s))\, ds)$$

$$- n^{-3/4} \tilde{Y}_{-1}(n^{3/2} \int_0^t \beta_{-1}^{(n)}(X_n(n^{1/2} s))\, ds) \tag{3.24}$$

$$- \int_0^t U_n(s)\,(U_n(s) + a)\,(U_n(s) - b)\, ds$$

and if $U_n(0) \to U(0)$ then $U_n \Rightarrow U$ where

$$U(t) = U(0) + \tilde{W}_1(t) - \tilde{W}_{-1}(t) - \int_0^t U(s)\,(U(s) + a)\,(U(s) - b)\, ds\ . \tag{3.25}$$

4. Processes with External Noise

Equations of the form (3.1) provide a straightforward way of modeling systems with external noise. We simply assume that the β_ℓ depend on a noise process ξ (taking values in $\mathbb{R}^m$ for example) and (3.1) becomes

$$X(t) = X(0) + \sum_\ell \ell\, Y_\ell(\int_0^t \beta_\ell(X(s), \xi(s))\, ds)\ . \tag{4.1}$$

The statement that the noise process is external we take to mean that ξ is independent of the Y_ℓ. For example in the logistic growth model we can replace λ, μ and γ in (3.6) and (3.7) by processes $\lambda(t)$, $\mu(t)$ and $\gamma(t)$. The noise process is then $\xi(t) = (\lambda(t), \mu(t), \gamma(t))$.

To obtain an analog of Theorem 3.1, let X_n satisfy

$$X_n(t) = X_n(0) + \sum_\ell n^{-1} \ell\, \tilde{Y}_\ell(n \int_0^t \beta_\ell(X_n(s), \xi(ns))\, ds)$$
$$+ \int_0^t F(X_n(s), \xi(ns))\, ds\ . \tag{4.2}$$

Note we have sped up ξ so we are thinking of rapidly fluctuating external noise. We assume ξ satisfies the following ergodic condition. There exists a probability measure ν such that

$$\lim_{T\to\infty} \frac{1}{T} \int_0^T f(\xi(s))\, ds = \int f(z)\, \nu(dz) \quad \text{a.s.} \tag{4.3}$$

for all bounded continuous f.

4.1 Theorem   Suppose for all compact $K \subset \mathbb{R}^d$

$$\sum_\ell |\ell| \sup_{x \in K} \sup_{z \in \mathbb{R}^m} \beta_\ell(x, z) < \infty \tag{4.4}$$

and there exists $M_K > 0$ such that

$$|F(x, z) - F(y, z)| \le M_K |x - y|\ , \quad x, y \in K, \quad z \in \mathbb{R}^m\ . \tag{4.5}$$

Suppose $\lim_{n\to\infty} X_n(0) = X(0)$, and X satisfies

$$X(t) = X(0) + \int_0^t \int F(X(s), z)\, \nu(dz)\, ds \equiv X(0) + \int_0^t G(X(s))\, ds \tag{4.6}$$

for all $t \geq 0$. Then

$$\lim_{n\to\infty} \sup_{s\leq t} | X_n(s) - X(s) | = 0 \quad \text{a.s.} \quad \text{for all } t > 0 . \tag{4.7}$$

We turn now to an analog of Theorem 3.2. Let X satisfy (4.6) and define $V_n = \sqrt{n}\,(X_n - X)$. Then

$$\begin{aligned} V_n(t) = V_n(0) &+ \sum_\ell \ell n^{-1/2} \tilde{Y}_\ell (n \int_0^t \beta_\ell (X_n(s), \xi(ns))\, ds) \\ &+ \int_0^t n^{1/2} (F(X_n(s), \xi(ns)) - F(X(s), \xi(ns)))\, ds \\ &+ \int_0^t n^{1/2} (F(X(s), \xi(ns)) - G(X(s)))\, ds . \end{aligned} \tag{4.8}$$

Let

$$U_n(t) \equiv \int_0^t n^{1/2} (F(X(s), \xi(ns)) - G(X(s)))\, ds \tag{4.9}$$

and note that U_n is independent of the Y_ℓ.

<u>4.2 Theorem</u> Suppose, in addition to the hypotheses of Theorem 4.1, $\beta_\ell(x,z)$, $\partial F(x,z)$ and $\partial G(x)$ are continuous; for all compact $K \subset \mathbb{R}^d$

$$\sum_\ell |\ell|^2 \sup_{x\in K} \sup_{z\in\mathbb{R}^m} \beta_\ell(x,z) < \infty \tag{4.10}$$

and

$$\sup_{x\in K} \sup_{z\in\mathbb{R}^m} |\partial F(x,z)| < \infty . \tag{4.11}$$

If $\lim_{n\to\infty} V_n(0) = V(0)$ and $U_n \Rightarrow U$, then $V_n \Rightarrow V$ where V satisfies

$$\begin{aligned} V(t) = V(0) &+ \sum \ell W_\ell (\int_0^t \int \beta_\ell (X(s), z)\, \nu(dz)\, ds) \\ &+ \int_0^t \partial G(X(s))\, V(s)\, ds \\ &+ U(t) . \end{aligned} \tag{4.12}$$

<u>Remark</u>. Note that U will be independent of the W_ℓ. In the limiting equation, U corresponds to the external noise and the W_ℓ correspond to the internal noise. Typically U will be a Gaussian process with independent increments.

In the next section we will give an example of a situation in which the hypothesis $U_n \Rightarrow U$ can be verified.

5. <u>Convergence of Generators</u>

It was observed in Section 2 that convergence of generators can be used to prove weak convergence. There are two approaches to justifying this. The first uses the fact that if X is a temporally homogeneous Markov process then

$$T(t)f(x) = E[f(X(t))|X(0) = x] \tag{5.1}$$

defines a semigroup of operators and applies the semigroup convergence theorems of TROTTER [11] and others to obtain convergence of the processes. The second approach uses the interpretation of (1.14) as the requirement that

$$M_f(t) = f(X(t)) - \int_0^t Af(X(s))\,ds \tag{5.2}$$

is a martingale for all $f \in \mathcal{D}(A)$, i.e.

$$E[M_f(t+h)|\mathcal{F}_t] = M_f(t), \qquad t, h \geq 0 . \tag{5.3}$$

The <u>martingale problem</u>, that is the problem of finding a process X for which (5.2) is a martingale for all $f \in \mathcal{D}(A)$ and showing that for each initial distribution this requirement uniquely determines the distribution of X, was introduced by STROOCK and VARADHAN (see [10]) in the study of diffusion processes. Its application here results from the fact that if X_n is a solution of the martingale problem for A_n, A_n converges to A, and $X_n \Rightarrow X$ then X will (under appropriate hypotheses) be a solution of the martingale problem for A.

We will not discuss the precise details of the general limit theorems here. See Chapter 11 of STROOCK and VARADHAN [10] and KURTZ [7, 8] for various formulations of these results. We will illustrate the kinds of calculations that are involved in the application of these theorems (as we already have in (2.4)).

We first consider the processes U_n given by (4.9). For simplicity we take the noise process ξ to be a two state Markov chain (say with values ± 1) having generator

$$Bf(z) = f(-z) - f(z) . \tag{5.4}$$

Note that for this choice of ξ, $\nu(\{1\}) = \nu(\{-1\}) = 1/2$ and hence $G(x) = 1/2\,(F(x,1) + F(x,-1))$.

We consider the triple $(X(t), U_n(t), \xi(nt))$ which is a Markov process whose generator is

$$\begin{aligned} A_n f(x,u,z) = {} & n^{1/2}(F(x,z) - G(x)) \cdot \partial_u f(x,u,z) \\ & + n(f(x,u,-z) - f(x,u,z)) \\ & + G(x) \cdot \partial_x f(x,u,z) \end{aligned} \tag{5.5}$$

for bounded, continuously differentiable f (say with compact support to avoid technicalities). Let $f(x,u)$ be twice continuously differentiable with compact support and set

$$f_n(x,u,z) = f(x,u) + \frac{1}{2} n^{-1/2}(F(x,z) - G(x)) \cdot \partial_u f(x,u) . \tag{5.6}$$

We observe that

$$\lim_{n \to \infty} \sup_{x,u,z} |f_n(x,u,z) - f(x,u)| = 0 \tag{5.7}$$

and

$$\begin{aligned} A_n f_n(x,u,z) = {} & n^{1/2}(F(x,z) - G(x)) \cdot \partial_u f(x,u) \\ & + \frac{1}{2} n^{1/2}(F(x,-z) - F(x,z)) \cdot \partial_u f(x,u) \\ & + \frac{1}{2} \sum_{i,j} (F_i(x,z) - G_i(x))(F_j(x,z) - G_j(x))\,\partial_{u_i}\partial_{u_j} f(x,u) \end{aligned} \tag{5.8}$$

$$+ G(x) \cdot \partial_x f(x,u) + \frac{1}{2} n^{-1/2} G(x) \cdot \partial_x [(F(x,z) - G(x)) \cdot \partial_u f(x,u)] .$$

Note that $F(x,z) - G(x) = \frac{1}{2}(F(x,z) - F(x,-z))$ so the sum of the first two terms is zero and the third is

$$\begin{aligned} &\frac{1}{q} \sum_{i,j} (F_i(x,1) - F_i(x,-1))(F_j(x,1) - F_j(x,-1)) \partial_{u_i} \partial_{u_j} f(x,u) \\ &\equiv \frac{1}{2} \sum_{i,j} a_{ij}(x) \partial_{u_i} \partial_{u_j} f(x,u) . \end{aligned} \tag{5.9}$$

It follows that

$$\lim_{n \to \infty} \sup_{x,u,z} |A_n f_n(x,u,z) - Af(x,u)| = 0 \tag{5.10}$$

where

$$Af(x,u) = \frac{1}{2} \sum_{i,j} a_{ij}(x) \partial_{u_i} \partial_{u_j} f(x,u) + G(x) \cdot \partial_x f(x,u) . \tag{5.11}$$

Taking (5.7) and (5.10) together, we see that A is the limit (in a generalized sense) of the A_n. This convergence is sufficient to conclude $(X, U_n) \Rightarrow (X, U)$ where (X, U) is a Markov process with generator A (assuming $U_n(0) \to U(0)$). Theorem 4.2 of KURTZ [8] again applies here. Of course X is deterministic while U can be represented by

$$U(t) = U(0) + \int_0^t \frac{1}{2}(F(X(s),1) - F(X(s),-1))\, dW(s) \tag{5.12}$$

where W is a standard Brownian motion. For a survey of more general results of this type see PAPANICOLAOU [9].

One important observation to make on the above calculation is that we do not require the usual type of operator convergence. For our purposes we will say $\lim_{n \to \infty} A_n = A$ if for every $f \in \mathcal{D}(A)$ there are $f_n \in \mathcal{D}(A_n)$ such that $\lim_{n \to \infty} f_n = f$ and $\lim_{n \to \infty} A_n f_n = Af$. In the above example, this convergence was uniform (in x, y and z). This uniformity is not necessary in general. See KURTZ [7].

6. Spatially Dependent Reaction Models

In this section we consider a way of passing from spatially dependent reaction models to the spatially homogeneous models considered in Section 3. For simplicity we consider only the reaction $A + B \to C$ taking place in a solution. The solution is in a container Γ which we will identify with $[0,d_1] \times [0,d_2] \times [0,d_3] \subset \mathbb{R}^3$. We can describe the state of the system by the vector $(k, \ell, x_1, x_2 \cdots x_k, y_1, y_2 \cdots y_\ell)$, where k is the number of molecules of reactant A and $x_1, x_2 \cdots x_k$ are their locations, $x_i \in \Gamma$, and ℓ is the number of molecules of reactant B with locations $y_1, y_2 \cdots y_\ell$. To specify the model we must specify how the molecules move and how they react. We assume that the molecules undergo Brownian motion with normal reflection at the boundary with the molecules moving independently up to the moment of reaction, and that the probability of a reaction between a molecule of A at x and a molecule of B at y in a time interval of length Δt is $\rho(x-y)\Delta t + o(\Delta t)$ for some nonnegative continuous ρ. Note, we can assume the process takes values in $\bigcup_{0 \le k, \ell \le K} \Gamma^k \times \Gamma^\ell$ for some finite K. For $(x_1, x_2 \cdots x_k) \in \Gamma^k$ we define $\eta_i(x_1, x_2 \cdots x_k)$ to be the element of Γ^{k-1} obtained by dropping x_i from $(x_1, x_2 \cdots x_k)$. The generator of the process has the form

$$
\begin{aligned}
Af(k,\ell,x_1\cdots x_k,y_1\cdots y_\ell) &= \sum_{i=1}^{k} \alpha\Delta_{x_i} f(k,\ell,x_1\cdots x_k,y_1\cdots y_\ell) \\
&+ \sum_{j=1}^{\ell} \beta\Delta_{y_j} f(k,\ell,x_1\cdots x_k,y_1\cdots y_\ell) \\
&+ \sum_{i=1}^{k}\sum_{j=1}^{\ell} \rho(x_i-y_j)(f(k-1,\ell-1,\eta_i(x_1\cdots x_k),\eta_j(y_1\cdots y_\ell)) \\
&- f(k,\ell,x_1\cdots x_k,y_1\cdots y_\ell))
\end{aligned} \tag{6.1}
$$

where Δ_{x_i} denotes the Laplacian in three dimensions acting on f as a function of x_i. The assumption of normal reflection means that we require the normal derivative of $f(k,\ell,x_1\cdots x_k,y_1\cdots y_\ell)$ at the boundary of $\Gamma^k \times \Gamma^\ell$ to vanish. (We assume, f $\Delta_{x_i} f$ and $\Delta_{y_i} f$ are all continuous.) To pass from this spatially dependent model to the spatially homogeneous model, we will speed up the diffusion rate. In particular let A_n denote the operator defined as in (6.1) but with α and β replaced by αn and βn. As in Section 5 we define a sequence

$$
f_n(k,\ell,x_1\cdots x_k,y_1\cdots y_\ell) = f(k,\ell) + n^{-1}h(k,\ell,x_1\cdots x_k,y_1\cdots y_\ell) \tag{6.2}
$$

and observe that

$$
\lim_{n\to\infty} \sup_{k,\ell,x_i,y_i} |f(k,\ell) - f_n(k,\ell,x_1\cdots x_k,y_1\cdots y_\ell)| = 0 , \tag{6.3}
$$

and $A_n f$ converges uniformly to

$$
\begin{aligned}
&\sum_{i=1}^{k}\sum_{j=1}^{\ell} \rho(x_i-y_j)(f(k-1,\ell-1) - f(k,\ell)) \\
&+ \sum_{i=1}^{k} \alpha\Delta_{x_i} h(k,\ell,x_1\cdots x_k,y_1\cdots y_k) \\
&+ \sum_{j=1}^{\ell} \beta\Delta_{y_j} h(k,\ell,x_1\cdots x_k,y_1\cdots y_k) .
\end{aligned} \tag{6.4}
$$

We claim there exists a function $g(x,y)$ with normal derivatives vanishing on the boundary of $\Gamma \times \Gamma$ satisfying

$$
\alpha\Delta_x g(x,y) + \beta\Delta_y g(x,y) = (\bar{\rho} - \rho(x-y))(f(k-1,\ell-1) - f(k,\ell)) \tag{6.5}
$$

where

$$\bar{\rho} = m(\Gamma)^{-2} \int_\Gamma\int_\Gamma \rho(x-y)\, m(dx)\, m(dy) \tag{6.6}$$

(m denoting Lebesgue measure). Setting

$$h(k,\ell,x_1\cdots x_k, y_1\cdots y_\ell) = \sum_{i=1}^{k} \sum_{j=1}^{\ell} g(x_i,y_j) \tag{6.7}$$

(6.4) becomes

$$\bar{\rho}\, k\ell (f(k-1,\ell-1) - f(k,\ell)), \qquad 0 \le k,\ell \le K, \tag{6.8}$$

which defines a generator for a finite Markov chain.

Let $(N_n(t), M_n(t))$ denote the numbers of molecules of A and of B at time t in the process with generator A_n (note (N_n, M_n) is not Markovian). If $N_n(0) = k_0$ and $M_n(0) = \ell_0$ for all n, then $(N_n, M_n) \Longrightarrow (N, M)$ where (N, M) is a Markov process with generator given by (6.8).

The above result would hold for more general mixing processes (besides Brownian diffusion). An abstract version of the result outlined here can be found in KURTZ [6].

7. Convergence of Stationary Distributions

Let $\{X_n\}$ be a sequence of Markov processes with values in $\mathbb{R}^d$ and set

$$T_n(t) f(x) = E[f(X_n(t)) \mid X_n(0) = x] \quad . \tag{7.1}$$

The techniques discussed above give conditions under which $X_n(0) = x_n \to x$ implies $X_n \Longrightarrow X$ where X is some Markov process with $X(0) = x$. In particular this implies

$$\lim_{n\to\infty} T_n(t) f(x_n) = T(t) f(x) \tag{7.2}$$

for all bounded continuous f (at least for all but countably many values of t) where

$$T(t) f(x) = E[f(X(t)) \mid X(0) = x] \quad . \tag{7.3}$$

With this observation in mind we note that if

$$\lim_{n\to\infty} T_n(t) f(x_n) = T(t) f(x) \tag{7.4}$$

for all x_n, x satisfying $\lim_{n\to\infty} x_n = x$, then

$$\lim_{n\to\infty} \sup_{x\in K} | T_n(t) f(x) - T(t) f(x)| = 0 \tag{7.5}$$

for all compact K.

Next note that ν_n is a stationary distribution for X_n if and only if

$$\int T_n(t) f \, d\nu_n = \int f \, d\nu_n \tag{7.6}$$

for all bounded continuous functions f. A sequence of probability measures $\{\nu_n\}$ on $\mathbb{R}^d$ is relatively compact in the weak topology (see BILLINGSLEY [1]) if and only if for every $\varepsilon > 0$ there exists a compact set K_ε such that

$$\inf_n \nu_n(K_\varepsilon) \geq 1 - \varepsilon , \tag{7.7}$$

and $\nu_n \Longrightarrow \nu$ (ν_n converges weakly to ν) if and only if

$$\lim_{n \to \infty} \int f \, d\nu_n = \int f \, d\nu \tag{7.8}$$

for every bounded continuous f.

7.1 Theorem Suppose $T(t)f$ is bounded and continuous whenever f is, and $\{T_n(t)\}$ and $T(t)$ satisfy (7.5) for all bounded continuous f, $t \geq 0$ and K compact. If ν_n is a stationary distribution for X_n (i.e. ν_n satisfies (7.6)) and $\nu_n \Longrightarrow \nu$, then ν is a stationary distribution for X.

Proof By (7.5) and (7.7) we have

$$\lim_{n \to \infty} \int (T_n(t) f(x) - T(t) f(x)) \, \nu_n(dx) = 0 \tag{7.9}$$

and hence for all bounded continuous f and $t \geq 0$

$$\begin{aligned} \int f \, d\nu &= \lim_{n \to \infty} \int f \, d\nu_n = \lim_{n \to \infty} \int T_n(t) f \, d\nu_n \\ &= \lim_{n \to \infty} \int T(t) f \, d\nu_n = \int T(t) f \, d\nu \end{aligned} \tag{7.10}$$

which implies ν is a stationary distribution for X.

References

1. Billingsley, Patrick. *Confergence of Probability Measures*. Wiley, New York, 1968.
2. Billingsley, Patrick. *Probability and Measure*. Wiley, New York, 1979.
3. Breiman, Leo. *Probability*. Addison-Wesley, Reading, Massachusetts, 1968.
4. Chung, Kai Lai. *Markov Chains with Stationary Transition Probabilities*, 2 ed., Grundlehren der mathematischen Wissenschaften, Vol. 104, Springer, Berlin, Heidelberg, New York, 1960.
5. Dynkin, E.B. *Markov Processes*, Grundlehren der mathematischen Wissenschaften, Vols. 121-122, Springer, Berlin, Heidelberg, New York, 1965; Vol. I.
6. Kurtz, Thomas G. A limit theorem for perturbed operator semigroups with applications to random evolutions. *J. Functional Anal. 12* (1973) 55-67.
7. Kurtz, Thomas G. Semigroups of conditioned shifts and approximation of Markov processes. *Ann. Probability 3* (1975) 618-642.
8. Kurtz, Thomas G. *Approximation of Population Processes*. SIAM, Philadelphia, 1981.
9. Papanicolaou, George C. Asymptotic analysis of transport processes. *Bull. Amer. Math. Soc. 81* (1975) 330-392.
10. Stroock, Daniel W. and S.R.S. Varadhan. *Multidimensional Diffusion Processes*, Grundlehren der mathematischen Wissenschaften, Vol. 233, Springer, Berlin, Heidelberg, New York, 1979.
11. Trotter, H.F. Approximation of semigroups of operators. *Pac. J. Math. 8* (1958) 887-919.

On the Asymptotic Behavior of Motions in Random Flows

G. Papanicolaou* and O. Pironeau
Courant Institute and University of Paris, Villetaneuse

Abstract

Given a random vector field we consider the trajectories of a moving particle through it and analyze various asymptotic limits.

1. Introduction and formulation

Let $v(x)$, $x \in \mathbb{R}^3$ be a random velocity field that is given with mean zero, is strictly stationary and has moments of all orders; it could for example be a gaussian random field. We want to consider motion of a particle in this random field which means to construct the trajectories $x(t;x)$ of the stochastic differential equation

$$\frac{dx(t)}{dt} = v(x(t)) , \qquad x(0) = x , \qquad x \in \mathbb{R}^3. \tag{1.1}$$

Equivalently we may consider the stochastic partial differential equation

$$\frac{\partial u(t,x)}{\partial t} - v(x)\cdot\nabla u(t,x) = 0 ,$$

$$u(0,x) = f(x) , \tag{1.2}$$

where ∇ is the gradient operator and $f(x)$ is a given function. Clearly (1.1) is the characteristic system for (1.2) and

$$u(t,x) = f(x(t;x)) . \tag{1.3}$$

In [1] the diffusion limit for (1.1) or (1.2) was analyzed when the random velocity field has a constant (or slowly varying) drift component $v_0 \neq 0$ and the random part is small. That is, when v in (1.1) is replaced by $v_0 + \varepsilon v(x)$ with $\varepsilon > 0$ a small parameter. It was proved there that the process defined by

$$x^\varepsilon(t) = x\left(\frac{t}{\varepsilon^2}\right) - \frac{v_0 t}{\varepsilon^2} \tag{1.4}$$

and satisfying

$$\frac{dx^\varepsilon(t)}{dt} = \frac{1}{\varepsilon}\, v\left(x^\varepsilon(t) + \frac{v_0 t}{\varepsilon^2}\right) , \qquad x^\varepsilon(0) = x ,$$

*Research supported by the Army Research Office under Grant No. DAAG29-78-G-0177.

converges weakly as $\varepsilon \to 0$ (as a measure on the space of continuous functions) to a diffusion Markov process which is essentially Brownian motion. Naturally it is necessary to assume that the vector field $v(x)$ has some strong mixing properties spelled out in [1].

It is important to note that the result of [1] is false if $v_0 = 0$. On the other hand, for the stochastic acceleration problem

$$\begin{aligned} \frac{dx^\varepsilon(t)}{dt} &= \frac{1}{\varepsilon^2} y^\varepsilon(t) , & x^\varepsilon(0) &= x \\ dy^\varepsilon(t) &= \frac{1}{\varepsilon} v(x^\varepsilon(t)) , & y^\varepsilon(0) &= y_0 \neq 0 , \end{aligned} \tag{1.6}$$

a diffusion limit for $y^\varepsilon(t)$ is expected on the grounds that in small time intervals (1.6) behaves like (1.5) with a v_0 that is not fixed but changes slowly and can get small but never zero. This is carried out in [2] and the actual details are much more involved than those of [1] so (1.6) will not be discussed further here.

In view of this the behavior of (1.5) for v_0 small or zero is bound to be very different from the case $v_0 \neq 0$. Returning to the P.D.E. formulation (1.2) we consider the case where external (molecular) diffusion is present and write

$$\begin{aligned} \frac{\partial u(t,x)}{\partial t} + v(x)\cdot\nabla u(t,x) &= \alpha\Delta u(t,x) , \qquad t > 0 \\ u(0,x) &= f(x) \end{aligned} \tag{1.7}$$

where $\alpha > 0$ is a constant (the molecular diffusivity). We restrict attention to flows in $\mathbb{R}^3$ (we comment on this below) that are incompressible

$$\nabla\cdot v(x) = 0 , \tag{1.8}$$

have mean zero

$$E\{v(x)\} = 0 \tag{1.9}$$

and are strictly stationary.

We are interested in the behavior of $u(t,x)$ as t tends to infinity. To obtain sensible results it is necessary to differentiate between length scales in (1.7), in addition to scaling the time. We shall assume that the length scale associated with the data f is large compared to that of the random field v. That is, the random velocity field is rapidly varying relative to the data. Replacing x by x/ε and t by t/ε^2 we see that the scaled problem to analyze as $\varepsilon \to 0$ is

$$\frac{\partial u^\varepsilon(t,x)}{\partial t} + \frac{1}{\varepsilon} v\left(\frac{x}{\varepsilon}\right)\cdot\nabla u^\varepsilon(t,x) = \alpha\Delta u^\varepsilon(t,x) , \qquad t > 0 \tag{1.10}$$

$$u^\varepsilon(0,x) = f(x)$$

with (1.8) and (1.9) holding.

The main result of this paper is this.

Theorem. If $f(x)$ is a smooth vector valued function on $\mathbb{R}^3$, $\alpha > 0$ and $v(x)$ is a strictly stationary random field satisfying (1.8) and (1.9) having moments of up to second order (along with ∇v) then there exists a positive constant matrix (γ_{ij}), called the turbulent diffusivity, such that if $\bar{u}(t,x)$ solves

$$\frac{\partial \bar{u}(t,x)}{\partial t} = \alpha \sum_{i,j} (\delta_{ij} + \gamma_{ij}) \frac{\partial^2 \bar{u}(t,x)}{\partial x_i \partial x_j} \tag{1.11}$$

$$\bar{u}(0,x) = f(x)$$

then

$$E \left\{ \int_{\mathbb{R}^3} |u^\varepsilon(t,x) - \bar{u}(t,x)|^2 dx \right\} \to 0 \quad \text{as} \quad \varepsilon \to 0. \tag{1.12}$$

This theorem is proved in the next section. We first make a few remarks about the proof and then about the meaning of the result. In section three we comment about some related problems.

The proof is based on the argument given in detail in [3] and the role of the dimension $d \geq 3$ of the space seems to be important; condition (1.8) is also important.

Our original intention in studying problems such as (1.10) was in obtaining some understanding of the Reynolds stress [4] in the theory of macroscopic flow interacting with homogeneous turbulence. The problem there is to account for the appearance of a coefficient very similar to γ in the theorem but in a context where randomness enters the equations (Navier-Stokes equations) in a more self-consistent fashion. We shall not make this statement precise here; it will be considered in a forthcoming paper. Other linear or nonlinear stochastic PDE amenable to our method are discussed in section 3.

There is a simple probabilistic interpretation of our theorem. Let $x(t)$ be the diffusion process solving the stochastic differential equation

$$dx(t) = v(x(t))dt + dw(t) , \qquad x(0) = x \tag{1.13}$$

where $w(t)$ is standard 3-dimensional Brownian motion and $v(x)$ is a random field, independent of $w(t)$, with the properties (1.8), (1.9) and stationary. The process $x(t)$ is a diffusion process in a random environment. The theorem says that $(x(t)-x)t^{-1/2}$ converges weakly as a diffusion and in mean square with respect to the random field v to a gaussian random vector with mean zero and covariance equal to $1 + \gamma$.

2. Proof of the theorem

It is a bit more convenient to work with the equivalent problem

$$-\alpha\Delta u^\varepsilon(x,\omega) + \frac{1}{\varepsilon} v(\frac{x}{\varepsilon},\omega)\cdot\nabla u^\varepsilon(x,\omega) + \lambda u^\varepsilon(x,\omega) = f(x) , \tag{2.1}$$

for $x \in R^3$ with $\lambda > 0$ fixed. Here ω belongs to Ω with $(\Omega,\mathcal{F},\mathcal{P})$ the probability space on which the random velocity field v is defined.

The first step in the analysis of (2.1) is to look for u^{ε} in the form $u_0(x) + \varepsilon u_1(x,x/\varepsilon,\omega) + \varepsilon^2 u_2 + \ldots$, where $u_j(x,y,\omega)$, $j \geq 1$, is for each x a stationary random field in y. This is the usual multiscaling analysis extended naturally to problems with random coefficients [cf. [5] for the case of periodic coefficients]. In many situations (for example in [1] and [2]) this will not work because u_1, u_2 etc. do not exist as stationary processes and in fact the implied ordering in $u_0 + \varepsilon u_1(x,\frac{x}{\varepsilon},\omega) + \varepsilon^2 u_2(x,\frac{x}{\varepsilon},\omega)$ is false because εu_1 is not small for ε small. It turns out however that for (2.1) and for the problem in [3] the expansion is sensible even though a stationary u_1 does not exist. This is because $\varepsilon u_1(x,\frac{x}{\varepsilon},\omega)$ is small as will now be made precise.

Inserting the expansion in (2.1), noting that the gradient operation becomes $\nabla_x + \varepsilon^{-1}\nabla_y$, where $y = x/\varepsilon$ is the fast variable, and collecting coefficients of powers of ε yields the following equations for $u_1, u_2, \ldots$

$$-\alpha\Delta_y u_1 + v(y)\cdot\nabla_y u_1 + v(y)\cdot\nabla_x u_0(x) = 0 , \tag{2.2}$$

$$-\alpha\Delta_y u_2 + v(y)\cdot\nabla_y u_2 + v(y)\cdot\nabla_x u_1 - 2\alpha\nabla_x\cdot\nabla_y u_1$$

$$-\alpha\Delta u_0 + \lambda u_0 = f , \ldots \tag{2.3}$$

We write u_1 in the form

$$u_1(x,y,\omega) = \sum_{j=1}^{3} \chi^j(y,\omega) \frac{\partial u_0(x)}{\partial x_j} . \tag{2.4}$$

Then χ^j must satisfy

$$-\alpha\Delta_y \chi^j(y) + v(y)\cdot\nabla_y \chi^j(y) + v^j(y) = 0 , \qquad j=1,2,3 . \tag{2.5}$$

Before continuing with (2.3) it is necessary to know what kind of solution (2.5) has. By adapting the method of [3] and using (i) the fact that we are in 3 (or more) dimensions and (ii) that $\nabla\cdot v = 0$ we can show that:

Eq. (2.5) has a unique generalized solution. $\chi^j(y,\omega)$, j=1,2,3, which is continuous (almost surely) $\chi^j(0,\omega) = 0$ but is not stationary. However $\nabla\chi^j(y,\omega)$ exists, has finite second moments and is stationary. (2.6)

Moreover

$$\frac{1}{|x|} \chi^j(x,\omega) \to 0 \quad \text{as} \quad |x| \to \infty \tag{2.7}$$

in mean square.

We note that condition (2.7) is just what is necessary in order that $\varepsilon u_1(x,x/\varepsilon,\omega)$ be $o(1)$ in mean square as $\varepsilon \to 0$. In the periodic case, of course, (2.5) has a periodic solution $\chi^j(y)$ so (2.7) is trivial and $\varepsilon u_1(x,x/\varepsilon,\omega)$ is $O(\varepsilon)$.

We use (2.4) now in (2.3) and try to solve for u_2 as a (random) function of y. Requiring that the rest of the terms have mean zero gives an operation for $u_0(x)$.

$$-\alpha \sum_{i,j=1}^{3} (\delta_{ij}+\gamma_{ij}) \frac{\partial^2 u_0(x)}{\partial x_i \partial x_j} + \lambda u_0(x) = f(x) . \tag{2.8}$$

where

$$\gamma_{ij} = E\{\nabla\chi^i(y)\cdot\nabla\chi^j(y)\} , \qquad i,j=1,2,3. \tag{2.9}$$

Note that eddy viscosity coefficients (v_{ij}) form a symmetric positive definite matrix.

Having constructed a formal expansion it is easy in the present case to complete the analysis by using this expansion. Set

$$z^\varepsilon = u^\varepsilon - u_0 - \varepsilon u_1 . \tag{2.10}$$

From (2.1), (2.4) and (2.8) one obtains an equation for z^ε, the error equation. Now the usual energy estimates, and one trick detailed in [2], explaining how to handle terms that are not at first glance small but have mean zero, give the statement (1.12) or its analog in the steady case (2.1).

3. Related problems

One can handle also the nonlinear problem

$$\frac{\partial u^\varepsilon}{\partial t} + (u^\varepsilon + \frac{1}{\varepsilon}v(\frac{x}{\varepsilon}))\cdot\nabla u^\varepsilon = -\nabla p^\varepsilon + \alpha\Delta u^\varepsilon , \qquad \nabla\cdot u^\varepsilon = 0, \tag{3.1}$$

$$u^\varepsilon(0,x) = f(x)$$

with v exactly as described in section one. The limit problem here is

$$\frac{\partial u}{\partial t} + u\cdot\nabla u = -\nabla p + \alpha \sum_{i,j=1}^{3} (\delta_{ij}+\gamma_{ij}) \frac{\partial u}{\partial x_i \partial x_j} , \qquad \nabla\cdot u = 0 \tag{3.2}$$

$$u(0,x) = f(x)$$

with the (γ_{ij}) determined as in section 2. Of course, part of the hypotheses here is that u of (3.2) has a nice solution.

It is perhaps clear now that from a model problem like (3.1) one can actually approach the Reynolds equation and its generalizations. This is not quite so, however, because there are a number of important obstacles that must be overcome.

References

1. H. Kesten and G. Papanicolaou, A limit theorem for turbulent diffusion, Comm. Math. Phys. 65 (1979), pp. 97-128.

2. H. Kesten and G. Papanicolaou, stochastic acceleration, Comm. Math. Phys., to appear.

3. G. Papanicolaou and S. Varadhan, Boundary value problems with rapidly oscillating coefficients, Proceedings of Conference on Random Fields, Esztergom 1979, published in Seria Colloquia Mathematica Societatis, János Bolyai, by North Holland.

4. A. Monin and A. Yaglom, Statistical Fluid Mechanics, Vol. I, MIT Press, 1971.

5. A. Bensoussan, J. L. Lions and G. Papanicolaou, Asymptotic analysis for periodic structures, North Holland, 1978.

Part III

Description of Internal Fluctuations

Some Aspects of Fluctuation Theory in Nonequilibrium Systems

G. Nicolis

Faculté des Sciences de l'Université Libre de Bruxelles, Campus Plaine, C.P. 226
B-1050 Bruxelles, Belgium

1. Introduction

Professor HAKEN has described the kind of transition phenomena that may take place when a nonlinear system is driven beyond a critical distance from thermodynamic equilibrium. From his discussion it was clear that, in the vicinity of transition points, the phenomenological analysis, based essentially on *bifurcation theory*, had to be supplemented with additional information pertaining to the behavior of the fluctuations. In this talk I would like to review the status of fluctuation theory for the particular, yet quite representative class of systems involving chemical reactions and diffusion. It is well known that the dynamics of these systems is amenable to a set of phenomenological laws of evolution for a limited number of macroscopic observables, typically the (appropriately scaled) composition variables $\bar{x}_i$ of the chemical species. Assuming Fickian diffusion in an ideal mixture and constant temperature throughout, we may write the evolution equations of these variables in the form

$$\frac{\partial \bar{x}_i}{\partial t} = v_i(\bar{x}_1, \ldots, \bar{x}_n; \lambda_1, \ldots, \lambda_m) + D_i \nabla^2 \bar{x}_i \quad (i = 1, \ldots, n) \tag{1.1}$$

D_i are diffusion coefficients, v_i the overall rate of change of $\bar{x}_i$ arising from the chemical reactions involving constituent i, and $\lambda_1, \ldots, \lambda_m$ stand for a set of control parameters.

The most common bifurcations associated with (1.1) are those leading (i) to multiple steady states without any change in spatial and temporal symmetries, (ii) to a time symmetry-breaking associated with the emergence of limit cycles, and (iii) to a space symmetry-breaking associated with pattern formation [1].

We would like now to incorporate fluctuations in the description of these transition phenomena. What are the options that we have in order to realize this program? As we saw in Professor PRIGOGINE's talk, stochastic behavior is an intrinsic property of large classes of physical systems. Fluctuation theory is thus an integral part of nonequilibrium statistical mechanics. The most natural option would therefore be to set up a description based on the kinetic equations for the distribution functions of the different degrees of freedom. Unfortunately, the complexity of this approach is such that this option has to be abandoned. Even near equilibrium, the statistical mechanics of reacting systems is far from being well understood. We have therefore to resort to other approaches, which will inevitably involve a phenomenological element.

An appealing option, which was adopted by several authors in the last years, is based on the analogy between nonequilibrium bifurcations and equilibrium critical phenomena. The reaction-diffusion dynamics is viewed as a Markovian process with *continuous realizations*. By adding white noise terms to (1.1) one can carry through much of the static and dynamic theory of critical phenomena [2,3]. This program was exhaustively carried out for the three types of bifurcation mentioned above in a series of papers by BORCKMANS, DEWEL, and WALGRAEF [4-6]. For a presentation of the ideas of these authors we refer to their contribution in this volume.

An alternative option is, however, open, namely, utilize at least at the beginning the maximum amount of specific information pertaining to the system. For a reaction-diffusion system we thus have to appeal explicitly to the mechanism of chemical reactions and diffusion. In particular, we view chemical reactions as *birth and death processes* corresponding to the appearance or disappearance of a small number of molecules at a time, and diffusion as a *random walk* between a volume element and its first neighbors. This is the description we shall adopt throughout this communication.

Let $\{X_{i\underline{r}}\}$ denote the number of particles of species i in a small volume ΔV centered at $\underline{r}$, where the partition in subvolumes ΔV covers the entire reaction space. Assuming, in agreement with the premises of the local description of irreversible processes, that the set $\{X_{i\underline{r}}\}$ defines a discrete Markov process, we can write the following equation for the multivariate probability function $P(\{X_{i\underline{r}}\},t)$

$$\begin{aligned}\frac{dP(\{X_{i\underline{r}}\},t)}{dt} &= \sum_{\underline{r}} L_{ch}(\underline{r})P + \sum_{\underline{r},\underline{\ell}} L_d(\underline{r},\underline{\ell})P \\ &= \sum_{r,\{X'_{i\underline{r}}\}} W(\{X'_{i\underline{r}}\}|\{X_{i\underline{r}}\})P(\{X'_{i\underline{r}}\},t) \\ &\quad + \sum_i \frac{\mathcal{D}_i}{2d} \sum_{\underline{r},\underline{\ell}} \Big\{(X_{i\underline{r}}+1)P(\ldots, X_{i\underline{r}}+1, \\ &\quad X_{i\underline{r}+\underline{\ell}}-1, \ldots, t) - X_{i\underline{r}}P(\{X_{i\underline{r}}\},t)\Big\} \ . \end{aligned} \qquad (1.2)$$

The chemical evolution operator $L_{ch}(\underline{r})$ is a local operator defined in each small space cell ΔV within our macroscopic system. It displays the transition rate W characteristic of the reaction kinetics. The diffusion operator $L_d(\underline{r},\underline{\ell})$ depends on all $\underline{r}$ as well as on their first neighbors $\underline{\ell}$. It is proportional to the jump frequencies $\mathcal{D}_i$ between adjacent cells and depends on the dimensionality d. An essential aspect of (1.2), which was already pointed out in the communication of Professor

KURTZ, is the *extensivity* of transition probabilities, expressed by the relation

$$W(\{X'_{i\underline{r}}\}|\{X_{i\underline{r}}\}) = \Delta V\, w\left[\left\{\frac{X_{i\underline{r}}}{\Delta V}\right\},\left\{\frac{X'_{i\underline{r}}}{\Delta V}\right\},\frac{1}{\Delta V}\right] . \tag{1.3}$$

In the sequel, we will be interested in the solutions of (1.2) in an infinite system, or a system subject to periodic boundary conditions.

What is the principal motivation for adopting this more *mechanistic description* rather than the critical dynamics approach? First, I believe that it is important to justify the premises and to understand the possible limitations of this latter formalism. In the multivariate master equation, (1.2), the partition in space cells must satisfy two requirements. In each cell ΔV one must have coherent fluctuations, and the jumps between cells must be assimilated to a random walk. This means that the linear dimension of ΔV should be, typically, of the order of the mean free path. As a result, (1.2) provides a very detailed description of the system. The question, how much of this information may turn out to be irrelevant near a bifurcation point, is a basic problem which is not clear even for equilibrium phase transitions. It is therefore desirable to have a formalism capable of handling this type of question.

A second motivation for adopting (1.2) is the possibility to incorporate properly and unambiguously the constraints imposed on our system by thermodynamics. This is achieved by the modelling of the transition rates W in terms of the frequencies of reactive collisions between molecules. Such information contains, as a byproduct, the properties of the random noise terms that are to be added to (1.1) in the critical dynamics approach. In the cases of interest, where contrary to what happens at equilibrium microscopic reversibility does not hold, it is important to have a way of controlling the assumptions made on these terms.

Finally, the description based on (1.2) enables us to understand the relative roles of reaction kinetics and diffusion in the properties of the fluctuations, without having to solve the equation and to construct an explicit representation of the probability function.

From the standpoint of probability theory, the use of (1.2) can also be of considerable interest. For instance, we can have some new insights on the status of a Markovian process with continuous realizations described by a Fokker-Planck type equation, and a jump process underlying the modelling adopted in (1.2). In short, we can see how the results described in the communication by Professor KURTZ can be extended to spatially distributed systems. Moreover, we are able to detect the mechanisms responsible for the gradual change in the character of the probability functions as the system approaches and then crosses a bifurcation point. For instance (cf. Sect.3) the probability function may switch from a (multi-) Gaussian to a form which does not belong to the class of stable distributions.

2. Some General Results

The solution of (1.2) confronts us with a formidable task. Even in the simplest case of one species we have a multivariate problem, because of the division in space cells; in each cell the number of states is countably infinite; the coefficients of the equation are generally nonlinear, because of the chemical kinetics; and there is no obvious perturbation parameter to be used. Indeed, the distance from bifurcation point appears in a very implicit manner. In addition, as we mentioned in the Introduction, the inverse of the volume element ΔV appearing in the extensivity condition, (1.3), cannot be used at the outset in a perturbation expansion: the only condition imposed is spatial coherence within each cell, and this in general will be violated if ΔV is taken to be very large.

For these reasons, it will be useful to review some general properties of the master equation which are model independent and whose validity does not require any uncontrollable approximation schemes.

2.1. Non-Poissonian Behavior From Chemical Kinetics

A first major result is that in a nonequilibrium open system, nonlinear chemical kinetics generates deviations from the Poissonian distribution [7]. This is somewhat unexpected, since we are dealing with ideal mixtures. It turns out that the deviation starts immediately with nonequilibrium (even in the linear range), and is proportional to the mass flux across the reaction chain.

2.2. Locally Poissonian Behavior From Diffusion

In contrast to this, one can easily check that the diffusion part of (1.2) tends to establish a locally Poissonian distribution. We are therefore witnessing an interesting competition between chemical kinetics and diffusion [8].

2.3. Origin of Spatial Correlations

Whenever deviations from the Poissonian are generated, the system can build up and sustain a spatial correlation function. The correlation length turns out to be related to the characteristic length over which a particle can diffuse before being transformed by chemical reactions [9].

2.4. Connection With Phenomenological Description

The deviations from the Poissonian are also responsible for differences between the structure of the phenomenological equations of evolution (cf. (1.1)) and the equations for the first moments $\langle X_{\underline{r}i} \rangle / \Delta V$ of the probability distribution. As it turns out, however [10], these deviations are weighted by the inverse of ΔV.

It follows that fluctuations of different scales will have different effects on the macroscopic evolution. Numerical simulations, based on molecular dynamics studies, confirm this conclusion [11].

3. Asymptotic Solution of the Multivariate Master Equation – Critical Behavior

We would now like to have more detailed information on the solutions of the multivariate master equation (1.2) by means of perturbation theory. As pointed out in section 2, ΔV cannot be taken as a large quantity *unless* spatial coherence is secured within each cell. Now, this latter requirement is compatible with a large ΔV in the following two instances. First, in the vicinity of a bifurcation point a system is expected to generate long range correlations if some additional conditions on the type of nonlinearity and the dimensionality are satisfied. One could thus envisage a partition into large cells, as long as their linear dimension does not exceed the correlation length. And second, well before bifurcation, when there exists a single asymptotically stable state, we expect the dynamics to become effectively linear. Therefore, the details of the partition into cells should be immaterial.

We now describe the consequences of considering the inverse cell size

$$\varepsilon = \Delta V^{-1} = \ell^{-d} \tag{3.1}$$

where ℓ is the linear dimension of the cells, as a perturbation parameter [12,13]. We write (1.2) in the generating function representation after extracting from P the macroscopic motion:

$$\psi = \prod_{\underline{r}} s_{\underline{r}}^{-\bar{X}_{\underline{r}}} F \equiv \sum_{\{X_{\underline{r}}\}} \prod_{\underline{r}} s_{\underline{r}}^{X_{\underline{r}}-\bar{X}_{\underline{r}}} P \tag{3.2}$$

where $\bar{X}_{\underline{r}}$ is the extensive variable associated to $\bar{x}_{\underline{r}}$ [cf. (1.1)]. The resulting equation for F or ψ becomes a partial differential equation wherein high order derivatives $\partial^N\psi/\partial s_1^{k_1} \dots \partial s_m^{k_m}$ (with $N = k_1 + \dots + k_m + \dots$) are multiplied by the increasingly smaller quantities ε^{-N}. This defines a *singular perturbation problem* associated with the formation of a *boundary layer* near $\{\delta_{\underline{r}} = 1\}$. Indeed, we expect F to be concentrated near this point as the maximum of the corresponding probability function $P(\{X_{\underline{r}}\},t)$ will drift to higher and higher values of $\{X_{\underline{r}}\}$ in the limit of $\varepsilon \to 0$. It is natural therefore to define scaled variables $\{\xi_{\underline{r}}\}$ through

$$\delta_{\underline{r}} = 1 + \varepsilon^a \xi_{\underline{r}} \tag{3.3}$$

and compute the scaling exponent a in such a way, that the equation for $\psi(\{1 + \varepsilon^a\xi_{\underline{r}}\},t) \equiv \psi_a(\{\xi_{\underline{r}}\},t)$ reduces to a regular perturbation problem. We do not go into the details of the algebra here, but simply summarize the main results for systems involving one variable and undergoing bifurcations leading to multiple steady states.

3.1. *Reduction to a Fokker-Planck Equation*

In the pretransitional region and up to the bifurcation point (1.1) yields the following equation for the probability $P(\{x_{\underline{r}}\},t)$ of the intensive variables $x_{\underline{r}} = X_{\underline{r}}/\Delta V$ in the limit of small ε

$$\frac{\partial P(\{x_{\underline{r}}\},t)}{\partial t} = -\sum_{\underline{r}} \frac{\partial^2}{\partial x_{\underline{r}}} \left\{ v(x_{\underline{r}}) + \frac{D}{2d} \sum_{\underline{\ell}} (x_{\underline{r}+\underline{\ell}} - x_{\underline{r}}) \right\} P$$

$$+ \frac{\varepsilon}{2} \sum_{\underline{r},\underline{r}'} \frac{\partial^2}{\partial x_{\underline{r}} \partial x_{\underline{r}'}} \left\{ Q(\bar{x}_{\underline{r}}) \delta^{kr}_{\underline{r},\underline{r}'} + \frac{\mathcal{D}}{2d} \sum_{\underline{\ell}} \left[(\bar{x}_{\underline{r}+\underline{\ell}} - \bar{x}_{\underline{r}}) \delta^{kr}_{\underline{r},\underline{r}'} \right.\right.$$

$$\left.\left. - (\delta^{kr}_{\underline{r}+\underline{\ell},\underline{r}'} - \delta^{kr}_{\underline{r},\underline{r}'})(\bar{x}_{\underline{r}} + \bar{x}_{\underline{r}'}) \right] \right\} P \quad . \tag{3.6}$$

This is a discretized form of a multi-variate Fokker-Planck equation with nonlinear drift and with constant (i.e., process-independent) diffusion. It is equivalent to the following Langevin equation

$$\frac{\partial x_{\underline{r}}}{\partial t} = v(x_{\underline{r}}) + \frac{\mathcal{D}}{2d} \sum_{\underline{\ell}} (x_{\underline{r}+\underline{\ell}} - x_{\underline{r}}) + \varepsilon^{1/2} F_{\underline{r}}(t) \tag{3.7}$$

where $F_{\underline{r}}(t)$ is a multi-Gaussian white noise defined by

$$\langle F_{\underline{r}}(t) F_{\underline{r}'}(t') \rangle = \left\{ Q(\bar{x}_{\underline{r}}) \delta^{kr}_{\underline{r},\underline{r}'} + \frac{\mathcal{D}}{2d} \sum_{\underline{\ell}} \left[(\bar{x}_{\underline{r}+\underline{\ell}} - \bar{x}_{\underline{r}}) \delta^{kr}_{\underline{r},\underline{r}'} \right.\right.$$

$$\left.\left. - (\delta^{kr}_{\underline{r}+\underline{\ell},\underline{r}'} - \delta^{kr}_{\underline{r},\underline{r}'})(\bar{x}_{\underline{r}} + \bar{x}_{\underline{r}'}) \right] \right\} \times \delta(t - t') \quad . \tag{3.8}$$

We recognize here the discretized version of the relations pointed out some time ago by GARDINER and GROSSMAN [14,15]. The point is that we have *deduced* these relations starting from the master equation, in which the notion of random force had not been built at the outset.

The reduction of (1.2) descriptive of a jump process, to (3.6) descriptive of a diffusion process, fails in the region of multiple steady states. As a matter of fact, one can show on simple models that the master equation and the Fokker-Planck equation give rise in this case to different stationary probability distributions. On the other hand, LEMARCHAND [16] was recently able to construct a partial differential equation for the logarithm of the probability distribution of the intensive variables χ, whose steady-state solution is identical to that of the master equation. This equation is different from the Fokker-Planck equation, but reduces to it if the additional assumption that $\partial P/\partial \chi$ is uniformly small is adopted. Whether this means that in the multiple steady state region the master equation should be equivalent to a more complex stochastic process with continuous realizations is not clear at the present time.

3.2. *Critical Behavior*

Near the bifurcation point the noise contribution arising from diffusion (last terms in (3.7,8) becomes negligible. The Fokker-Planck equation can then be solved exactly at the steady state. The result reads

$$P_{St}(\{x_{\underline{r}}\}) \sim \exp\left\{-\frac{2\varepsilon^{-1}}{Q(\bar{x}_{st})}\sum_{\underline{r}}\left[-\int\{dx_{\underline{r}}\}v(\{x_{\underline{r}}\}) + \frac{\mathcal{D}}{8d}\sum_{\underline{\ell}}(x_{\underline{r}+\underline{\ell}} - x_{\underline{r}})^2\right]\right\} \quad . \tag{3.9}$$

This equation establishes the link between the master equation approach and the theory of equilibrium critical phenomena. Indeed, it features the exponential of the *Landau-Ginzburg functional* and can therefore be studied by renormalization group methods [2]. As it is well known, the result of this analysis is a critical dimensionality d_c and the concomitant appearance of non-classical exponents describing the divergence of variances, correlation functions, and so forth for $d < d_c$.

The meaning of a critical dimensionality in the context of reaction-diffusion systems is quite clear, and can be understood by direct inspection of the master equation (1.2), or the associated Fokker-Planck equation (3.6). Near a bifurcation point the reaction kinetics is slowed down, because the derivative of the rate function $\partial v(x_{\underline{r}})/\partial x_{\underline{r}}$ tends to zero. On the other hand, in a large system (ΔV large or ε small) the chracteristic time of diffusion is also very long even if the diffusion coefficient is large. The critical dimensionally corresponds to the situation where there is matching between these two time scales. For $d > d_c$ diffusion is very effective in homogeneizing the system quasi-instantaneously. Hence, mean-field theory should be applicable. For $d < d_c$ however, spatially inhomogeneous fluctuations subsist and induce non-classical behavior.

For a system involving cubic nonlinearities, like the Schlögl model [17], the above arguments give $d_c = 4$. Equation (3.9) reduces to the structure characteristic of the Ising model

$$P_{St}(\{x_{\underline{r}}\}) \sim \exp\left\{-\frac{2\varepsilon^{-1}}{Q(\bar{x}_{st})}\sum_{\underline{r}}\left[\frac{\lambda-\lambda_c}{2}(x_{\underline{r}} - \bar{x}_{St})^2 + \frac{\mu}{4}(x_{\underline{r}} - \bar{x}_{st})^4 + \frac{\mathcal{D}}{8d}\sum_{\underline{\ell}}(x_{\underline{r}+\underline{\ell}} - x_{\underline{r}})^2\right]\right\} \tag{3.10}$$

where $\lambda - \lambda_c$ is the distance from the bifurcation point. As we anticipated in the Introduction, this form does *not* belong to the class of stable distributions (this remains true even in the mean-field limit). Moreover, the underlying stochastic process is not a process with independent increments. These dramatic changes reflect the *coherence* associated with bifurcation.

The above results can be extended to the case of degenerate bifurcations. One can also obtain an approximate evaluation of the critical exponents directly from the master equation, without having to evaluate explicitly the probability function [13].

3.3. *Mean-Field Theory*

The asymptotic method of solution of the master equation was also applied to a number of situations where the generalized detailed balance condition does not hold. So far, most of this work is limited to the mean-field approximation, in which spatially inhomogeneous fluctuations are discarded. One case studied refers to the spectrum of the master operator below and at the bifurcation point. Our method allows us to recover, straightforwardly, the results of MATSUO [18]. Another case is the reduction of certain bivariate problems to problems involving only one variable. In particular, the mechanism and the limitations of the adiabatic elimination procedure [19] can be followed in detail.

A very interesting problem is the stochastic description of the onset of limit cycle behavior in systems involving two variables. The mean-field master equation can again be reduced to a Fokker-Planck equation in the limit $\varepsilon \to 0$. By switching to polar coordinates and by scaling the radial variable in agreement with the general procedure described at the beginning of the present section, one can separate the radial and angular variables to the dominant order in ε and construct the explicit form of the stationary probability distribution (see also ref. [20]). Further work on limit cycles using the critical dynamics approach is reported in the paper by WALGRAEF et al. in this volume.

4. Conclusions and Perspectives

We have seen that it is possible to develop a systematic perturbation approach to nonequilibrium transition phenomena. We have been led to some striking analogies with equilibrium phase transitions.

However, as pointed out also in Professor HAKEN's communication, one should be fully aware of some basic differences. For instance, the arguments on critical dimensionality developed in the preceding section show that the source of non classical behavior is completely different in the two types of phenomena.

Several challenging problems remain open. The extension of the perturbative approach to include inhomogeneous fluctuations in systems involving two or more variables is necessary to understand the mechanism of symmetry-breaking bifurcations. Perhaps the most exciting question is the development of a perturbative approach capable of handling the initial value problem associated with the multi-variate master or Fokker-Planck equations. Despite significant progress achieved by SUZUKI and VAN KAMPEN among others [21-23], it is still not possible to see in detail the mechanisms by which an initially Gaussian distribution peaked on an unstable or a marginally stable state is finally attracted to a distribution belonging to a quite different class of functions, such as the form displayed in (3.10).

References

1. G. Nicolis, I. Prigogine: *Self-Organization in Nonequilibrium Systems* (Wiley, New York 1977)
2. S. Ma: *Modern Theory of Critical Phenomena* (W.A. Benjamin, Reading, MASS 1976)
3. P. Hohenberg, B. Halperin: Rev. Mod. Phys. *49*, 435 (1977)
4. G. Dewel, D. Walgraef, P. Borckmans: Z. Physik B*28*, 235 (1977)
5. D. Walgraef, G. Dewel, P. Borckmans: Phys. Rev. A*21*, 397 (1980)
6. D. Walgraef, G. Dewel, P. Borckmans: Adv. Chem. Phys., to be published
7. G. Nicolis, I. Prigogine: Proc. Nat. Acad. Sci. (U.S.A.) *68*, 2102 (1971)
8. C. Van den Broeck, J. Houard, M. Malek-Mansour: Physica *101*A, 167 (1980)
9. M. Malek-Mansour: Ph.D. Dissertation, University of Brussels (1979)
10. M. Malek-Mansour, G. Nicolis: J. Stat. Phys. *13*, 197 (1975)
11. J. Boissonade: Pys. Lett. *74*A, 285 (1979)
12. G. Nicolis, M. Malek-Mansour: J. Stat. Phys. *22*, 495 (1980)
13. M. Malek-Mansour, C. Van den Broeck, G. Nicolis, J.W. Turner: Annals of Physics, in press
14. C. Gardiner: J. Stat. Phys. *15*, 451 (1976)
15. S. Grossman: J. Chem. Phys. *65*, 2007 (1976)
16. H. Lemarchand: Physica *101*A, 518 (1980)
17. F. Schlögl: Z. Physik *253*, 147 (1972)
18. K. Matsuo: J. Stat. Phys. *16*, 169 (1977)
19. H. Haken: *Synergetics*, 2 ed., Springer Series in Synergetics, Vol.1 (Springer, Berlin, Heidelberg, New York 1978)
20. J.W. Turner: In *Dynamics of Synergetic Systems*, Proc. Int. Symposium, Bielefeld, FRG, Sept. 4-8, 1979. Springer Series in Synergetics, Vol.6 (Springer, Berlin, Heidelberg, New York 1980)
21. M. Suzuki: J. Stat. Phys. *16*, 11 (1977)
22. M. Suzuki: In Proc. XVIIth Solvay Conference on Physics (Wiley, New York 1980)
23. N.G. Van Kampen: J. Stat. Phys. *17*, 71 (1977)

Aspects of Classical and Quantum Theory of Stochastic Bistable Systems

C.W. Gardiner

Physics Department, University of Waikato, Hamilton, New Zealand

1. Introduction

I want to talk today about *bistable systems*, by which I mean systems which can exist in strictly 2 macroscopic configurations, for example a chemical molecule with 2 isomeric states, an electronic flip-flop device such as a tunnel diode in a suitable circuit, or an optical system which may exist in a state of high light intensity, or low light intensity. I do not mean to discuss subjects like liquid-vapour systems, in which there are 2 distinct physical states, but many configurations, depending on the shape of the boundary between the two states. I will review here today A) the classical description of bistable systems by Markov processes, concentrating mainly on a description in terms of Fokker Planck equations, and extending the usual methods into several dimensions and B) Quantum mechanical bistable systems viewed as Quantum Markov processes, with an emphasis on optical systems.

2. Classical Description of one variable Bistable Systems by Markov Processes

(a) Fokker Planck Equation:

One variable Systems have been treated by many authors, both as jump processes [4,5] and as diffusion processes [1,2,3,5]. Since similar behaviour appears in both descriptions, I will confine myself to the mathematically more tractable diffusion description. VAN KAMPEN'S paper [5] illustrates this similarity.

To my knowledge, KRAMERS [1] was the first to write down a Fokker-Planck equation for a bistable system, in the form

$$\partial_t P(x,t) = \partial_x[U'(x)P(x,t)] + D\partial_x{}^2 P(x,t) \tag{2.1}$$

where $U(x)$ is a double well potential of the form illustrated in Fig.1. The questions which it is considered of interest to pose about such a system are

i) Given the potential $U(x)$, what is the stationary distribution function $P_s(x)$? In the case of the one variable system (2.1) we know

$$P_s(x) = \mathcal{N} \exp[-U(x)/D] \tag{2.2}$$

In multivariable systems this question cannot always be answered.

ii) If the particle is one well (say around a) how long does it stay there? Although an essentially definitive answer to this question was given by KRAMERS [1], and has been elaborated by other [1,2,3,5], one still sees papers claiming to solve this problem. I will review KRAMERS method in a formulation which makes the assumption and their importance clear, and which enables extensions to multivariable systems possible.

iii) The relaxation of a probability distribution initially concentrated at the central maximum of the potential is still a subject of study [6,7,8]. The most careful treatment of this problem is by CAROLI, CAROLI AND ROULET [7,8]. I will not discuss this problem here.

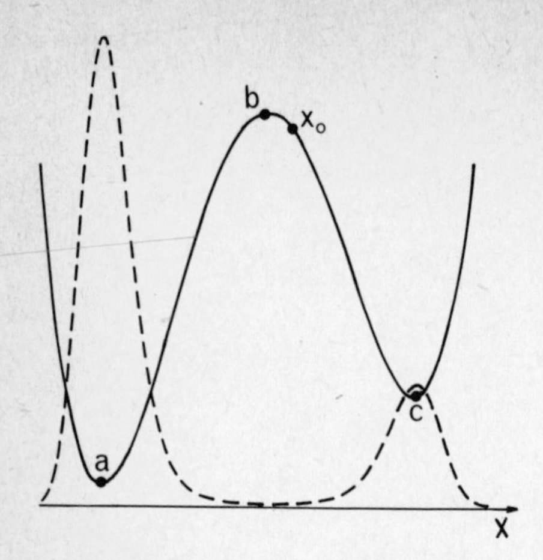

Fig.1. Double well potential *(full curve)* and stationary distribution function *(dashed line)*. Points a,c are local minima of the potential, and b is a local maximum. The point x_0 is close to the local maximum

(b) KRAMER'S Method Reformulated

Using the notation of Fig.1, define

$$M(x,t) = \int_{-\infty}^{x} dx' P(x',t) \tag{2.3}$$

$$N_a(t) = 1 - N_c(t) = M(b,t) \tag{2.4}$$

$$N_0(t) = (c-a)P(x_0,t) \tag{2.5}$$

and define also the corresponding stationary quantities by

$$n_a = 1 - n_c = \int_{-\infty}^{b} P_s(x)dx \tag{2.6}$$

$$n_0 = (c-a)P_s(x_0)\,. \tag{2.7}$$

We now solve as follows: from the Fokker-Planck equation (2.1) follows

$$\dot{M}(x,t) = DP_s(x)\,\frac{\partial}{\partial x}\,\{P(x,t)/P_s(x)\} \tag{2.8}$$

which is integrated to give

$$\int_{a}^{x_0} dx\dot{M}(x,t)/P_s(x) = D[P(x_0,t)/P_s(x_0) - P(a,t)/P_s(a)]\,. \tag{2.9}$$

Assumption I: We are considering only *long time* behaviour, so that $P(x,t)$ has attained the stationary shape in each well, but the relevant weights of the two peaks have not reached the stationary value. Quantitatively this means we can set

$$P(x,t) = P_s(x)N_a(t)/n_a \;(x<b); \;\; = P_s(x)N_c(t)/n_c \;(x>b)\,. \tag{2.10}$$

Substituting in the right hand side of the relaxation equation (2.9), for $x_0 < b$ we find

$$\kappa(x_0)\dot{N}_a(t) = D[N_0(t)/n_0 - N_a(t)/n_a] \tag{2.11a}$$

$$\mu(x_0)\dot{N}_c(t) = D[N_0(t)/n_0 - N_c(t)/n_c] \tag{2.11b}$$

with $$\kappa(x_0) = \int_a^{x_0} dx P_s(x)^{-1}[1 - \psi(x)] \qquad (2.12a)$$

and $$\mu(x_0) = \int_{x_0}^{c} dx P_s(x)^{-1}[1 - \psi(x)] \qquad (2.12b)$$

and $$\psi(x) = n_a^{-1} \int_x^b P_s(z)dz \quad (x < b); \quad = n_c^{-1} \int_b^x P_s(z)dz \quad (x > b). \qquad (2.13)$$

Notice that $\kappa(x_0) + \mu(x_0)$ is independent of x_0.

Assumption II: $P_S(x)$ must have a sharp minimum at x=b - thus the two states are well separated. As VAN KAMPEN [5] points out, the very concept of a two state system requires this. This means that i) $P_S(x)^{-1}$ has a sharp maximum at x=b and ii) $\psi(x)$ is extremely small in the vicinity of x=b. In this case one can write approximately

$$\kappa(x_0) \simeq \int_a^{x_0} dx P_s(x)^{-1}. \qquad (2.14)$$

(c) Three State Interpretation

Eqs. (2.11) correspond to a process symbolically able to be written as

$$N_a \leftrightharpoons N_0 \leftrightharpoons N_c \qquad (2.15)$$

except that there is no equation for N_0. However, by noting that $N_a + N_c = 1$, we find that

$$N_0(t) = n_0[\mu(x_0)N_a(t) + \kappa(x_0)N_c(t)]/[\kappa(x_0) + \mu(x_0)]. \qquad (2.16)$$

This is the same solution as would be obtained by adiabatically eliminating the variable $N_0(t)$ from the differential equation for $N_0(t)$

$$\dot{N}_0(t) = D\left\{N_a(t)/[n_a\kappa(x_0)] + N_c(t)/[n_c\mu(x_0)] - N_0(t)[\{n_0\kappa(x_0)\}^{-1} + \{n_0\mu(x_0)\}^{-1}]\right\} \qquad (2.17)$$

which is implied by the reaction scheme (2.5) and the other 2 equations (2.11). Such a procedure is valid when the rate constant for $N_0(t)$ is much larger than the other two, and, it can be seen that the ratio is of the order of n_a/n_0, which is very large, since n_a is of the order of 1, and n_0 is proportional to the very small probability of being at x_0.

This three state interpretation is essentially the transition state theory of EYRING [9].

(d) Elimination of Intermediate States

Eliminating $N_0(t)$ from (2.11) we get

$$-\dot{N}_a(t) = N_c(t) = r_a N_a(t) - r_c N_c(t) \qquad (2.18)$$

with $$r_a = D\left\{ n_a \int_a^c dx P_s(s)^{-1}[1 - \psi(x)]\right\}^{-1} \qquad (2.19a)$$

$$r_c = D\left\{n_c \int_a^c dxP_s(x)^{-1}[1 - \psi(x)]\right\}^{-1} \tag{2.19b}$$

where it is noted that r_a and r_c are independent of x_0: thus the precise choice of x_0 does not affect the interpeak relaxation.

(e) Splitting Probabilities

VAN KAMPEN [5] has considered the probability that the system, started at x_0, will reach the vicinity of a or c. The picture (2.15) indicates that the ratio of these two probabilities, π_a and π_c, must be the ratio of the rates at which N_0 decays respectively to N_a and N_c. Thus we deduce

$$\pi_a = \mu(x_0)/[\kappa(x_0) + \mu(x_0)]; \quad \pi_c = \kappa(x_0)/[\kappa(x_0) + \mu(x_0)] \tag{2.20}$$

which is in agreement with his results if we make Assumption II.

(f) The Escape Probability per unit time for a particle initially near a, to reach x_0, is the decay rate of $N_a(t)$ under the condition that an absorbing barrier is at x_0. This means that in (2.9) we set $P(x_0,t) = 0$ [but note that $P_s(x)$ is defined by (2.2)]. Similar reasoning gives us

$$\dot{N}_a(t) = -DN_a(t)/[n_a\kappa(x_0)] \tag{2.21}$$

so that the mean first passage time is given by

$$\tau_a = n_a D^{-1} \int_a^{x_0} dxP_s(x)^{-1}[1 - \psi(x)]. \tag{2.22}$$

This result is exact for $x_0 < b$, and both it and the exact result differ very little from the approximation obtained from Assumption II, which sets $\psi(x) = 0$, and is valid for x_0 near b.

3. Multivariable Systems

The multidimensional case was first treated by LANDAUER AND SWANSON [10] and restated by LANGER [11], in both cases, for situation in which a particle moves in a potential. In the following we generalise and formalise LANDAUER AND SWANSON'S method to non-potential situations.

(a) Fokker Planck Equation

We consider a completely general Fokker-Planck equation in ℓ dimensions.

$$\partial_t P = \nabla.[-\underset{\sim}{v}(\underset{\sim}{x})P + \varepsilon \underset{\approx}{D}(\underset{\sim}{x}).\nabla P] \tag{3.1}$$

whose stationary solution is to be called $P_s(\underset{\sim}{x})$, and can only be exhibited explicitly if (3.1) satisfies potential conditions. We assume that $P_s(\underset{\sim}{x})$ has two well defined maxima at $\underset{\sim}{a}$ and $\underset{\sim}{c}$, and well defined saddlepoint at $\underset{\sim}{b}$ (See Fig.2). We assume that the value at the saddlepoint is very much smaller than the values at $\underset{\sim}{a}$ and $\underset{\sim}{c}$. We introduce a family of $(\ell - 1)$ dimensional planes $\underset{\sim}{S}(w)$, where w is a parameter which labels the planes. We choose $\underset{\sim}{S}(a)$ to pass through $\underset{\sim}{a}$, $\underset{\sim}{S}(b)$ through $\underset{\sim}{b}$, and $\underset{\sim}{S}(c)$ through $\underset{\sim}{c}$. The planes $\underset{\sim}{S}(w)$ are assumed to be oriented in such a way that $P_s(\underset{\sim}{x})$ has a unique maximum when restricted to any one of them. We define, similarly to the previous treatment,

$$M[\underset{\sim}{S}(w)] = \int_{L(w)} d\underset{\sim}{x}P(\underset{\sim}{x}) \tag{3.2}$$

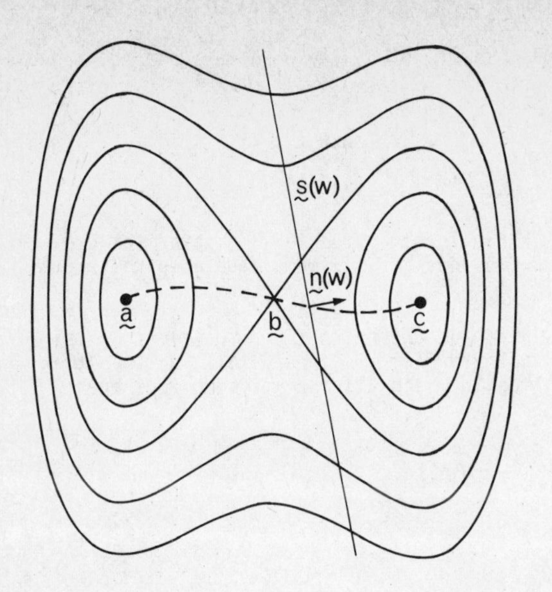

Fig.2. Contours of the stationary distribution function $P_S(x)$. The plane S(w) is oriented so that $P_S(x)$ has a unique maximum there and the curve $\underset{\sim}{x} = \underset{\sim}{u}(w)$ is the locus of these maxima *(dashed line)*

where L(w) is the region of space to the left of the plane $\underset{\sim}{S}(w)$, then

$$\dot{M}[\underset{\sim}{S}(w)] = \int_{\underset{\sim}{S}(w)} d\underset{\sim}{S}.[-\underset{\sim}{v}(\underset{\sim}{x})P + \varepsilon \underset{\approx}{D}(x).\nabla P]. \tag{3.3}$$

(b) Current in Stationary State is defined by

$$\underset{\sim}{J}_s = -\underset{\sim}{v}(\underset{\sim}{x})P_s + \underset{\approx}{D}(\underset{\sim}{x}).\nabla P_s. \tag{3.4}$$

Assumption I: We exclude cases in which finite currents $\underset{\sim}{J}_S$ occur where P_S is very small. Because of $\nabla.\underset{\sim}{J}_S = 0$, we can write

$$\underset{\sim}{J}_s = -\varepsilon\nabla.(\underset{\approx}{A}P_s) \tag{3.5}$$

where $\underset{\approx}{A}$ is an antisymmetric tensor. *We require that $\underset{\approx}{A}$ be of the same order of magnitude as $\underset{\approx}{D}(x)$, or smaller.*

(c) Relaxation Equations are derived in two stages. Define a quantity $\beta(\underset{\sim}{x})$ by

$$\beta(\underset{\sim}{x}) = P(\underset{\sim}{x},t)/P_s(\underset{\sim}{x}) \simeq N_a(t)/n_a \quad (\underset{\sim}{x} \text{ near } \underset{\sim}{a}) \tag{3.6}$$

$$\simeq N_c(t)/n_c \quad (\underset{\sim}{x} \text{ near } \underset{\sim}{c}).$$

This is the assumption that all relaxation within peaks has ceased. Substitute now in (3.3), integrate by parts discarding terms at infinity, and we obtain

$$\dot{M}[\underset{\sim}{S}(w)] = \varepsilon \int_{\underset{\sim}{S}(w)} d\underset{\sim}{S}.[\underset{\approx}{\mathcal{D}}(\underset{\sim}{x}).\nabla\beta]P_s(\underset{\sim}{x}) \tag{3.7}$$

with

$$\underset{\approx}{\mathcal{D}}(\underset{\approx}{x}) = \underset{\approx}{D}(\underset{\sim}{x}) + \underset{\approx}{A}(\underset{\sim}{x}). \tag{3.8}$$

Assumption II: $P_s(\underset{\sim}{x})$ is sharply singly peaked on $\underset{\sim}{S}(w)$, so we may make the approximate evaluation

$$\dot{M}[\underset{\sim}{S}(w)] = \varepsilon\left\{[\underset{\sim}{n}(w).\underset{\approx}{\mathscr{D}}(\underset{\sim}{x}).\nabla\beta]_{\underset{\sim}{u}(w)} + \delta(w)\right\} \left|\iint_{\underset{\sim}{S}(w)} d\underset{\sim}{S}P_s(\underset{\sim}{x})\right| \tag{3.9}$$

where $\delta(w)$ is expected to be very much smaller than the term in square brackets. Here $\underset{\sim}{u}(w)$ is the position at which $P_s(x)$ has its maximum value when restricted to $\underset{\sim}{S}(w)$.

Assumption III: The direction of $\underset{\sim}{n}(w)$ can be chosen so that $\underset{\approx}{\mathscr{D}}^T(x).\underset{\sim}{n}(w)$ is parallel to the tangent at w to the curve $\underset{\sim}{x} = \underset{\sim}{u}(w)$ - without violating the other assumptions. (Note: Examples can be given in which all three assumptions are satisfied). Hence

$$\varepsilon\underset{\approx}{\mathscr{D}}^T[\underset{\sim}{u}(w)].\underset{\sim}{n}(w) = d(w)\partial_w\underset{\sim}{u}(w). \tag{3.10}$$

Defining now

$$p(w) = \left|\iint_{\underset{\sim}{S}(w)} d\underset{\sim}{S}P_s(\underset{\sim}{x})\right| \tag{3.11}$$

which is (up to a slowly varying factor) the probability density for the particle to be on the plane $\underset{\sim}{S}(w)$, and is expected to have a two peaked shape, with maxima at $w = a$ and $w = c$, and a minimum at $w = b$.

Assumption IV: These are assumed to be sharp maxima and minimum. Neglecting $\delta(w)$ and making the choice (3.11), and noting

$$\partial_w\underset{\sim}{u}(w).\nabla\beta[\underset{\sim}{u}(w)] = \partial_w\beta[\underset{\sim}{u}(w)], \tag{3.12}$$

we find

$$\int_a^{w_0} dw\{\dot{M}[\underset{\sim}{S}(w)]/[p(w)d(w)]\} = \beta(w_0) - \beta(a) \tag{3.13}$$

and using the sharp peaked nature of $p(w)^{-1}$, (3.13) can now be approximated by taking the value at the peak, using (3.6) and,

$$N(a,t) = M[\underset{\sim}{S}(b),t] \tag{3.14}$$

as well as defining

$$\kappa(w_0) = \int_a^{w_0} [p(w)]^{-1}dw \tag{3.15a}$$

$$\mu(w_0) = \int_{w_0}^{b} [p(w)]^{-1}dw, \tag{3.15b}$$

to obtain the relaxation equations

$$\kappa(w_0)\dot{N}_a(t) = d(w_0)[N_0(t)/n_0 - N_a(t)/n_a] \tag{3.16a}$$

$$\mu(w_0)\dot{N}_c(t) = d(w_0)[N_0(t)/n_0 - N_c(t)/n_c] \tag{3.16b}$$

which are of exactly the same form as those in the one variable case, and all the same interpretations can be made.

(d) Range of Validity of Assumptions

This matter is still under investigation, and will be published elsewhere. It can be shown that, under the condition that Gaussian approximations can be made to most integrals, all conditions are always satisfied if $J_s = 0$ (i.e. the system satisfied potential conditions), D is non-singular, and under similar conditions for two dimensional systems even when $J_s \neq 0$, all conditions are satisfied. At the time of writing the question is open for higher dimensional systems.

4. Other Methods

The most elegant methods for potential systems are thos of CAROLI et al. [7,8] which use VAN KAMPEN'S [12] transformation of a Fokker-Planck equation into a Schrodinger equation, to which W.K.B. methods are applied. This enables a treatment of the medium term relaxation to be given, but has not yet been adapted to non potential situations. A more mathematical method based on asymptotic solutions of singularly perturbed Fokker-Planck equations which has been presented by SCHUSS and MATKOWSKY [13] gives a rigorous basis to Kramer's method, and is probably equivalent to that of CAROLI et al. Again, this has been carried out only for the potential case. SCHENZLE and BRAND [14] have shown that, if there is a potential solution, the eigenvalue problem may be reduced to a variational problem, which can be used numerically, in principle, to estimate all eigenvalues.

5. Quantum Systems

(a) Quantum Stochastics is now about 15 years old, and has not yet found a place in the mathematical literature. Basic reference are by HAKEN [15] HAAKE [16] and LOUISELL [17]. Recent technical advances by the author and coworkers [18,19,20] have introduced the concept of stochastic processes in the complex plane as a useful technique, and there are some optically bistable systems which can only be handled with these techniques.

(b) Fokker-Planck equations are derived in quantum systems by an expansion of the quantum density matrix in coherent states, first introduced by GLAUBER [21]. When first developed, although it was realised that non-positive semidefinite diffusion matrices could arise in these Fokker-Planck equations, the small deviation from positive semidefiniteness was considered negligible. In certain strongly nonlinear systems this is not physically justifiable, for example in the following Fokker-Planck equation for an optically bistable system (DRUMMOND and WALLS [22])

$$\partial_t P(\alpha,\beta,t) = \Big\{\partial_\alpha[\kappa\alpha + 2\chi\alpha^2\beta - \varepsilon] - \chi\partial_\alpha{}^2\alpha^2 + \partial_\beta[\kappa^*\beta + 2\chi^*\alpha\,\beta^2 - \varepsilon^*] - \chi^*\partial_\beta{}^2\beta^2 + \Gamma\partial_\alpha\partial_\beta\Big\} P(\alpha,\beta,t). \tag{5.1}$$

This is clearly not an ordinary Fokker-Planck equation, since the diffusion matrix is not even real, and certainly not positive definite. However, a stochastic interpretation is possible. It has been shown by DRUMMOND and GARDINER [18] that it is equivalent to the pair of Ito stochastic differential equations for the complex quantities α and β.

$$\begin{pmatrix} d\alpha \\ d\beta \end{pmatrix} = \begin{pmatrix} \varepsilon - \kappa\alpha - 2\chi\alpha^2\beta \\ \varepsilon^* - \kappa^*\beta - 2\chi^*\alpha\beta^2 \end{pmatrix} dt + \begin{pmatrix} -2\chi\alpha^2 & \Gamma \\ \Gamma & -2\chi^*\beta^2 \end{pmatrix}^{\frac{1}{2}} \cdot \begin{pmatrix} dW_1(t) \\ dW_2(t) \end{pmatrix} \tag{5.2}$$

in the sense that all physically observable quantities are the same. These are mean values of *analytic functions* of α and β. The Fokker-Planck equation corresponding to (5.2) is a four variable equation for the real and imaginary parts of α and β, and hence also predicts values for the unobservable non-analytic functions of α and β. Using equations like (5.2), all the usual techniques for calculating spectra, etc, by linearising can be used, but since such techniques are local only, general features of bistability are not easy to investigate.

The system (5.2) is under certain conditions deterministically bistable - for by setting $n = \alpha\beta$, and choosing the particular values of the parameters, corresponding to dispersive bistability,

$$\kappa = \kappa' + i\kappa'' \qquad (\kappa > 0) \tag{5.3a}$$

$$\chi = i\chi'' \qquad (\chi'' \text{ real}) \tag{5.3b}$$

$$\varepsilon = iE \qquad (E \text{ real}) \tag{5.3c}$$

we find a solution to the deterministic equation must satisfy

$$E^2 = n[(\kappa')^2 + (2\chi''n + \kappa'')^2] \tag{5.4}$$

and

$$\alpha = \beta^* = iE/[\kappa' + i(\kappa'' + 2\chi''n] \tag{5.5}$$

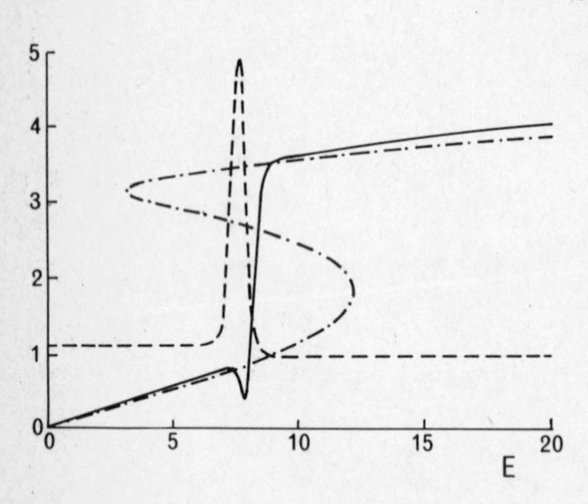

Fig.3. Chain curve: deterministic solution for $|\alpha|$ of the stochastic system eq.(5.2). *Full curve:* $|\langle\alpha\rangle|$ as obtained by $P_S(x)$ from eq.(5.8). *Dashed curve:* $g^2(0)$, all plotted as a function of E

For $\chi''\kappa'' < 0$, and $\kappa'' > 3(\kappa')^2$ (5.4) represents an S shaped curve when E is plotted against n, and there are regions where n has three distinct real positive roots (see Fig.3). Linearised stability analysis shows that only the central solution is locally unstable. Stationary solutions of the Fokker-Planck equations corresponding to (5.2) are not given by potential conditions. However, it is possible to solve (5.1) in the stationary state by a potential method if $\Gamma = 0$, which corresponds to zero thermal noise. In this case we find

$$P_s(\alpha,\beta) \propto \alpha^{c-2}\beta^{d-2}\exp[\mu/\alpha + \mu^*/\beta + 2\alpha\beta] \tag{5.6}$$

with

$$c = d^* = \kappa/\chi, \qquad \mu = \varepsilon/\chi . \tag{5.7}$$

This solution is not normalisable on any real range of α and β; the ranges of α and β must be taken as contours around the origin in the respective complex planes. Moments are given by

$$\langle\alpha^r \beta^s\rangle = \mathcal{N}^{-1} \oint\oint d\alpha d\beta\, \alpha^r \beta^s P_s(\alpha,\beta), \tag{5.8}$$

where $\mathcal{N}$ is the normalisation integral. In Fig.3 are also illustrated $|\langle\alpha\rangle|$ as well as the quantity

$$g^2(0) = \langle(\alpha\beta)^2\rangle/\langle\alpha\beta\rangle^2 \tag{5.9}$$

which is a measure of photon statistics. Notice that $|\langle\alpha\rangle|$ exhibits an interference phenomenon near the transition from one branch to the other, which is characteristic of a vector situation.

(c) Lifetime of the Metastable States

This can in principle be treated by applying the generalisation of Kramer's method to a Fokker-Planck equation corresponding to the equations (5.2). This has not been done yet, and may not succeed, since the diffusion matrix is singular. Furthermore, at the very least, an approximate estimate of the 4 variable stationary distribution function is needed, and this does not at present exist, nor indeed does any estimate of the tensor potential $\underset{\approx}{A}$.

ACKNOWLEDGEMENTS

I wish to thank the New Zealand University Grants Committee, and the Zentrum für Interdisziplinäre Forschung, Bielefeld, for financial assistance for this project.

In developing the ideas here, I have been greatly assisted by conversations with N.G. van Kampen, S. Chaturvedi, G. Nicolis, D.F. Walls, P.D. Drummond and K.J. McNeil.

REFERENCES

1. H.A. Kramers, Physica 7, 284 (1940) = Collected Scientific Papers (Nth Holland, 1956 p. 784.
2. S. Chandrasekhar, Rev. Mod. Phys. 15, 1 (1943).
3. R. Landauer, J. Appl. Physics 33, 7 (1962).
4. I. Matheson, D.F. Walls, and C.W. Gardiner, J. Stat. Phys. 12, 21 (1975).
5. N.G. van Kampen, Suppl. Prog. Theor. Physics. 64, 289 (1978).
6. M. Suzuki, Suppl. Prog. Theor. Physics. 64, 402 (1978).
7. B. Caroli, C. Caroli, and B. Roulet, J. Sat. Phys. 21, 415 (1979).
8. B. Caroli, C. Caroli, B. Roulet, and J.F. Gouet, J. Stat. Phys. 22, 515 (1980).
9. H. Eyring, J. Chem. Phys. 3, (1935).
10. R. Landauer and J.A. Swanson, Phys. Rev. 121, 1668 (1961).
11. J.S. Langer, Ann. Phys. (N.Y.) 54, 258 (1961).
12. N.G. van Kampen, J. Stat. Phys. 17, 71 (1977).
13. Z. Schuss and B. Matkowsky, SIAM, J. Appl. Math. 35, 401 (1979).
14. A. Schenzle and H. Brand, Optics. Comm. 31, 401 (1979).
15. H. Haken, "Laser Theory", in Light and Matter Ic, ed. by L. Genzel, Encyclopedia of Physics, Vol. XXV/2c, Springer, Berlin, Heidelberg, New York, 1970
16. F. Haake, "Statistical Treatment of Open Systems by Generalized Master Equations", in Quantum Statistics in Optics and Solid-State Physics, Springer Tracts in Modern Physics, Vol. 66, Springer, Berlin, Heidelberg, New York, 1973.
17. W.H. Louisell, Quantum Statistical Properties of Radiation, Wiley (N.Y.), 1973.
18. P.D. Drummond and C.W. Gardiner, J. Phys. A13, 2353 (1980).
19. C.W. Gardiner and S. Chaturvedi, J. Stat. Phys. 17, 429 (1977).
20. S. Chaturvedi and C.W. Gardiner, J. Stat. Phys. 18, 501 (1978).
21. R.J. Glauber, Phys. Rev. 130, 2529 (1963).
22. P.D. Drummond and D.F. Walls, J. Phys. A13, 725 (1980).

Postscript: After presentation of this paper, Dr. M. SAN MIGUEL brought to my attention the fact that the paper of J.S. LANGER [11] does in fact consider certain particular Fokker-Planck equations for which there is a non-vanishing stationary current. LANGER computes the decay time of metastable states by a method reminiscent of that used in section 3 of this paper, which thus represents a previous extension of KRAMERS' method to non-potential systems.

Kinetics of Phase Separation

Kurt Binder

Institut für Festkörperforschung der Kernforschungsanlage Jülich,
D-5170 Jülich, Fed. Rep. of Germany

1. Introduction

This talk considers the kinetics of phase transformations, induced by a sudden change of external parameters of the system. In particular, we are concerned with a system which is suddenly cooled from the disordered phase to a state below the critical temperature of an order-disorder transition (Fig.1a). Then the initially disordered system has to separate into ordered domains of macroscopic size. Alternatively, we are concerned with binary systems (AB) with a miscibility gap (Fig.1b). "Quenching" the system at time t=0 from a state in the one-phase region to a state below the coexistence curve again leads to phase separation: equilibrium would require macroscopic regions of both phases with concentrations $c^{(1)}_{coex}, c^{(2)}_{coex}$. The system develops towards this equilibrium through states far from equilibrium. These states become more and more inhomogeneous as time goes on. The final equilibrium (inhomogeneous at macroscopic scales) is reached only for $t \to \infty$ in the thermodynamic limit $N \to \infty$.

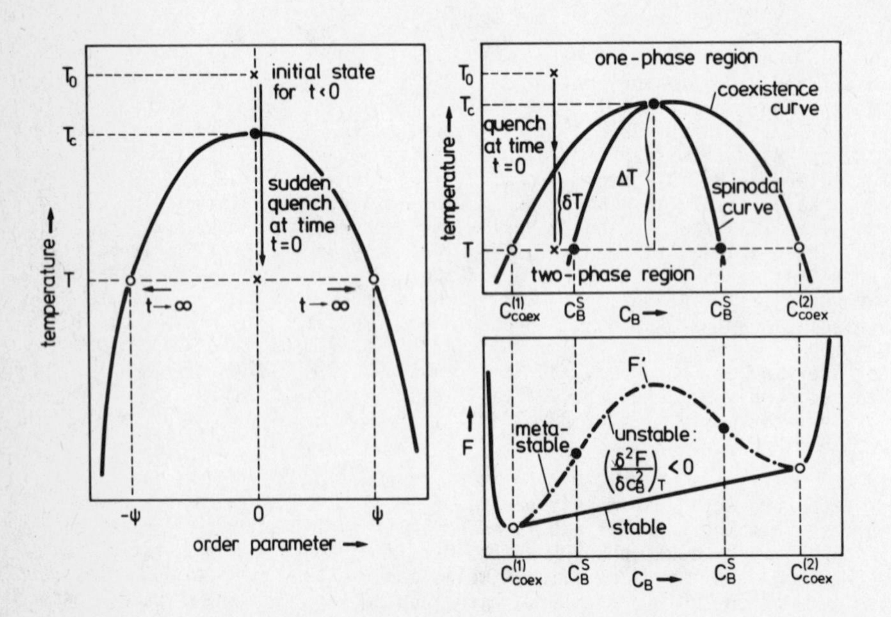

Fig.1 a) Order parameter of a phase transition plotted vs. temperature; b) Phase diagram of a binary system; c) Free energy vs composition at temperature T (schematic)

The free energy (Fig.1c) decreases with concentration c_B in the one phase region (entropy of mixing!), while it varies linearly with c_B in the two-phase region (there the amount of the two coexisting phases changes). The most naive theory on kinetics starts by introducing a free energy F' of one-phase states within the two-phase region. The regime where F'>F but still $(\partial^2F'/\partial c_B^2)_T>0$ is called "metastable", while the regime where $(\partial^2F'/\partial c_B^2)_T<0$ is called "unstable". The locus of inflection points $(\partial^2F'/\partial c_B^2)_T=0$ is called "spinodal curve" $c_B^s(T)$. On this basis, one considers two different transformation mechanisms (Fig.2): "spinodal decomposition" in the unstable regime is the growth of long-wavelength (delocalized) concentration fluctuations, in the metastable regime "nucleation" of (localized) microdomains of the other phase starts the transformation.

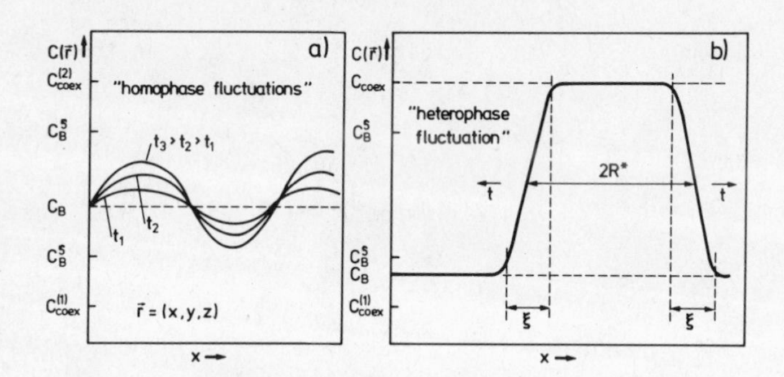

Fig.2 Unstable fluctuations in the two-phase regime

In this talk, the state of the art of theories on nucleation and spinodal decomposition will be summarized. Emphasis is on the regime in the vicinity of the critical point T_c, as there the theories can be cast in a scaled universal form (analogous to dynamic scaling for the decay of fluctuations close to equilibrium). We will argue that there is no sharp distinction between two mechanisms, but rather a gradual transition from one to the other (and hence the spinodal in Fig.1 has not much meaning). A unified description which naturally incorporates this fact is presented on the basis of the "cluster dynamics"-approach. In this way, one can also understand to some extent the asymptotic power laws describing the domain growth, as well as the scaling behavior during the late stages. Finally the main unsolved problems will be spelled out.

2. Cahn-Hilliard Theory [1]

The local concentration $c(\vec{r},t)$ of the binary system satisfies the conservation law $\partial c(\vec{r},t)/\partial t+\nabla\vec{j}(r,t)=0$. The concentration current $\vec{j}$ is assumed proportional to the gradient of the local chemical potential-*difference* $\mu(\vec{r})$, $\vec{j}=-M\nabla\mu(\vec{r})$, where M is a mobility. To find $\mu(\vec{r})$, the free energy of the inhomogeneous system is described as

$$F_{inh}=\int d^3r\{f[c(\vec{r})]+K[\nabla c(\vec{r})]^2\}, \quad \mu(\vec{r}) = \frac{\delta F}{\delta c(\vec{r})} = \left(\frac{\partial f}{\partial c}\right)_T - K\nabla^2 c \, , \tag{2.1}$$

where the gradient term describes the additional energy of inhomogeneities (domain walls, etc.), and f is the density of F' (Fig.1c), e.g. $f=f_0+A(c-c_B^{crit})^2+B(c-c_B^{crit})^4+\ldots$, A<0, B>0. From these assumptions results the Cahn-Hilliard equation

$$\partial c(\vec{r},t)/\partial t = M\nabla^2[(\partial f(c)/\partial c)_T - K\nabla^2 c]. \tag{2.2}$$

Because of the nonlinearity it has no analytic solution. Thus one assumes, in the initial stages of unmixing, that $c(\vec{r})$ is not much different from c_B (Fig.2a); hence Eq. (2.2) is linearized at c_B

$$\partial(c-c_B)/\partial t = M\nabla^2[(\partial^2 f/\partial c^2)_T\big|_{c_B} - K\nabla^2](c-c_B). \tag{2.3}$$

By fourier transforming $\{\delta c(q,t) \equiv \int d^3r \exp(i\vec{q}\cdot\vec{r})[c(\vec{r},t)-c_B]\}$ one readily finds the equal-time structure factor

$$S(\vec{q},t) \equiv \langle\delta c(-\vec{q},t)\delta c(\vec{q},t)\rangle_T = \langle\delta c(-\vec{q})\delta c(\vec{q})\rangle_{T_0} \exp[2R(\vec{q})t] \ , \qquad (2.4)$$

where the "amplification factor" $R(\vec{q}) \equiv Mq^2[(\partial^2 f/\partial c^2)_T \big|_{c_B} + Kq^2]$. Note that a similar treatment applies to the case of the order-disorder transition (Fig.1a), one has to replace the (conserved) order parameter $c-c_B^{crit}$ by the (nonconserved) ψ, and Mq^2 by a kinetic rate factor Γ.

From Eq. (2.4) it is clear that the only thermal fluctuations included so far are those of the initial state. For metastable states $R(\vec{q})<0$ for all q, hence all fluctuations die out [only homophase fluctuations (Fig.2a) are included in the linearized Eq. (2.3)]. In unstable states $R(\vec{q})>0$ for $0<q<q_c \equiv [-(\partial^2 f/\partial c^2)_T \big|_{c_B}/K]^{1/2}$: fluctuations with wavevectors in this regime increase with time, until the neglected nonlinear terms become important.

To account for fluctuations in the final state, one must add a random force term [2,3] $\nabla^2\eta(\vec{r},t)$ to Eqs. (2.2), (2.3), where

$$\langle\eta(\vec{r},t)\eta(\vec{r}',t')\rangle_T = \langle\eta^2\rangle_T\ \delta(\vec{r}-\vec{r}')\delta(t-t'),\ \langle\eta^2\rangle_T = k_BTM \ . \qquad (2.5)$$

The nonlinear Cahn-Hilliard equation with fluctuations is expected to describe phase separation satisfactorily. However, a solution exists in the linearized case only [2],

$$S(\vec{q},t) = S_{T_0}(\vec{q})\exp[2R(\vec{q})t] + S_T(\vec{q})\{1-\exp[2R(\vec{q})t]\} \ , \qquad (2.6)$$

$$S_T(\vec{q}) = k_BT/[(\partial^2 f/\partial c^2)_T \big|_{c_B} + Kq^2] = (k_BT/K)(\xi^{-2}+q^2)^{-1} \ , \qquad (2.7)$$

$\xi = \sqrt{K/(\partial^2 f/\partial c^2)_T \big|_{c_B}}$ being the correlation length of concentration fluctuations. This approximation still does not distinguish stable and metastable states. Both ξ and the critical wavelength $\lambda_c = 2\pi/q_c$ diverge (according to this theory) at the spinodal (Fig.3a). Also the radius R^* of the "critical cluster" (Fig.2b) having the size to just overcome the nucleation barrier ΔF^* diverges there, although ΔF^* there becomes zero (Fig.3b). In this framework, the spinodal is a line of critical points, and one also has "critical slowing down" of fluctuations there.

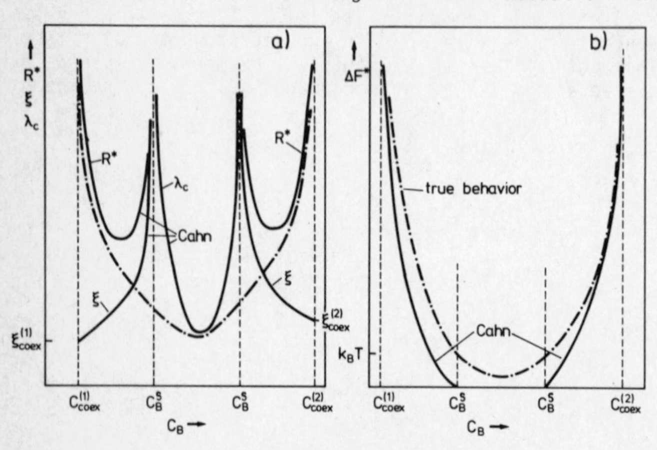

Fig.3 Characteristic lengths (a) and energy barrier (b) plotted vs. concentration (schematic)

While the spinodal is well defined for infinitely long-ranged (infinitely weak) attractive forces (where nucleation is suppressed and metastable states have infinite lifetime), this singularity (Fig.3a) is rounded off for short-range interactions [5,6]. Clearly, for $c_B \gtrsim c_B^s$ the Cahn-Hilliard critical modes ($\lambda > \lambda_c$) produce a very slow rate of transformation only (due to the above critical slowing down). But also fluctuations of finite amplitude and smaller linear dimension (cf. Fig.2b) can form here, involving nonzero but small energy barriers ($\sim k_B T$) [Fig.3b, dash-dotted curve]. These fluctuations also are unstable and transform the system by a much quicker rate, and hence the slow Cahn-Hilliard modes have no chance to come into play. The situation differs from equilibrium, where all fluctuations decay, and hence the slowest ones dominate the late stages of relaxation: in the nonequilibrium relaxation from one state to another the quickest mechanism dominates.

In addition, the position of the spinodal cannot be defined without ambiguity [6]. Consider e.g. a lattice model of binary mixtures: $c_i=1$ if the i-th lattice site is taken by a B-atom, $c_i=0$ for an A-atom. The concentration field $c(\vec{r})$ postulated by continuum theory is found by coarse-graining over a volume V_α [a = lattice spacing, d = dimensionality]

$$c(\vec{r}) = (1/V_\alpha) \sum_{i \in V_\alpha} c_i, \qquad a^d \ll V_\alpha \ll \xi^d , \tag{2.8}$$

$\vec{r}$ being the center of gravity of V_α. The inequality is needed in order for the system being homogeneous within a cell; $c(\vec{r})$ is then slowly varying from one cell to the next. Since $\xi \gg a$ holds only close to the critical point (T_c, c_B^{crit}), the continuum theory makes sense in the critical region only.

Now f(c) means the free-energy density of a cell V_α, and not of the whole volume {choosing $V_\alpha \gg \xi^d$ equilibrium consists of a finite B-rich cluster with surrounding A-rich background [7], i.e. c would no longer be homogeneous, contrary to the assumption about f(c)}. As a result, f(c) depends on the (somewhat arbitrary!) choice of V_α, and so does the location of the inflection point c_B^s: there is no well-defined spinodal curve, but rather a spinodal region. Of course, $S(\vec{q},t)$ cannot depend on V_α, which is introduced only for computational convenience.

While this theory is unsatisfactory for phase separation, it is useful for quenches from one state to another one in the one-phase region {then Eq. (2.6) can be derived more generally from master equations [8]}. Recent experiments [9] agree with theory [8] quantitatively.

3. Theory of Langer-Baron-Miller [10] and Billotet-Binder [11-13]

This theory *approximately* includes both fluctuations and nonlinear terms. Eq. (2.2) with random forces [Eq. (2.5)] still contains information on critical behavior, which is in itself a formidable problem. Thus one does not use Eq. (2.8) but rather $V_\alpha = (\xi/\alpha)^d$, α being a constant of order unity, and one assumes static critical behavior to be known. This restricts the validity of the theory to the regime [12] [β, γ, ν... critical exponents]

$$q\xi \lesssim 1, \quad (c_B - c_B^{crit})/(1-T/T_c)^\beta \gtrsim 1 . \tag{3.1}$$

Then a crucial approximation is made: one derives exact equations of motion for the probability distributions $\rho_1[c(\vec{r})]$, $\rho_2[c(\vec{r}_1), c(\vec{r}_2)]$. As expected, the equation of motion for ρ_1 couples to ρ_2, the equation for ρ_2 couples to ρ_3, etc. A "closed" equation for ρ_1 is then obtained by the decoupling $[\delta c(\vec{r}) = c(\vec{r}) - c_B]$

$$\rho_2\, c(\vec{r}_1), c(\vec{r}_2)] = \rho_1[c(\vec{r}_1)]\rho_1[c(\vec{r}_2)]\left\{1 + \frac{\langle \delta c(\vec{r}_1)\delta c(\vec{r}_2)\rangle_T}{\langle(\delta c)^2\rangle_T}\, \frac{\delta c(\vec{r}_1)\delta c(\vec{r}_2)}{\langle(\delta c)^2\rangle_T}\right\} . \tag{3.2}$$

If there were no correlation between concentrations at $\vec{r}_1$ and $\vec{r}_2$ ρ_2 would be a simple product of one-point probabilities. Thus the correction is put proportional to $\langle\delta c(\vec{r}_1)\delta c(\vec{r}_2)\rangle_T$. The calculation yields [10]

$$dS(\vec{q},t)/dt = -2Mq^2\{[A(t) + Kq^2]\ S(\vec{q},t) - k_BT\}, \tag{3.3}$$

A(t) itself depending on $S(\vec{q},t)$ selfconsistently. Thus the expression $(\partial^2 f/\partial c^2)_T \big|_{c_B}$ in Cook's equation of motion [2] is replaced by A(t).

The main results of this theory are: (i) $S(\vec{q},t)$ increases with time only linearly or weaker [10], in agreement with computer simulations [14]: thus there is *no regime of times where linearized Cahn-Hilliard theory holds*. (ii) The position $q_{max}(t)$ where $S(\vec{q},t)$ is maximal shifts to $q\to 0$ for $t\to\infty$, in qualitative agreement with simulation [14] and experiment [15]. (iii) The structure factor satisfies dynamic scaling [$\epsilon=1-T/T_c$],

$$S_{T,c_B}(\vec{q},t) = \epsilon^{-\gamma}\ \tilde{S}[q\epsilon^{-\nu},\ t\epsilon^{z\nu},\ (c_B-c_B^{crit})\epsilon^{-\beta}] \tag{3.4}$$

bearing out the concept [16] that dynamic scaling can apply to relaxation far from equilibrium. The scaling function $\tilde{S}$ is universal [12] - i.e. the same for all systems in a "class" apart from amplitude factors for the arguments of $\tilde{S}$. The dynamic exponent $z\approx 2$ for nonconserved order parameter, while $z=4-\eta$ or $z\tilde{=}3$ for solid or liquid mixtures, respectively [17]. (iv) The theory is easily extended to the two-time structure factor $S(\vec{q},t_1,t_2)$ [13], which describes the decay of fluctuations during the separation process. For large q one finds (Fig.4) $S(\vec{q},t_1,t_2)\approx S_{eq.}(\vec{q},t_2-t_1)$

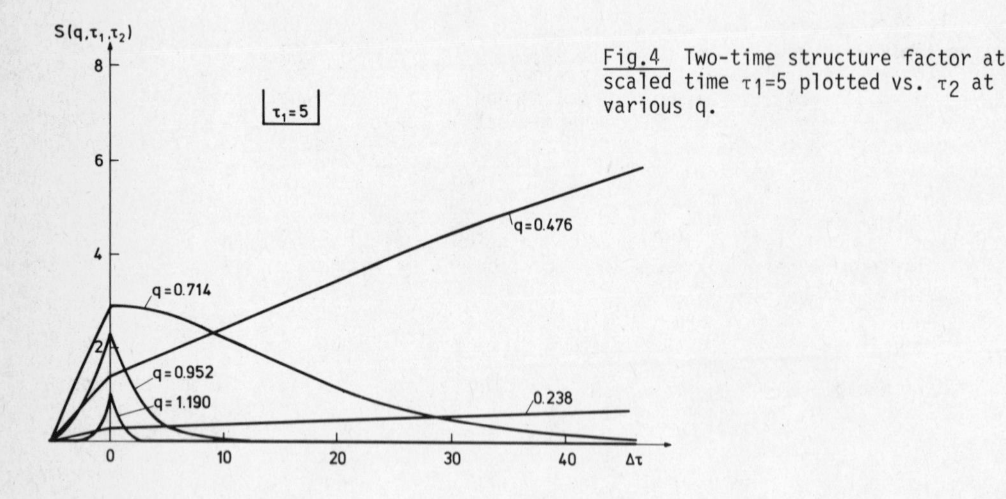

Fig.4 Two-time structure factor at scaled time τ_1=5 plotted vs. τ_2 at various q.

$= S_T(\vec{q})\exp[-D_Tq^2(t_2-t_1)]$, D_T being the diffusivity, while for small q [$q\lesssim q_{max}(t_1)$] $S(\vec{q},t_1,t_2)$ deviates from equilibrium strongly even for large times t_1 after the quench. (v) Unfortunately, the theory still exhibits metastable states of infinite lifetime (Fig.5) [11,12]. The "spinodal" is located in between Cahn's spinodal and the coexistence curve. It is clearly an artefact of the approximation, as its precise location depends on the arbitrary parameter α. The structure factor there first also develops a peak but later on saturates at the structure factor of the metastable state: nucleation is not yet included in the theory.

(vi) In the later stages of the separation, the structure factor can be expressed in scaled form [8,11,12] (Fig.6a)

$$S(\vec{q},t) = \varphi[q_{max}(t)]\tilde{S}[q/q_{max}(t)]\ . \tag{3.5}$$

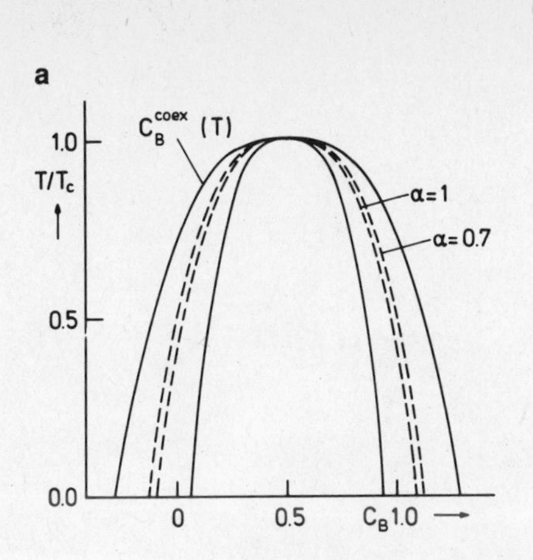

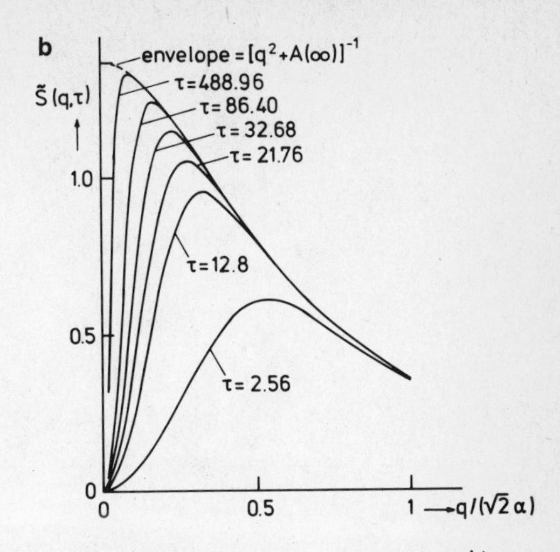

Fig.5 a) Phase diagram [11] and b) Scaled structure factor [11] for $(c_B-c_B^{crit})/c_{coex}^{(1)}-c_B^{crit})=0.88$.

The results for the nonconserved case [12], $q_{max}(t)\propto t^{-1/2}$, $\varphi\propto q_{max}^{-3}\propto t^{3/2}$, are in good agreement with simulation [18] (Fig.6b), while for very late stages $S(q,t\to\infty)\to q^{-2}$, $\varphi\to 0$, which is an unphysical result {Note that here $S(\vec{q},t)$ is always maximal at q=0, Fig.6b, and q_{max} is found from the maximum of $q^2S(\vec{q},t)$, Fig.6a}. For solid mixtures, one finds [8] $q_{max}(t)\propto t^{-1/4}$, which is not expected to be correct [8,11], one rather expects to find Lifshitz-Slyozov behavior [19] $q_{max}(t)\propto t^{-1/3}$. However, Eq. (3.5) is found to be valid both for simulations [20] and experiment on fluid mixtures [21] (Fig.7a). Neither the present theory [10-13] nor its extension to fluid mixtures [22] can calculate the scaling function $\tilde{S}$ reliably, as the behavior of Eq. (3.4) is unreliable for $q_{max}(t)\xi<<1$, the theory is valid for early stages only [10]. For late stages one does not expect the predicted q^{-2} decay (Fig.7b) but rather a decay governed by two lengths $L(t)\propto q_{max}^{-1}(t),\xi$ (Fig.7c). The "background" $\propto[q^2+\xi^{-2}]^{-1}$ represents the scattering from fluctuations inside the growing domains, while the extra scattering [Eq. (3.5)] is due to their walls.

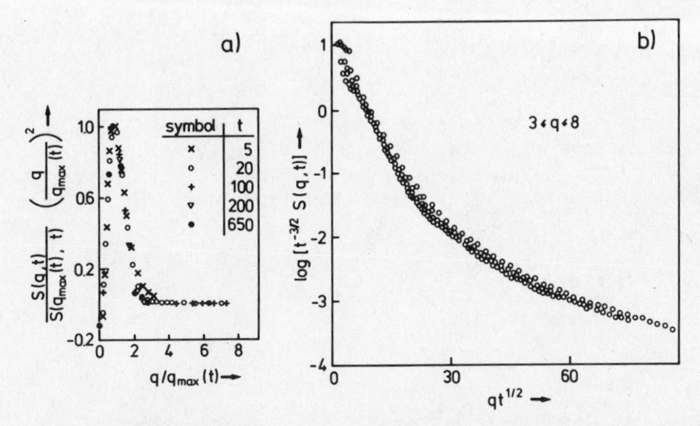

Fig.6 Scaling plot for structure factor [12] (a) and corresponding simulation [18] (b)

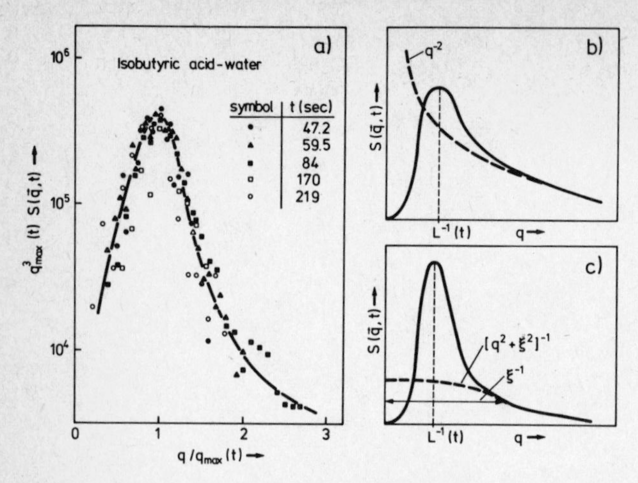

Fig.7 Scaling plot for the scattering from a fluid mixture [21] (a) Schematic variation of the structure factor at late times according to Eq. (3.3) (b) and presumable correct behavior (c) [11,12].

4. Nucleation Theory

To be specific, we consider a solid alloy and the probability $P(\vec{X},t)$ of a state $\vec{X}=(c_1,c_2,\ldots,c_N)$. Atomic diffusion jumps are described by a transition probability per unit time [τ = characteristic time, $\delta\mathcal{H}$ = energy change]

$$W(c_i\to c_j, c_j\to c_i)=\tau^{-1}\exp(-\delta\mathcal{H}/k_BT)/[1+\exp(-\delta\mathcal{H}/k_BT)]. \tag{4.1}$$

Then $P(\vec{X},t)$ follows a master equation,

$$dP(\vec{X},t)/dt=-\sum_{\vec{X}'} W(\vec{X}\to\vec{X}')P(\vec{X},t)+\sum_{\vec{X}'} W(\vec{X}'\to\vec{X})P(\vec{X},t)\ . \tag{4.2}$$

Eq. (4.2) can be used to derive Cahn's theory [1] (or the theories of Cook [2] or Langer [10]) by meanfield assumptions [23] or variants thereof [8]. Here we rather introduce a description of the state $\vec{X}$ by coordinates $\{\ell_i,\{\alpha_i\}\}$ of clusters of B-atoms (Fig.8a). The number of B-atoms in the i-th cluster is ℓ_i, and $\{\alpha_i\}$ denotes other degrees of freedom. They are not considered explicitly, but rather the average cluster concentration $\bar{n}_\ell(t)=\langle n_\ell^{\{\alpha\}}\rangle_{\{\alpha\}}$. In principle, one can derive reaction rates $C_{\ell,\ell'}^{\{\alpha\},\{\alpha'\}}$ for all kinds of cluster reactions from Eq. (4.2). An exact kinetic equation for $\bar{n}_\ell(t)$ would involve terms like $\langle C_{\ell,\ell'}^{\{\alpha\},\{\alpha'\}} n_\ell^{\{\alpha\}} n_{\ell'}^{\{\alpha'\}}\rangle_{\{\alpha\},\{\alpha'\}}$. We now introduce a meanfield-like factorization, replacing such terms by $C_{\ell,\ell'}\bar{n}_\ell(t)\bar{n}_{\ell'}(t)$. Fluctuations of cluster properties to some extent are included by a "multi-cluster-coordinate-theory [24] but this is beyond our scope here. The full kinetic equations, involving both cluster coagulation $\{\propto C_{\ell,\ell'}\}$ and splitting $\{\propto S_{\ell,\ell'}\}$, are [8]

$$\frac{d}{dt}\bar{n}_\ell(t)=\sum_{\ell'=1}^{\infty} S_{\ell+\ell',\ell'}\bar{n}_{\ell+\ell'}(t)-\frac{1}{2}\sum_{\ell'=1}^{\ell-1} S_{\ell,\ell'}\bar{n}_\ell(t)+\frac{1}{2}\sum_{\ell'=1}^{\ell-1} C_{\ell-\ell',\ell'}\bar{n}_{\ell'}(t)\bar{n}_{\ell-\ell'}(t)-\sum_{\ell'=1}^{\infty} C_{\ell,\ell'}\bar{n}_{\ell'}(t). \tag{4.3}$$

Eq. (4.3) is still consistent with the conservation of total concentration, $c_B = \Sigma\ell n_\ell(t)$=const. It can also be applied in equilibrium, where detailed balance holds between splitting and coagulation, $S_{\ell+\ell',\ell'}n_{\ell+\ell'}=C_{\ell,\ell'}n_\ell n_{\ell'}\equiv W(\ell,\ell')$ [n_ℓ=equilibrium

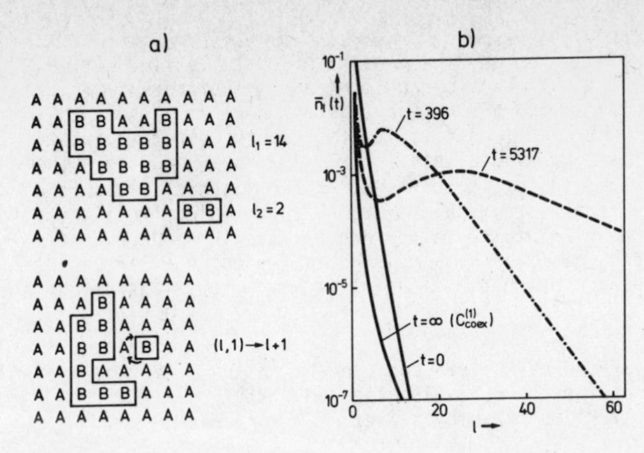

Fig.8 a) "Clusters" defined from contours around groups of B-atoms and their reactions; b) Cluster size distribution resulting from numerical solution of Eq. (4.3) [24]

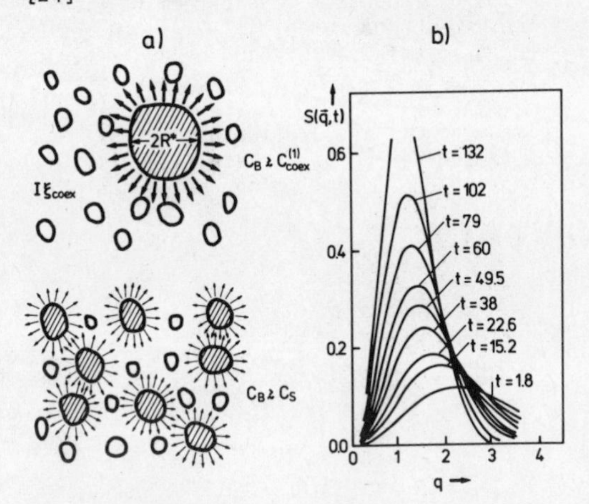

Fig.9 a) Schematic "snapshot pictures" of fluctuations of an unmixing system in the metastable (above) or unstable regime (below) [unstable fluctuations are shaded] b) $S(\vec{q},t)$ plotted vs. q for various times after the quench from infinite temperature to $T=0.6\ T_c$ for a 3 dim.-Ising model at $c_B=0.1$. All units contain scale factors [11]

cluster concentration]. With plausible assumptions on the reaction matrix $W(\ell,\ell')$ and on n_ℓ one can integrate Eq. (4.3) numerically, specifying as initial state $\bar{n}_\ell(0)$ e.g. the cluster concentration of random mixing. The final equilibrium ($t\to\infty$) is then given by n_ℓ (which describes fluctuations in A-rich phase of concentration $c^{(1)}_{coex}$), while the excess $\Sigma\ell[\bar{n}_\ell(0)-n_\ell]$ is redistributed into macroscopic B-rich domains (of concentration $c^{(2)}_{coex}$), i.e. occurs at cluster size $\ell\to\infty$. For intermediate times, one finds [25] a nonmonotonic cluster size distribution (Fig.8b): a minimum occurs at the "critical size" ℓ^* of nucleation, while the broad maximum at larger sizes is due to the growing supercritical clusters which have already been nucleated. For $\ell<\ell^*$ the cluster size distribution is basically the equilibrium size distribution of a slightly supersaturated gas. As time goes on, the peak shifts to larger and larger sizes, and at the same time the supersaturation is diminished, until for $t\to\infty$ the peak has shifted to $\ell\to\infty$. This picture of phase separation emerges also very clearly from the computer simulations [14].

Now the *gradual transition from nucleation to spinodal decomposition* is readily understood [11,24]. In the metastable regime, the density of unstable fluctuations ("critical clusters") is very small (Fig.9a), because the energy barrier $\Delta F^* >> k_BT$. In the spinodal regime, on the other hand $\Delta F^* \approx k_BT$, and hence there is a high density of unstable fluctuations: ΔF^* here no longer limits the growth, but rather the conservation of concentration: near growing clusters $c(\vec{r})$ decreases, and there then no other cluster can grow. Due to this "excluded volume"-interaction of clusters a nearly periodic variation of concentration results - roughly equivalent to a wave-packet of Cahn's concentration waves. But the latter are not growing independently, rather they are strongly interacting. It is more reasonable to consider large clusters as objects nearly independent of each other (apart from the reactions in Eq. (4.3)) rather than these waves.

For a more quantitative comparison of the present concept to the other approaches, we relate $S(\vec{q},t)$ to $\bar{n}_\ell(t)$ introducing the conditional probability $g_\ell(\vec{r})$ that the site $\vec{r}'+\vec{r}$ is taken by a B-atom if $\vec{r}'$ is taken by a B-atom of an ℓ-cluster,

$$S(\vec{q},t) = \sum_{\ell=1}^{\infty} \ell\bar{n}_\ell(t) d^3r\{g_\ell(\vec{r}) - c_B\} \exp(i\vec{q}\cdot\vec{r}). \tag{4.4}$$

Simple assumptions for $g_\ell(\vec{r})$ yield $S(\vec{q},t)$ [Fig.9b] in qualitative agreement with experiment [15], simulation [14] and the theory of Langer et al. [10-13]. In contrast to the latter, the present theory reduces to Lifshitz-Slyozov behavior [19] at late times and incorporates nucleation and growth also.

Now we briefly sketch the analytic treatment of Eq. (4.3). Clusters are divided in two classes: for $\ell<\ell_c$ equilibrium with a time-dependent supersaturation is assumed $\{$i.e., $\bar{n}_\ell(t)=n_\ell \exp[-\Delta\mu(t)\ell/k_BT]\}$. For $\ell>\ell_c$ domain splitting can be neglected, and sums replaced by integrals. Hence $[R_\ell=(1/n_\ell)\sum_{\ell'=1}^{\ell_c} \ell'^2 W(\ell,\ell')]$

$$\frac{\partial\bar{n}_\ell}{\partial t} = \frac{\partial}{\partial\ell}\left\{R_\ell n_\ell[\Delta\mu(t)]\frac{\partial}{\partial\ell}\frac{\bar{n}_\ell(t)}{n_\ell[\Delta\mu(t)]}\right\} + \frac{1}{2}\int_{\ell_c}^{\ell} d\ell'\, W(\ell-\ell',\ell')\,\frac{\bar{n}_{\ell'}(t)}{n_{\ell'}}\,\frac{\bar{n}_{\ell-\ell'}(t)}{n_{\ell-\ell'}}$$

$$- [\bar{n}_\ell(t)/n_\ell]\int_{\ell_c}^{\infty} d\ell' W(\ell,\ell')\bar{n}_{\ell'}(t)/n_{\ell'}\ , \tag{4.5}$$

and $\Delta\mu(t)$ is obtained from $\bar{n}_\ell(t)$ implicitly via the conservation law. Eq. (4.5) relates several theories: (i) for equilibrium fluctuations, coagulation terms are neglected. One obtains the correlation function as expected [D_ℓ=cluster diffusivity]

$$S_T(\vec{q},t)=S_T(\vec{q})\exp[-D_Tq^2t],\quad D_T=\Sigma\ell^2D_\ell n_\ell/\Sigma\ell^2n_\ell. \tag{4.6}$$

Applying scaling assumptions for n_ℓ, D_ℓ close to T_c, one obtains [8] $D_T \propto \epsilon^\gamma$ as expected [17]. However, then one must use no longer the "contour description" (Fig.8a) but must define clusters by the excess order parameter of local concentration fluctuations [26] to avoid difficulties with percolation. (ii) Standard nucleation theory [24] follows from neglecting coagulation, for $\Delta\mu$ constant (and negative). The nucleation rate J (number of supercritical clusters per unit volume and second) then becomes $J=[\int d\ell/\{R_\ell n_\ell\}]^{-1}$. Near T_c one again derives a scaling law $J=\varepsilon^{2-\alpha+\nu z}\mathfrak{J}(\delta T/\Delta T)$ [cf. Fig.1] [24], but the calculation of $\mathfrak{J}$ is still hampered by imprecise knowledge of the (effective) surface free energy f_ℓ of clusters $[n_\ell\equiv\exp(-f_\ell/k_BT)]$ [7,24]. (iii) Lifshitz-Slyozov theory [19] results by also neglecting coagulation but taking the time-dependence of $\Delta\mu(t)$ properly into account. For large ℓ one finds [8] a *scaling solution* $\bar{n}_\ell(t)=t^{-2x}\tilde{n}(\ell t^{-x})$ with $x=d/3$, i.e. the law $L(t) \propto t^{1/3}$. On this basis, one also expects scaling of $S(\vec{q},t)$ for $t\to\infty$ [Eq. (3.5)]. (iv) The opposite approximation, valid at intermediate times and c_B

not too close to $c_{coex}^{(1)}$, consists in neglecting the nucleation term, taking only coagulation into account. For large ℓ again the above scaling solution applies, but x is different: for solid mixtures and low T, $x=d/(d+3)$, a law suggested [7] to explain simulations [14]. For liquids, Stokes law $D_\ell \propto \ell^{(1-2/d)}$ implies x=1.[8] However, hydrodynamic interactions lead to much quicker coalescence [27] $L(t) \propto t$. Finally, for nonconserved order parameter [24,26] $L(t) \propto t^{1/2}$.

5. Conclusions

In this talk it was stated that (i) a linearized theory (exponential growth of fluctuations) fails completely for phase separation; (ii) a sharp spinodal separating metastable from unstable states does not exist, the transition from nucleation to spinodal decomposition is gradual; (iii) approximate nonlinear theories à la Langer et al. are satisfactory only outside of the "spinodal region" and for early stages of the separation only (iv) nucleation theory is still hampered by the lack of reliable methods to compute the surface free energy of clusters; (v) the "cluster dynamics"-approach to spinodal decomposition and domain growth is qualitatively satisfactory, but many quantitative problems remain: fluctuations in cluster properties; correlations between clusters; hydrodynamic interactions in fluids [27]; percolation phenomena [28] at compositions near c_B^{crit}; etc. Hence one is still far from a reliable theoretical understanding of the experiments.

References

1. J.W. Cahn, Trans. Metall. Soc. AIME 242, 166 (1968), and references therein
2. H.E. Cook, Acta Met. 18, 297 (1970)
3. J.S. Langer, Ann. Phys. (N.Y.) 65, 53 (1971)
4. O. Penrose and J.L. Lebowitz, J. Statist. Phys. 3, 211 (1971)
5. K. Binder, Phys. Rev. B8, 3423 (1973)
6. J.S. Langer, physica 73, 61 (1974)
7. K. Binder and M.H. Kalos, J. Statist. Phys. 22, 363 (1980)
8. K. Binder, Phys. Rev. B15, 4425 (1977)
9. N.-C. Wong and C.M. Knobler, Phys. Rev. Lett. 43, 1733 (1979)
10. J.S. Langer, M. Baron, and H.D. Miller, Phys. Rev. A11, 1417 (1975)
11. K. Binder, C. Billotet, and P. Mirold, Z. Physik B30, 183 (1978)
12. C. Billotet and K. Binder, Z. Physik B32, 195 (1979)
13. C. Billotet and K. Binder, physica (1980)
14. See e.g. J. Marro, A.B. Bortz, M.H. Kalos, and J.L. Lebowitz, Phys. Rev. B12, 2000 (1975)
15. See e.g. V. Gerold and J. Kostorz, Appl. Cryst. 11, 376 (1978)
16. K. Binder and E. Stoll, Phys. Rev. Lett. 31, 47 (1973)
17. B.I. Halperin and P.C. Hohenberg, Revs. Modern Phys. 49, 435 (1977)
18. M.K. Phani, J.L. Lebowitz, M.H. Kalos, and O. Penrose, Phys. Rev. Lett. 45, 366 (1980)
19. I.M. Lifshitz and V.V. Slyozov, J. Phys. Chem. Solids 19, 35 (1961)
20. J. Marro, J.L. Lebowitz, and M.H. Kalos, Phys. Rev. Lett. 43, 282 (1979)
21. Y.C. Chou and W.I. Goldburg, preprint
22. K. Kawasaki and T. Ohta, Progr. Theor. Phys. 59, 362 (1978)
23. K. Binder, Z. Physik 267, 313 (1974)
24. K. Binder and D. Stauffer, Adv. Phys. 25, 343 (1976); K. Binder, J. de phys. 41, C4-51 (1980)
25. P. Mirold and K. Binder, Acta Met. 25, 1435 (1977)
26. R. Kretschmer, K. Binder and D. Stauffer, J. Stat. Phys. 15, 267 (1976)
27. E. Siggia, Phys. Rev. A20, 595 (1979)
28. K. Binder, Solid State Comm. 34, 191 (1980)

Chemical Instabilities and Broken Symmetry: The Hard Mode Case

D. Walgraef*, G. Dewel*, and P. Borckmans*
Chimie-Physique II, C.P. 231, Université Libre de Bruxelles
B-1050 Bruxelles, Belgium

Introduction

Chemical instabilities are interesting for the theorist for several reasons. The most important to our purposes are the following. These instabilities occur far out of thermal equilibrium and ordered states may appear through space or time symmetry breaking. In the vicinity of the transition points anomalies in the static and dynamical behaviour of the fluctuations, correlation functions, ..., occur. These anomalies are very similar to critical slowing down, divergence of correlation lengths, existence of long range fluctuations, which are well known to play an essential role in the understanding of equilibrium phase transitions. So despite the loss of important properties like detailed balance or fluctuation-dissipation theorems, far from equilibrium chemical instabilities show important similarities with ordinary equilibrium phase transitions [1-3]. Moreover, near the instability the behaviour of the system may be thought to be governed by a few modes only which evolve on long time scales and for these modes detailed balance may be recovered [4]. Great progress has been made these last years in this domain, namely by the use of renormalization group theories to justify scaling properties, universality concepts and to calculate critical indices also by the derivation of exact relations or theorems associated to the breaking of continuous symmetries in the ordered phases [3,7]. As our understanding of the real world lives on analogies, it is then an attractive idea to apply these powerful concepts to get a deeper understanding of non equilibrium phase transitions.

We are, however, aware that although such analogies are helpful they are seldom complete and their eventual limitations must be discussed.

In this work we illustrate the application of these methods to the study of the hard mode instability leading to chemical oscillations. Such a program has already been carried out for other types of instabilities, namely the transition between multiple steady states or the pattern selection in systems spontaneously developing spatial ordering [5,6].

---

* Chercheur qualifié au Fonds National de la Recherche Scientifique

The open non-linear chemical systems we consider are assumed to be in mechanical and isothermal equilibrium. They may be described by non-linear reaction-diffusion equations for the local concentrations of the intermediate species. These deterministic rate equations are unable to give a complete description of the macroscopic system. Because of the complexity of microscopic events, of molecular motions and also of the feeding mechanisms, fluctuations are to be taken into account. It is well known that they play a major role near instability points. These effects are introduced in the description by adding appropriate noise terms to the rate equations. We will, however, restrict ourselves to internal noise and consider the pump parameters as mere constants.

These noise terms, coming from chemical and diffusive processes which are local and rapid, quickly loose the memory of their earlier behaviour and are correlated over short times and distances. So we will assume as most of the physicists dealing with this kind of problem that the noise is Gaussian and white. We further admit that the system remains in a state of local equilibrium throughout the transition region so that local fluctuation-dissipation holds. The correlation functions may then be obtained by integration over the local fluctuating forces using the response function associated to the rate equations.

For small constraints the system admits one stable homogeneous solution. When the constraints are increased this reference state usually becomes unstable and different types of bifurcation may appear. In this paper we will consider only the hard mode instability leading to time periodic solutions of the limit cycle type. The unstable modes play the role of order parameter and evolve on a much longer time scale than the others. The rapid modes may be adiabatically eliminated and the description of the system near the instability depends only on the slow mode dynamics [2].

The Model

To pursue our analysis let us now discuss a concrete example. We consider the Brusselator model [1]. Using standard notations, the rate equations for the local concentrations are in Fourier transforms

$$\partial_t X_q = A - (B + 1 - q^2 D_x) X_q + L^{-d} \sum_{k,k'} X_k X_{k'} Y_{q-k-k'}$$

$$\partial_t Y_q = B X_q - q^2 D_y Y_q - L^{-d} \sum_{k,k'} X_k X_{k'} Y_{q-k-k'} \quad , \tag{1}$$

A is a constant and B plays the role of the pump parameter. The inhomogeneous fluctuations around the uniform steady state X = A, Y = B/A obey the following kinetic equation

$$\partial_t \begin{pmatrix} x_q \\ y_q \end{pmatrix} = \begin{pmatrix} B-1-q^2D_x & A^2 \\ -B & -A^2-q^2D_y \end{pmatrix} \begin{pmatrix} x_q \\ y_q \end{pmatrix} + \Bigg(L^{-d/2} \sum_k x_k$$

$$\cdot \left(\frac{B}{A} x_{q-k} + 2Ay_{q-k}\right) + L^{-d} \sum_{k,k'} x_k x_{k'} y_{q-k-k'} \Bigg) \cdot \begin{pmatrix} +1 \\ -1 \end{pmatrix} . \tag{2}$$

From the characteristic equations for the eigenvalues of the linear evolution matrix one deduces that, when,

$$(D_x/D_y)^{1/2} > [(1 + A^2)^{1/2} - 1] \cdot A^{-1}$$

a hard mode transition to chemical oscillations is the first to occur at $B = B_c = 1 + A^2$, $q = 0$.

Around this instability point the eigenfrequencies may be written

$$\omega_q = \frac{1}{2} \{B - B_c - q^2(D_x + D_y) \pm 2iA[1 + q^2(D_x - D_y)$$

$$- 4A^{-2}((B - B_c) - q^2(D_x - D_y))^2]^{1/2}\} . \tag{3}$$

The corresponding eigenvectors play the role of a two component order parameter. They are written, at the lowest order in q^2 and $B - B_c$

$$T_q = \frac{i - A}{2i} \left[\frac{A + i}{A} x_q + y_q\right] , \quad S_q = T^*_q . \tag{4}$$

The associate non-linear Langevin equation may easily be obtained in the rotating wave approximation (valid only near the instability point $-|B - B_c| \ll A$ - and for time much greater than A^{-1}) as

$$\partial_\tau \sigma_q = -[r + cq^2]\sigma_q - uL^{-d} \sum_{k,k'} \sigma_k \sigma^*_{k'} \sigma_{q-k-k'} + \eta_k \tag{5}$$

with

$$\sigma_q = T_q e^{ia\tau} , \quad \tau = B_c t , \quad r = r_1 + ir_2 , \quad c = c_1 + ic_2 , \quad u = u_1 + iu_2$$

$$r_1 = (B_c - B)/B_c , \quad r_2 = (B - B_c)^2/2AB_c , \quad c_1 = (D_x + D_y)/B_c , \quad c_2 = (D_x - D_y)/2AB_c$$

$$u_1 = (2 + A^2)/2A^2B_c , \quad u_2 = (4 - 7A^2 + 4A^4)/3A^3B_c$$

$$\langle\eta_q(\tau)\eta^*_{q'}(\tau')\rangle = 2\left[\Gamma_c + q^2\Gamma_d\right]\delta_{q,-q'}\delta(\tau - \tau') \tag{6}$$

where Γ_c is the chemical noise and Γ_d the diffusive noise which is irrelevant to the long wavelength behaviour.

As the instability occurs at $q = 0$, the new phase will be homogeneous and as a first step of our description let us study the homogeneous problem which corresponds to the zero dimensional regime. Then (5) is equivalent to the TDGL model

$$\partial_\tau \sigma_0 = -\frac{\delta F_0}{\delta \sigma_0^*} - i[r_2\sigma_0 + u_2|\sigma_0|^2\sigma_0] + \eta_0 \tag{7}$$

where the Ginzburg-Landau potential is

$$F_0 = \frac{1}{\Gamma_0}\left[r_1|\sigma_0|^2 + \frac{u_1}{2}|\sigma_0|^4\right] \tag{8}$$

the distribution function for the fluctuations being

$$P \simeq \exp - F_0 \quad . \tag{9}$$

A first manifestation of universality lies in the fact that F_0 is the standard homogeneous potential for a system with a two component order parameter such as the XY model [7], the superfluid [8] or the single mode laser [9]. Its shape, a Gaussian well as long as r_1 is positive, becomes craterlike for r_1 negative, around the orbit of the limit cycle corresponding to the stationary solution of (7). This was already known from a master equation description of this problem [10].

While the static behaviour of the system belongs at this level of description to the n = 2 universality class the dynamics will include specific effects due to the divergence free current $V_0 = i[r_2\sigma_0 + u_2|\sigma_0|^2\sigma_0]$. These effects already occur in the mean field or Landau theory where the actual value $\bar{\sigma}_0$ of the order parameter maximizes the distribution function.

Mean Field Theory

Writing $\bar{\sigma}_0(\tau) = R(\tau)\exp i\phi(\tau)$ one obtains

$$R^2(\tau) = \begin{cases} - r_1R^2(0)[u_1R^2(0) - \left(r_1 + u_1R^2(0)\right)\exp 2r_1\tau]^{-1} & r_1 \neq 0 \\ R^2(0)[1 + 2u_1R^2(0)\tau]^{-1} & r_1 = 0 \end{cases} \tag{10}$$

leading to the stationary values

$$R^2(\tau \to \infty) = \bar{R}^2 = \begin{cases} 0 & r_1 \geqq 0 \\ |r_1|/u_1 & r_1 < 0 \end{cases} \tag{11}$$

characteristic of a mean field second order transition.

On the other hand the phase behaviour is given by

$$\phi(\tau) = \begin{cases} \phi(0) - \frac{u_2}{2u_1}\ln\left[1 + u_1R^2(0)/r_1\right] - r_2\tau & r_1 > 0 \\ \phi(0) - \frac{u_2}{u_1}\ln\frac{R(0)}{\bar{R}} - (r_2 + u_2\bar{R}^2)\tau & r_1 < 0 \quad . \end{cases} \tag{12}$$

We see that the specific aspects of the Brusselator lead to a frequency shift in the ordered phase ($r_1 < 0$) and a phase shift due to any perturbation even of the radius of the limit cycle only and of amplitude $\phi(0) - (u_2/u_1)\ \ln\ [R/(0)/\bar{R}]$. The corresponding isochrons, which are the locus of all the points in the ($\mathrm{Re}\sigma$, $\mathrm{Im}\sigma$) plane which come into phase synchrony with one another on the limit cycle for $t \to \infty$ [11] are thus logarithmic and show a 2π discontinuity for $R_0(k) = \bar{R}\ \exp(u_1/u_2)(\phi_0 + 2k\pi)$.

The static correlation functions with respect to the rotating frame may be calculated from the probability distribution giving identical results to those obtained for the single mode laser [12]. Let us recall that far from the transition point and for intermediate time scales ($t \lesssim A^{-1}$) important deviations from this behaviour may occur as the rotating wave approximation is no more justified as discussed for instance by GROSSMAN et al. [13]. Nevertheless the transition appears to be second order with a phase symmetry breaking. As in the case of equilibrium phase transitions of this kind one expects that the inhomogeneous fluctuations may drastically affect the behaviour of the system for space dimensionality less than four. Moreover the symmetry which is spontaneously broken is continuous and one expects the existence of symmetry restoring mechanisms associated to GOLDSTONE modes. Let us note that in the zero dimensional regime the phase diffusion is the symmetry restoring mechanism. To discuss these problems we have to take into account the full inhomogeneous problem.

Inhomogeneous Fluctuations

Going along the same lines as in the zero dimensional description, (7,8) is equivalent to the following TDGL model

$$\partial_t \sigma_k = -\Gamma \frac{\delta F}{\delta \sigma^*_{-k}} + V_k + \eta_k \quad . \tag{13}$$

The generalized potential and the divergence-free current are such that

$$\sum_k \left(\frac{\delta}{\delta \sigma_k} V_k + \frac{\delta}{\delta \sigma^*_k} V^*_k \right) \exp - F = 0 \tag{14}$$

and

$$F = \frac{1}{\Gamma} \left[\sum_k (r_1 + c_1 k^2) |\sigma_k|^2 + \frac{1}{2L^d} \sum_{k,k',k''} u(k,k',k'') \sigma_k \sigma^*_{k'} \sigma_{k''} \sigma^*_{-k-k'-k''} \right]$$

$$u(k,k',k'') = u_1 + i(c_1 u_2 - c_2 u_1) \beta(k,k',k'') \cdot [\alpha(k,k',k'') + i c_2 \beta(k,k',k'')]^{-1}$$

$$\beta(k,k',k'') = k^2 - k'^2 + k''^2 - (k + k' + k'')^2$$

$$\alpha(k,k',k'') = 4r_1 + c_1 [k^2 + k'^2 + k''^2 + (k + k' + k'')^2] \tag{15}$$

$$V_k = i(r_2 + c_2 k^2) \sigma_k + L^{-d} \sum_{k',k''} \left[\frac{i(c_1 u_2 - c_2 u_1) \beta(k,k',k'')}{\alpha(k,k',k'') + i c_2 \beta(k,k',k'')} - i u_2 \right] \sigma_{k'} \sigma_{k''} \sigma^*_{k-k'-k''} \quad . \tag{16}$$

It may easily be checked that (14) is verified up to terms of order six in σ and this is consistent with the approximations made up to now. Moreover only the space independent part of the noise is taken into account as the only relevant one to the critical behaviour of the system. One immediately sees that for $c_2 = 0$, $u_2 = 0$ one recovers the standard $n = 2$ GINZBURG-LANDAU potential. Moreover when $c_2, u_2 \neq 0$ the k dependent terms in u have a critical dimensionality of two and are irrelevant to the critical behaviour of $d > 2$ systems. So three dimensional systems have the static critical indices of the $n = 2$ model obtained by the renormalization group approach as for example

$$|\sigma_0| \simeq (B - B_c^*)^\beta \quad \beta = \frac{1}{2} - \frac{3\varepsilon}{20} \quad (\varepsilon = 4 - d)$$

$$\langle\sigma^*_{-k}\sigma_k\rangle \cong \Gamma/c_1k^2 + \xi_1^{-2} \quad \xi_1 \simeq |B - B_c^*|^{-\nu_1} \quad \nu_1 = \frac{1}{2} + \frac{\varepsilon}{10} \ . \tag{17}$$

B_c^* is the true instability point as the mean field instability point is shifted by the fluctuations. These results are in complete agreement with HENTSCHEL's [14] who directly applied the dynamical renormalization group to (5). Moreover he showed that in the disordered phase the dynamical correlation functions behave as

$$\langle\sigma^*_{-k}(t)\sigma_k(0)\rangle \simeq 2\,\langle\sigma^*_{-k}\sigma_k\rangle \exp - \Gamma t/\tau(k)$$

where

$$\tau^{-1}(k) = c_1k^2 + \xi_1^{-2} + i(c_2k^2 + c_2\xi_1^{-2} + \xi_2^{-2}) \tag{18}$$

with

$$\xi_2 \simeq |B - B_c^*|^{-\nu_2} \quad \nu_2 = \frac{1}{2} \ .$$

Despite the spontaneous symmetry breaking in the ordered phase, the order parameter can rotate homogeneously in the ($\mathrm{Re}\sigma$, $\mathrm{Im}\sigma$) plane without change in the potential as a consequence of degeneracy. Consequently the small wavenumber fluctuations are not weak. The existence of such long range fluctuations in systems with broken continuous symmetries is well known in quantum field theory as the GOLDSTONE theorem. Using the standard methods to study the response of the system to a long wavelength inhomogeneous phase shift we may derive the following BOGOLUBOV inequality for the transverse correlation function [15]

$$g_\perp(k) = \langle|\sigma_k - \sigma_k^*|^2\rangle \geq |\sigma_0|^2/[\mathrm{coeff.}]\cdot k^2 \tag{19}$$

which is the confirmation of the result obtained in the quasi-linear approximation [16].

Equation (19) implies the destruction of long range order for $d \leq 2$ infinite systems as in this case the mean square of the order parameter fluctuations become infinitely large. This contradiction may be removed if the radius of the limit cycle vanishes as well. However, for $d = 2$ this divergence being of logarithmic type is weak. It may thus be unessential for systems of finite size at sufficiently high values of B. Indeed the reduced mean square fluctuations of the order parameter behave as

$$\int d^d q \langle \sigma_q \sigma^*_{-q} \rangle / |\sigma_0^2| \simeq \frac{u_1 \Gamma B_c}{B - B_c} \ln \frac{L}{\Lambda} \tag{20}$$

L being the linear size of the system. This quantity is small for $L < L_c = \Lambda \exp(B - B_c)/u_1 \Gamma B_c$ so that finite systems become ordered for sufficiently high values of B. Furthermore the following discussion shows the evidence for some kind of transition even in two dimensional systems [17]. For low values of B the systems exhibits short range order as the correlation function of the order parameter behaves as

$$|\langle \sigma(r')\sigma^*(r) \rangle| \simeq \frac{\Gamma}{|r - r'|^{1/2}} \exp - \frac{|r - r'|}{c_1 \xi_1} . \tag{21}$$

For sufficiently great values of B, macroscopically large parts of the system may become ordered and the order parameter looks like $\sigma(r) = \bar{R} \exp i\phi(r)$, $\bar{R}$ being the mean field value of $|\sigma(r)|$ while ϕ changes slowly as a consequence of the local phase symmetry breaking.

The potential then becomes

$$F = F_0 + \frac{\bar{R}^2 c_1}{2\Gamma} \int d^d r (\nabla\phi)^2 . \tag{22}$$

The order parameter correlation functions may then be written as

$$|\langle \sigma(r')\sigma^*(r) \rangle| = \bar{R}^2 \exp - \frac{1}{2} \langle [\phi(r') - \phi(r)]^2 \rangle \tag{23}$$

and

$$\langle [\phi(r') - \phi(r)]^2 \rangle = \begin{cases} \dfrac{2\Gamma}{\bar{R}^2 c_1} \displaystyle\int^{\Lambda} \frac{dq}{q} [1 - J_0(q|r - r'|)] & d = 2 \\ \dfrac{4\Gamma}{\bar{R}^2 c_1} \displaystyle\int^{\Lambda} \frac{dq}{q^2} [1 - \cos q|r - r'|] & d = 1 \end{cases} \tag{24}$$

leading at large distance to

$$|\langle \sigma(r')\sigma^*(r) \rangle| \simeq \begin{cases} \bar{R}^2 \left(\dfrac{|r - r'|}{\Lambda} \right)^{-(2\Gamma/\bar{R}^2 c_1)} & d = 2 \\ \bar{R}^2 \exp - \dfrac{2\pi\Gamma}{\bar{R}^2 c_1} |r - r'| & d = 1 . \end{cases} \tag{25}$$

For one dimensional systems there is short range order for any finite values of R while for two dimensional systems there is no definite correlation length as correlations decrease in a power like manner with distance. This quasi-long range order is well known in equilibrium XY or superfluid systems.

The transitions between the two regimes is of topological nature as extensively discussed in [18-20]. The next section is devoted to the applicability of these concepts to the system we consider. For $B > B_c$ the kinetic equation may be written in terms of the radius and phase fluctuations

$$\sigma(r,t) = [\bar{R} + \rho(r,t)]\exp i[\omega t + \phi(r,t)]$$

$$\omega = r_2 + u_2\bar{R}^2$$

$$\partial_t\rho = -(2u_1\bar{R}^2 - c_1\nabla^2)\rho - \bar{R}c_2\nabla^2\phi - 3u_1\bar{R}\rho^2 - 2c_2(\nabla\rho)(\nabla\phi) - c_1\bar{R}(\nabla\phi)^2 + \eta_\rho \tag{26}$$

$$\partial_t\phi = -\left(2u_2\bar{R} - \frac{c_2}{\bar{R}}\nabla^2\right)\rho + c_1\nabla^2\phi - u_2\rho^2 + \frac{2c_1}{\bar{R}}(\nabla\rho)(\nabla\phi) - c_2(\nabla\phi)^2 + \eta_\phi \quad .$$

The stability analysis shows that the behaviour of the system depends on the sign of $u_1c_1 + u_2c_2$. When $u_1c_1 + u_2c_2 > 0$ the limit cycle is unstable against inhomogeneous fluctuations and this may lead to phase turbulence [21]. In similar equilibrium systems the phase and radius fluctuations are decoupled ($u_2 = c_2 = 0$) and the phase is the only symmetry restoring variable. In our case phase and radius fluctuations are coupled on short distances and are both symmetry restoring. On the contrary they are decoupled for long distances $[|r - r'| > L_0 = (c_2u_2/u_1r_1)^{1/2}]$. On this length scale, the radius fluctuations may be adiabatically eliminated and we obtain another TDGL description of the system which may be written at the lowest order in the non-linearity

$$\partial_t\phi = \left(c_1 + \frac{u_2c_2}{u_1}\right)\nabla^2\phi - \left(c_2 - \frac{u_2c_1}{u_1}\right)(\nabla\phi)^2 + \eta_\phi$$

$$= -\frac{\delta F}{\delta\phi} - \left(c_2 - \frac{u_2c_2}{u_1}\right)(\nabla\phi)^2 + \eta_\phi$$

$$F = \frac{\bar{R}^2}{2\Gamma}\left(c_1 + \frac{u_2c_2}{u_1}\right)\int_{L_0} d^dr(\nabla\phi)^2 + F_0$$

$$\langle\eta_\phi(r,t)\eta_\phi(r',t')\rangle = \frac{2\Gamma}{\bar{R}^2}\,\delta(r - r')\delta(t - t') \tag{27}$$

where the regular part F_0 of F contains the short range phase and radius fluctuations ($L < L_0$). Again the static properties will be the same as for the XY model or superfluid films while the dynamics will show specific effects. As a consequence the results obtained by KOSTERLITZ and THOULESS for the topological phase transition in the XY model may be extended to our system. Let us recall the main points of their theory [18]. The whole story is based on the fact that vortices could play an important role at finite temperatures. Vortices are isolated points where the phase is not continuous or differentiable anymore because the order parameter vanishes at these points. They are characterized by a net circulation which has to be quantized in units for 2π as exp $i\phi$ must be single valued

$$\oint \vec{\nabla}\phi \, d\vec{r} = 2\pi q \qquad q = 0, \pm 1, \pm 2, \ldots \tag{28}$$

where the contour surrounds a vortex of charge q. Consequently $|\vec{\nabla}\phi|$ falls off like r^{-1} far from the vortex while the energy of an isolated vortex diverges logarithmically with the system size according to (28). This divergence may be removed if one considers a pair of vortices of opposite charge. Because of cancellation from the two vortices charges $\vec{\nabla}\phi$ falls off faster than r^{-1} for distances large compared to the vortex separation, the energy of a pair grows logarithmically with this separation and remains finite as long as the vortices are bounded. It is essential to note that pairs of vortices do not destroy the quasi long range order while isolated vortices do. A transition from quasi-long-range order to short range order may thus occur if the vortex pairs dissociate to form a "vortex plasma". In the model discussed here F_0, the non-singular part of F may contain strong fluctuations of the phase but also of the radius of the order parameter. In particular vortices may exist and we may transpose the first argument KOSTERLITZ-THOULESS gave for the existence of a phase transition. The potential associated to an isolated vortex may be estimated as

$$F_{(1)} = F_c + 2\pi \int_{L_0}^{L} dr\, r \frac{\bar{R}^2}{2u_1\Gamma} (u_1c_1 + u_2c_2)(\nabla\phi)^2$$

$$F_c = c + \frac{\pi\bar{R}^2}{u_1\Gamma} (u_1c_1 + u_2c_2) \ln \frac{L}{L_0} \tag{29}$$

where F_c is the core correction coming from regions where $r < L_0$ which causes no trouble as it remains finite. On the other hand the potential associated to a pair of vortices of opposite charge ($q_1 = -q_2 = q$) located at (r_1 , r_2) may be written

$$F_{(2)} = F_c + \frac{\pi\bar{R}^2}{u_1\Gamma} (u_1c_1 + u_2c_2) \ln \frac{|r_1 - r_2|}{L_0} \quad . \tag{30}$$

The probability density for an isolated vortex is then proportional to $\exp\text{-}F_{(1)}$ and for large L we have

$$P_{(1)} \simeq (L_0/L)^{(\pi\bar{R}^2/u_1\Gamma)\,(u_1c_1+u_2c_2)} \tag{31}$$

and the total probability of finding an isolated vortex in the system is given by

$$P \simeq L^2 P_{(1)} \simeq L^{2-(\pi\bar{R}^2/u_1\Gamma)(u_1c_1+u_2c_2)} \to 0 \tag{32}$$

if $\pi\bar{R}^2(u_1c_1 + u_2c_2)/u_1\Gamma > 2$ when $L \to \infty$. When this inequality is not satisfied free vortices should appear and the quasi-long-range order should be destroyed. This topological phase transition occurs at $B_1 = B_c[1 + 2u_1^2\Gamma/\pi(u_1c_1 + u_2c_2)]$.

The detailed argument is of course more complicated as it must include the interactions between vortices [18,19]. This may, however, be done as a mere transition from the XY model to ours. The specific aspects of the Brusselator in this description lie in its dynamics and we will now discuss the effects due to the presence of vortices. In particular let us study the dynamics of the fluctuations of some characteristic phase configurations $\theta_0(r)$ which minimize the potential (27), i.e., for which $\nabla^2\theta_0 = 0$. Writing the phase as $\theta(r,t) = \theta_0(r) + (u_1c_1 + u_2c_2)/(u_1c_2 - u_2c_1)$ $\ln \psi(r,t)$ and introducing the imaginary time $\tau = -i(u_1c_1 + u_2c_2)\, t$ one has $(r > L_0)$

$$i\partial_\tau\psi(r,\tau) = -\nabla^2\psi(r,\tau) - \left(\frac{u_1c_2 - u_2c_1}{u_1c_1 + u_2c_2}\right)^2 (\nabla\theta_0)^2\psi(r,\tau) \quad . \tag{33}$$

This equation is identical to the Schrödinger equation for a particle in an attractive potential. The asymptotic space-time behaviour of ψ is then given by the contribution of the bound states if any [22]. Moreover the dominant term corresponds to the ground state and ψ behaves like $\exp[iK\tau - \sqrt{K}r]$ which is a general property of bound state eigenfunctions, $-K$ being the ground state eigenvalue, determined by the short range behaviour of the system $(r \leq L_0)$ [23]. Let us turn out to specific examples. The phase fluctuations may be of "spin-wave" type or of vortex type. In the first case we may have for example $\theta_0 = \alpha \ln|r - r_i|/L_0$ which occurs with a probability $P \simeq \exp - (\pi\bar{R}^2/u_1\Gamma)(u_1c_1 + u_2c_2)\alpha^2 \ln L/L_0$.

The potential in (33) is $V_\alpha = \left((u_1c_2 - u_2c_1)/(u_1c_1 + u_2c_2)\right)^2\alpha^2|r - r_i|^{-2}$. One obtains for the phase a target pattern defined by

$$\theta(r,t) = \alpha \ln|r - r_i| + \frac{u_1c_1 + u_2c_2}{u_1c_2 - u_2c_1}\,[(u_1c_1 + u_2c_2)K_\alpha(t - t_0) - \sqrt{K_\alpha}|r - r_i|] \tag{34}$$

K_α being the maximum eigenvalue of the operator $(\nabla^2 + \nabla_\alpha)$. As α may vary continuously there is always a non-vanishing probability to have such isolated patterns or pairs of such patterns. The frequency shift varies with α and the corresponding frequency is higher than the frequency of the bulk oscillation.

Let us also note that the target centers $\{r_i\}$ are distributed according to the statistics defined by $\exp-(\bar{R}^2/2u_1\Gamma)(u_1c_1 + u_2c_2) \int d^dr(\nabla\theta_0)^2$.

When the system has non-vanishing local vorticity the situation is somewhat different. Let us consider for example a pair of vortices of charge $+q$ at r_i and $-q$ at r_j. In this case $\theta_0 = q[\mathrm{tg}^{-1}(y - y_i/x - x_i) - \mathrm{tg}^{-1}(y - y_j/x - x_j)]$ and $(\nabla\theta_0)^2 = q^2|r_i - r_j|^2 \cdot |r - r_i|^{-2} \cdot |r - r_j|^{-2}$.

This potential varies like r^{-2} near the vortex cores and like r^{-4} far away from them and the bound state eigenfunctions may be expressed as the superposition of one vortex eigenfunctions. The ground state corresponds to the "molecular" structure

$$\psi_{as} \simeq e^{(u_1c_1 + u_2c_2)K_q(t-t_{0_i})-\sqrt{K_q}|r-r_i|} + e^{(u_1c_1+u_2c_2)K_q(t-t_{0_j})-\sqrt{K_q}|r-r_j|} \quad ; (35)$$

this leads to

$$\theta(r,t) \simeq q\left(t_g^{-1}\frac{y - y_i}{x - x_i} - t_g^{-1}\frac{y - y_j}{x - x_j}\right) + \frac{u_1c_1 + u_2c_2}{u_1c_2 - u_2c_1}\,\mathrm{Max}[0, f_i, f_j]$$

$$f_k = (u_1c_1 + u_2c_2)K_q(t - t_{0_k}) - \sqrt{K_q}|r - r_k| \quad . \qquad (36)$$

The isoconcentration lines of the order parameter then form pairs of Archimedian spirals. As q is an integer the most probable patterns correspond to $q = 1$ and have the same frequency shift. When isolated vortices become the most probable, dissociation of spiral molecules occurs corresponding to the destruction of the quasi-long-range order.

The situations described above present strong analogies with experimental ones namely the spatio-temporal organization observed in the ZABOTINSKII system [24,25]. After a perfect stirring, target patterns arise, the center of which appear at unpredictable times and locations. They are out of phase with the bulk, the corresponding periods being variable but shorter than that of the bulk oscillation. Mainly they are not due to impurities or dust as their density depends on the concentration of the reacting products. When the stirring is not perfect spirals may appear. They grow usually by pairs, they are also out of phase with the bulk but their frequency seems to be constant. However, transition like phenomena from a target or spiral molecules behaviour to an isolated spiral behaviour have not yet been reported. This may be due to the size of the experimental systems as such a transition might not occur in too small a system. Indeed, the above description in terms of the potential (27) is valid for $L^2 > L_0^2 = c_2u_2/u_1r_1$, while the eventual topological transition takes place at $r_1 \simeq 4\Gamma u_1^2/(u_1c_1 + u_2c_2)$. This would only be possible in systems such that

$$L^2 > \frac{c_2u_2}{u_1} \cdot \frac{u_1c_1 + u_2c_2}{4\Gamma u_1^2} \quad . \qquad (37)$$

Otherwise other mechanisms due to a stronger coupling between phase and radius fluctuations would destroy the quasi-long-range order.

So despite the fact that our description bares strong resemblance with experimental situations, more quantitative experimental data on systems with various sizes and concentrations are highly desirable to test the proposed analogy developed here between the spatiotemporal organization of oscillatory chemical systems and the topological order of two-dimensional XY models or superfluid films.

References

1. G. Nicolis, I. Prigogine: Self-Organization in Non-Equilibrium Systems (Wiley, New York 1977)
2. H. Haken: Synergetics, 2ed., Springer Series in Synergetics, Vol.1 (Springer, Berlin, Heidelberg, New York 1978)
3. Proceedings of the XVIIth Solvay Conference on Physics: Order and Fluctuations in Equilibrium and Non-Equilibrium Statistical Mechanics (Wiley, New York 1981, ed. by G. Nicolis, G. Dewel, J. Turner
4. R. Graham: Phys. Rev. A*10*, 1762 (1974)
5. G. Dewel, D. Walgraef, P. Borckmans: Z. Physik B*28*, 235 (1977)
   D. Walgraef, G. Dewel, P. Borckmans: Phys. Rev. A*21*, 397 (1980)
6. D. Walgraef, G. Dewel, P. Borckmans: Adv. in Chem. Phys., submitted
7. S.K. Ma: Modern Theory of Critical Phenomena (Benjamin, New York 1976)
8. P.C. Hohenberg, B.I. Halperin: Rev. Mod. Phys. *49*, 435 (1977)
9. V. Degiorgio, M. Scully: Phys. Rev. A*2*, 1170 (1970)
10. J.W. Turner: "Stationary and Time Dependent Solutions of Master Equations in Several Variables", in *Dynamics of Synergetics Systems*, ed. by H. Haken (Springer, Berlin, Heidelberg, New York 1980)
11. A. Winfree: J. Math. Biol. *1*, 73 (1974)
12. R. Graham: in *Fluctuations, Instabilities and Phase Transitions*, ed. by T. Riste (Plenum, New York 1975)
13. P. Schranner, S. Grossmann, P. Richter: Z. Physik B*35*, 363 (1979)
14. H. Hentschel: Z. Physik B*31*, 401 (1978)
15. N.N. Bogolubov: Lectures in Quantum Statistics (Mac Donald, London 1971) Vol.2
16. H. Mashiyama, A. Ito, T. Ohta: Progr. Theor. Phys. *54*, 1050 (1975)
17. V.L. Pokrovskii: Adv. in Phys. *28*, 597 (1979)
18. J.K. Kosterlitz, D. Thouless: J. Phys. C*6*, 1181 (1973)
19. B.I. Halperin: in *Physics of Low Dimension Systems*, Proceedings of the Kyoto Summer Institute, ed. by Y. Nagaoka, S. Hikami (Publication Office, Progress in Theoretical Physics, Kyoto 1979)
20. D. Nelson: To appear in the Proceedings of the 1980 Summer School on Fundamental Problems in Statistical Mechanics, Enschede, Netherlands
21. Y. Kuramoto: Prog. Theor. Physics *56*, 679 (1976)
22. Y. Kuramoto, T. Yamada: Prog. Theor. Phys. *56*, 724 (1976)
23. P. Ortoleva: J. Chem. Phys. *69*, 300 (1978)
24. T. Yamada, Y. Kuramoto: Prog. Theor. Phys. *55*, 2035 (1976)
24. A.T. Winfree: Science *175*, 634 (1972)
25. M.L. Smoes: "Chemical Waves in the Oscillatory Zhabotinskii System. A Transition from Temporal to Spatio-Temporal Organization", in *Dynamics of Synergetic Systems*, ed. by H. Haken (Springer, Berlin, Heidelberg, New York 1980)

Part IV

Long-Term Behavior of Stochastic Systems

Asymptotic Behavior of Several Dimensional Diffusions *

R.N. Bhattacharya

Department of Mathematics, University of Arizona,
Tucson, AZ 85721, USA

1. Definitions, Preliminaries

This is primarily an expository survey of results, recent and past, on criteria for recurrence, ergodicity and the validity of the central limit theorem for multidimensional diffusions.

Let $p(t,x,dy)$ be a transition probability function on R^k. This means that (i) $p(t,x,dy)$ is a nonnegative measure on the Borel sigma field of R^k for each $t \geq 0$ and each $x \in R^k$, $0 < p(t,x,R^k) \leq 1$, (ii) $x \to p(t,x,B)$ is Borel measurable for each $t \geq 0$ and each Borel set B, (iii) the semigroup property holds:

$$p(t + s,x,B) = \int p(s,z,B)p(t,x,dz), \tag{1.1}$$

or, equivalently, $T_{t+s} = T_t T_s$ where T_t is the transition operator: $T_t f(x) = \int f(y)p(t,x,dy)$ acting on the class $B(R^k)$ of all real bounded Borel measurable functions on R^k. Assume that this semigroup has the Feller property: $T_t(C_b(R^k)) \subset C_b(R^k)$, where $C_b(R^k)$ is the set of all real bounded continuous functions on R^k. Let $C' = \{f \in C_b(R^k): \|T_t f - f\| \to 0 \text{ as } t \downarrow 0\}$, where $\|\cdot\|$ is sup norm. The infinitesimal generator A of the semigroup $\{T_t\}$ is defined by $Af = \lim_{t\downarrow 0} (T_t f - f)/t$ on the set $\mathcal{D}(A)$ on which this limit exists in sup norm and belongs to C'. Under mild regularity conditions on p one can realize a family of probability measures $\{P_x : x \in R^k\}$ on a space Ω' of right continuous functions on $[0,\infty)$ into $R^k \cup \{'\infty'\}$ ('∞' is the point at infinity in the one point compactification of R^k) having the Markov property: if $X(t)$, $t \geq 0$, are the coordinate random variables (i.e., $X(t)(\omega) = \omega(t)$ for $\omega \in \Omega'$), F_t is the smallest sigma field on Ω' with respect to which $X(s)$, $0 \leq s \leq t$, are measurable maps into $R^k \cup \{'\infty'\}$, then the conditional distribution of $X(t + s)$ given F_s is given by

$$P_x(X(t + s) \in B|F_s) = p(t,X(s),B), \tag{1.2}$$

B being an arbitrary Borel subset of $R^k \cup \{'\infty'\}$. Here p has been extended to $R^k \cup \{'\infty'\}$ by $p(t,x,\{'\infty'\}) = 1 - p(t,x,R^k)$ for $x \in R^k$, $p(t,'\infty',\{'\infty'\}) = 1$. One may also show that the strong Markov property holds: if τ is any stopping time (i.e., τ is a function on Ω' into $[0,\infty]$ such that $\{\tau_t < t\} \in F_t$ for all $t \geq 0$), $F_{\tau+} = \bigcap_{\varepsilon>0} F_{\tau+\varepsilon}$ ($F_{\tau+\varepsilon}$ is the sigma field generated by the random variables $X((\tau+\varepsilon) \wedge t), t \geq 0$), and X_τ^+ is the shifted process defined by $X_\tau^+(s)(\omega) = \omega(\tau(\omega)+s)$, then the conditional distribution of X_τ^+ given $F_{\tau+}$ (the past upto time τ) is simply $P_{X(\tau)}$ on the set $\{\tau < \infty\}$. An important property of the measures P_x is that for every $f \in \mathcal{D}(A)$ the process $\{Y_t = f(X(t)) - \int_0^t Af(X(s)ds : t \geq 0\}$ is a martingale: $E(Y_{t+s}|F_s) = Y_s$. This follows from

$$
\begin{aligned}
E(Y_{t+s}|F_s) &= E[f(X(t+s))|F_s] - \int_0^s Af(X(u))du - \int_s^{t+s} E[Af(X(u))|F_s]du \\
&= T_t f(X(s)) - \int_0^s Af(x(u))du - \int_s^{t+s} T_{u-s}Af(X(s))du \\
&= T_t f(X(s)) - \int_0^s Af(X(u))du - \int_0^t T_u Af(X(s))du \\
&= f(X(s)) - \int_0^s Af(X(u))du, \qquad (1.3)
\end{aligned}
$$

since $T_t f(y) - f(y) = \int_0^t T_u Af(y)du$. In the so called "martingale problem" of STROOCK and VARADHAN [16] one goes in the reverse direction by constructing the measures P_x (and, therefore, p) such that $f(X(t)) - \int_0^t Af(X(s))ds$, $t \geq 0$, are martingales for a large enough class of functions on which an appropriate operator A is defined. From now on we shall only consider elliptic differential operators A and use the letter L (instead of A) to emphasize that construction of the measures in not via semigroup theory, but by an extension of Itô's theory of stochastic differential equations. Let $L = \frac{1}{2}\sum a_{ij}(x)(\partial^2/\partial x_i \partial x_j) + \sum b_i(x)(\partial/\partial x_i)$, where $((a_{ij}(x))$ is a symmetric positive definite matrix for each x and $a_{ij}(\cdot)$ are continuous functions on R^k, $b_k(\cdot)$ are Borel measurable real functions which are bounded on compact subsets of R^k. We shall replace Ω' by Ω — the set of all continuous functions on $[0,\infty)$ into $R^k \cup \{'\infty'\}$. However, we let $X(t)$ continue to denote coordinate random variables (on Ω) and F_t the corresponding sigmafields; also let F denote the smallest sigma field with respect to which all the $X(t)$, $t \geq 0$, are measurable. Here is a basic result due to STROOCK and VARADHAN [16] (Theorems 4.2.1, 7.2.1, Corollary 10.1.4, Exercise 6.7.5). A transition probability p has the strong Feller property if $x \to p(t,x,dy)$ is continuous in variation norm, for $t > 0$.

Theorem 1.1. *In addition to the above assumptions on the coefficients of* L *assume that they are bounded and that the smallest eigenvalue of* $((a_{ij}(x)))$ *is bounded away from zero. Then for each* x *in* R^k *there exists a unique probability measure* P_x *on* (Ω,F) *such that* (i) $P_x(X(0)=x)=1$, (ii) *for every real* f *on* R^k *whose derivatives upto the second order are all bounded and continuous, the process* $\{f(X(t)) - \int_0^t Lf(X(s))ds : t \geq 0\}$ *is a martingale under* P_x.

Further, the measures P_x _define a strong Markov process whose transition probability has the strong Feller property_: $T_t(B(R^k)) \subset C_b(R^k)$. _Finally, the support of_ P_x _is the set of all continuous functions on_ $[0,\infty)$ _into_ R^k _starting at_ x.

In case $a(\cdot)$, $a(\cdot)^{-1}$, $b(\cdot)$ are unbounded one may construct a unique probability measure $P_{x,N}$ corresponding to some L_N whose coefficients agree with those of L on the ball $\bar{B}_N = \{|x| \leq N\}$, and piece these together as $N \to \infty$. In this general case, there is a possibility of _explosion_: if $\zeta(\omega) = \inf\{t \geq 0 : \omega(t) = '\infty'\}$ is the _explosion time_, then $P_x(\zeta < \infty)$ may be positive. The measure P_x is such that $P_x(X(t) \neq '\infty'$ for some $t \geq \zeta) = 0$. Thus the Markovian particle is absorbed at $'\infty'$ after explosion. This leads to the possibility of nonuniqueness in the Cauchy problem or the initial value problem for $\partial/\partial t - L$. KHAS'MINSKII [9] provided a criterion for nonexplosion and another for explosion. Proofs may be found in McKEAN [12] and STROOCK and VARADHAN [16]. Necessary and sufficient conditions for nonexplosion in the one-dimensional case were obtained earlier by Feller (see McKEAN [12]). The family $\{\Omega, F, P_x : x \in R^k\}$ is the _diffusion generated by_ L.

2. Recurrence, Invariant Measures, and the Ergodic Theorem

We begin by recalling a basic result of Doeblin in the theory of discrete parameter Markov processes (see DOOB [7], Chapter V). Let S be a set with a sigmafield $A(S)$ on it. A (one-step) transition probability $q(x,dy)$ satisfies _Doeblin's condition_ if there is a finite nonzero measure ϕ on $A(S)$, a number $\varepsilon > 0$ and an integer $m \geq 1$ such that if $\phi(A) \leq \varepsilon$ for some A, then $q^{(m)}(x,A) \leq 1 - \varepsilon$ for all $x \in S$. Here $q^{(m)}$ is the mth iterate of q. If, in addition, there does not exist x, x' such that $q^{(n)}(x,dy)$ and $q^{(n)}(x',dy)$ are mutually singular for every n, then there exists a unique _invariant probability measure_ $\bar{q}$: $\int q(x,A)\bar{q}(dx) = \bar{q}(A)$ for all $A \in A(S)$. Also, $q^{(n)}(x,dy)$ _converges to_ $\bar{q}(dy)$ _in variation norm uniformly in_ x _and exponentially fast as_ $n \to \infty$. Let $Q_{\bar{q}}$ be the measure on the product space $(S^{\mathbb{Z}^+}, A(S^{\mathbb{Z}^+}))$ defined by:

$$Q_{\bar{q}}(X_0 \in A) = \bar{q}(A), \quad Q_{\bar{q}}(X_{n+n'} \in A | G_n) = q^{(n')}(X_n, A),$$

where X_0, X_1, ... are the coordinate maps and G_n is the sigmafield generated by $X_0, \ldots, X_n$. Clearly $Q_{\bar{q}}$ is invariant under time shift. The following result is then standard. The first part is a trivial application of the ergodic theorem, while the second part follows, e.g., from Theorem 20.1 in BILLINGSLEY [4].

Theorem 2.1. _Let_ q _satisfy the above hypothesis_. (a) _If_ f _is a real measurable function on_ S _which is integrable with respect to_ $\bar{q}$, _then_

$$Q_{\bar{q}}(n^{-1}(f(X_0) + f(X_1) + \ldots + f(X_{n-1})) \to \int f d\bar{q} \text{ as } n \to \infty) = 1. \tag{2.1}$$

(b) _If_ $\int f d\bar{q} = 0$, $\int f^2 d\bar{q} = 1$, _then_

$$Q_{\bar{q}}(n^{-1/2}[f(X_0) + f(X_1) + \dots + f(X_{n-1})] \leq y) \to \Phi_{\sigma^2}(y) \tag{2.2}$$

uniformly in $y \in R^1$, as $n \to \infty$. _Here_ Φ_{σ^2} _is the Gaussian distribution function with mean zero and variance_ σ^2 _given by_

$$\sigma^2 = \int f^2(x)\bar{q}(dx) + 2\sum_{n=1}^{\infty}\int f(x)[\int f(y)q^{(n)}(x,dy)]\bar{q}(dx). \tag{2.3}$$

Remark 2.1. The _ergodic coefficient_ of Kolmogorov is defined for the n-step transition probability by

$$\alpha^{(n)} = 1 - \sup_{x,x'} \|q^{(n)}(x,dy) - q^{(n)}(x',dy)\|,$$

where $\|\mu\| = \sup\{|\mu(B)|:B \in A(S)\}$. DOBRUSHIN [6] has shown that $\sigma^2 > 0$, if $\alpha^{(n)} > 0$ for some positive integer n. But, as stated above, under the hypothesis of Theorem 2.1, $q^{(n)}(x,dy)$ converges in norm to $\bar{q}(dy)$ uniformly in x, so that $\alpha^{(n)} \to 1$. Hence $\sigma^2 > 0$. From the fact that $q^{(n)}(x,dy)$ converges to $\bar{q}(dy)$ exponentially fast, it is clear that $\sigma^2 < \infty$.

Remark 2.2. Both statements (2.1), (2.2) hold for every initial distribution μ (i.e., with Q_μ replacing $Q_{\bar{q}}$). At the same time one may prove a functional central limit theorem instead of (2.2) (see BILLINGSLEY [4], Theorem 20.1). Similarly, the ergodic theorem in part (a) may be strengthened to a Strassen type almost sure invariance principle under the hypothesis of part (b).

For a continuous parameter Markov process with a transition probability $p(t,x,dy)$ the analogue of Theorem 2.1 concerning the asymptotic behavior of $\int_0^T f(X(s))ds$ for large T may be stated and proved in the same manner, provided, of course, $p(t_0,x,dy)$ satisfies the same assumptions as q does, for some $t_0 > 0$. This is typically the case when one considers nonsingular diffusions (without killing) on a compact and connected manifold without boundary or on a compact connected manifold with boundary and reflecting boundary conditions. However, for ergodic diffusions on R^k such assumptions are rarely satisfied. An effective tool in studying asymptotic behavior in this case is the method of renewal, which is used extensively for denumerable Markov chains in CHUNG [5], and by MARUYAMA and TANAKA [11] and KHAS'MINSKII [9] for multidimensional recurrent diffusions. Since complete results on one dimensional diffusions are known (see MANDL [10]), from now on we shall only consider diffusions on R^k, $k \geq 2$, unless otherwise specified.

Consider the diffusion generated by an elliptic operator L whose coefficients satisfy regularity conditions stated in Section 1 (but no growth conditions are imposed). A point $x \in R^k$ is said to be recurrent for the diffusion if for every nonempty open subset U of R^k one has

$$P_x(X(s) \in U \text{ for some } s > 0) = 1. \tag{2.4}$$

A point $x \in R^k$ is transient for the diffusion if

$$P_x(\lim_{s \uparrow \zeta} |X(s)| = \infty). \tag{2.5}$$

Recall ζ is the explosion time. A diffusion is said to be recurrent (transient) if all points of R^k are recurrent (transient). It may be noted that multidimensional diffusions cannot be 'point recurrent' i.e., in (2.5) U cannot be replaced by singletons $\{y\}$ (this follows, e.g., from inequality (3.11) of Section 3). Although this is an important difference between denumerable Markov chains and one dimensional diffusions on the one hand and multidimensional diffusions on the other, the following dichotomy still holds (see the author's article [1]): A diffusion is either recurrent or transient.

Perhaps of greater importance are effective criteria for recurrence and transience (Theorem 3.2). But before deriving them let us outline some important asymptotic properties of recurrent diffusions which were obtained by MARUYAMA and TANAKA [11], and KHAS'MINSKII [9]. Let $\tau_F = \inf\{t \geq 0 : X(t) \in F\}$ denote the first hitting time of a set F. Choose $z \in R^k$ and positive numbers $r_0 < r_1$, and write $B(z:r) = \{x : |x - z| < r\}$. Consider the successive hitting times

$$\begin{aligned} \eta_1 &= \tau_{\partial B(z:r_0)}, \quad \eta_{2i} = \inf\{t > \eta_{2i-1} : |X(t) - z| = r_1\}, \\ \eta_{2i+1} &= \inf\{t > \eta_{2i} : |X(t) - z| = r_0\} \qquad (i = 1,2, \ldots). \end{aligned} \tag{2.6}$$

For a recurrent diffusion η's are a.s. (P_x) finite, and, because of the strong Markov property, the sequence of random variables $Y_i = X(\eta_{2i})$ $(i = 1,2, \ldots)$ form a Markov chain on the state space $B(z:r_1)$ with transition probability q given by

$$q(a,B) = P_a(X(\eta_2) \in B) \qquad a \in \partial B(z:r_1). \tag{2.7}$$

Now, applying the strong Markov property again,

$$P_a(X(\eta_2) \in B) = E_a[P_a(X(\eta_2) \in B | F_{\eta_1 +})] = E_a H(X(\eta_1), B), \tag{2.8}$$

where $H(b,dy)$ is the hitting measure on $\partial B(z:r_1)$, i.e.,

$$H(b,B) = P_b(X(\tau_{\partial B(z:r_1)}) \in B). \tag{2.9}$$

Also, $b \to g(b) = H(b,B)$ is L-harmonic on $B(z:r_1)$, i.e., $g(b) = E_b g(X(\tau_{\partial U}))$ - for every nonempty open set whose (compact) closure is contained in $B(z:r_1)$. The function $b \to H(b,B)$ is, therefore, continuous on $B(z:r_1)$ (see the author's article [1]). Suppose now that for some $a \in \partial B(z:r_1)$ and some Borel subset B of $\partial B(z:r_1)$, the probability $q(a,B)$ is positive. Then (2.8) implies $H(b,B)$ is positive for some $b \in \partial B(z:r_0)$, and the maximum principle implies that $b \to H(b,B)$ is strictly positive on $B(z:r_1)$. Together with the property of continuity this leads to the conclusion that $H(b,B)$ is bounded away from zero on $\partial B(z:r_0)$, which in turn implies (in view of (2.8)) that $q(a',B)$ is bounded away from zero on $\partial B(z:r_1)$. From this Doeblin's condition follows immediately by contradiction, if one takes for $\phi(dy)$ the measure $q(a_0,dy)$ for some a_0 (see MARUYAMA and TANAKA [11] or the author's lecture notes [2]). Also, the measures $q(a,dy)$, $a \in \partial B(z:r_1)$, are clearly absolutly continuous with respect to each other. The conclusions of Theorem 2.1 now hold. Let $\bar{q}(dy)$ denote the invariant probability measure for q.

For every nonnegative Borel measurable function f on R^k which is bounded on compacts, the sequence

$$X_i(f) = \int_{\eta_{2i}}^{\eta_{2i+2}} f(X(s))ds \qquad (i = 0,1,2, \ldots), \qquad \eta_0 = 0, \tag{2.10}$$

is a stationary sequence of random variables under $P_{\bar{q}}$. This sequence is easily seen to be ϕ-mixing with a mixing coefficient decaying exponentially fast. In particular, it is ergodic. Hence

$$E_{\bar{q}} X_0(f) = \lim_{n\to\infty} \frac{1}{n}\sum_{i=0}^{n-1} X_i(f) \qquad \text{a.s. } P_{\bar{q}}, \tag{2.11}$$

where $P_{\bar{q}}(C) = \int P_x(C)\bar{q}(dx)$ for $C \in F$. This limit is finite if f is bounded and has compact support, as may be shown by using Proposition 3.1 and Theorem 3.2 below. Taking $f = I_B$, the indicator function of the set B, one gets a Borel measure m on R^k.

$$m(B) = E_{\bar{q}} X_0(I_B). \tag{2.12}$$

This is the unique sigma-finite invariant measure (upto a scalar multiple) for the recurrent diffusion. Let $N(T) = \sup\{n \geq 1 : \eta_{2n} \leq T\}$. Then (2.11) leads to

$$\lim_{T\to\infty} \frac{1}{N(T)} \int_0^T f(X(s))ds = \int f(y)m(dy) \qquad \text{a.s. } P_\mu, \tag{2.13}$$

where μ is an arbitrary initial distribution, and f is nonnegative or bounded with compact support. If g is another such function and $\int g(y)m(dy) \neq 0$, then one gets (MARUYAMA and TANAKA [11], KHAS'MINSKII [9])

$$\lim_{T\to\infty} \frac{\int_0^T f(X(s))ds}{\int_0^T g(X(s))ds} = \frac{\int f(y)m(dy)}{\int g(y)m(dy)} \qquad (\text{a.s. } P_\mu). \tag{2.14}$$

A recurrent diffusion is said to be positive recurrent if the invariant measure m is finite. This is the case if $E_{\bar{q}}\eta_2 < \infty$. Else it is null recurrent. The ergodic theorem implies that if the diffusion is positive recurrent, then (2.11) holds for all m-integrable f, and in particular for $f \equiv 1$. The latter says: $\eta_{2n}/n \to E_{\bar{q}}\eta_2$ a.s. $P_{\bar{q}}$ (or P_μ for arbitrary μ) as $n \to \infty$. Hence $N(T)/T \to (E_{\bar{q}}\eta_2)^{-1}$ a.s. as $T \to \infty$. Let us normalize the measure m (by dividing by $E_{\bar{q}}\eta_2$) to make it a probability measure but continue to denote the latter by m. One now has the desired ergodic theorem: for a positive recurrent diffusion one has

$$\lim_{T\to\infty} \frac{1}{T}\int_0^T f(X(s))ds = \int f dm \qquad \text{a.s. } P_\mu, \tag{2.15}$$

whatever the initial distribution μ (MARUYAMA and TANAKA [11]).

The above results easily imply the weak convergence of the transition probability $p(t,x,dy)$ to $m(dy)$ as $t \to \infty$ for all x. One can, however, prove the following stronger result for all $x \in R^k$ (BHATTACHARYA and RAMASUBRAMANIAN [3]):

$$\|p(t,x,dy) - m(dy)\| \to 0 \quad \text{as} \quad t \to \infty, \tag{2.16}$$

where $\|\cdot\|$ denotes variation norm.

3. Criteria for Transience, Recurrence, and Ergodicity. The Central Limit Theorem

Consider the Dirichlet problem (for L) in an annulus $A = \{r_0 < |x - z| < r_1\}$:

$$\begin{aligned} &Lh(x) = 0 \quad \text{for} \quad r_0 < |x - z| < r_1, \quad h(x) \to f(a) \\ &\text{as} \quad x \to a \quad \text{for} \quad a \in \partial A, \end{aligned} \tag{3.1}$$

f being a continuous function prescribed on ∂A. If, in addition to the hypothesis on L introduced in Section 1, the coefficients of L are locally Hölder continuous, then there exists a unique solution h of (3.1). This solution is given by

$$h(x) = E_x f(X(\tau_{\partial A})), \quad x \in \bar{A} = A \cup \partial A. \tag{3.2}$$

Now let $r_1 = N$, $f(a) = 0$ if $|a - z| = N$ and $f(a) = 1$ if $|a - z| = r_0$. Then the solution is

$$h_N(x) = P_x(\eta'_{r_0} < \eta'_N), \qquad \eta'_r = \inf\{t \geq 0: |X(t) - z| = r\}. \tag{3.3}$$

Since $\eta'_N \uparrow \zeta$ (the explosion time) as $N \uparrow \infty$, and $\{\eta'_{r_0} < \zeta\}$ is equivalent (a.s.) to $\{\eta'_{r_0} < \infty\}$, one has

$$h(x) \equiv \lim_{N\to\infty} h_N(x) = P_x(\eta'_{r_0} < \infty). \tag{3.4}$$

It is simple to show that recurrence means that this last probability is identically one, while transience means that it is strictly less than one for $|x - z| > r_0$. Thus one way to prove recurrence is to prove that $h_N(x) \to 1$ as $N \to \infty$. In general, it is not possible to solve (3.1) explicitly. In an interesting special case, however, this Dirichlet problem (with a radial boundary function f) reduces to a one dimensional two-point boundary value problem and, therefore, admits an explicit solution. To see when this happens (with origin at z) write

$$\begin{aligned} A_z(x) &= \sum_{i,j} a_{ij}(x)(x_i - z_i)(x_j - z_j)/|x - z|^2, \\ B(x) &= \sum_i a_{ii}(x), \qquad C_z(x) = 2\sum_i (x_i - z_i)b_i(x). \end{aligned} \tag{3.5}$$

If F is a smooth function on $(0,\infty)$, then

$$LF(|x - z|) = \tfrac{1}{2}A_z(x)F''(|x - z|) + \frac{F'(|x-z|)}{2|x-z|}(B(x) - A_z(x) + C_z(x)). \tag{3.6}$$

Hence $LF(|x - z|)$ is a function of $|x - z|$ if and only if $A_z(x)$ and $B(x) + C_z(x)$ are. This has an interesting probabilistic interpretation: L <u>transforms all smooth radial</u> (with respect ot center z) <u>functions into radial functions if and only if</u> $\{|X(t) - z|:t \geq 0\}$ <u>is a one dimensional diffusion on the state space</u> $(0,\infty)$ <u>having generator</u> $L_r = \frac{1}{2}\alpha_z(r)\frac{d^2}{dr^2} + \frac{1}{2}\,\frac{\gamma_z(r)-\alpha_z(r)}{r}\,\frac{d}{dr}$, <u>where</u> $\alpha_z(|x - z|) = A_z(x)$, $\gamma_z(|x - z|) = B(x) + C_z(x)$. In this case $h_N(x) = g(|x - z|)$ is obtained by solving

$$\begin{aligned} &\tfrac{1}{2}\alpha_z(r)g''(r) + \tfrac{1}{2}\,\frac{\gamma_z(r)-\alpha_z(r)}{r}g'(r) = 0 \quad \text{for } r_0 < r < N \\ &g(r_0) = 1, \quad g(N) = 0. \end{aligned} \tag{3.7}$$

Letting $N \uparrow \infty$, a necessary and sufficient condition for recurrence is obtained. In the general case, one attempts to obtain radial functions which are superharmonic or subharmonic to provide lower and upper bounds to h_N. Analogously, to obtain criteria for positive and null recurrence one derives radial upper and lower bounds to the function $x \to E_x(\eta'_{r_0} \wedge \eta'_N)$ which is the solution to the Dirichlet problem:

$$
\begin{aligned}
&Lh(x) = -1 \quad \text{for} \quad r_0 < |x - z| < N, \\
&h(x) = 0 \quad \text{if} \quad |x - z| = r_0 \ \text{or} \ N.
\end{aligned}
\tag{3.8}
$$

To express these bounds write

$$
\begin{aligned}
&\bar{\alpha}(r) = \sup_{|x-z|=r} A_z(x), \qquad \underline{\alpha}(r) = \inf_{|x-z|=r} A_z(x), \\
&\bar{\beta}(r) = \sup_{|x-z|=r} (B(x) - A_z(x) + C_z(x))/A_z(x), \\
&\underline{\beta}(r) = \inf_{|x-z|=r} (B(x) - A_z(X) + C_z(x))/A_z(x), \\
&\underline{I}(r) = \int_{r_0}^{r} \frac{\underline{\beta}(u)}{u} du, \qquad \bar{I}(r) = \int_{r_0}^{r} \frac{\bar{\beta}(u)}{u} du.
\end{aligned}
\tag{3.9}
$$

Let $0 < c < d$, and let h be a continuous nonnegative function on $(0,\infty)$. Define the functions

$$
\begin{aligned}
&\underline{\psi}(r) = \int_c^r \exp\{-\underline{I}(u)\} du, \qquad \bar{\psi}(r) = \int_c^r \exp\{-\bar{I}(u)\} du, \\
&\underline{F}_1(r) = \int_c^r \exp\{-\underline{I}(u)\} \left(\int_c^u \frac{h(v)}{\underline{\alpha}(v)} \exp\{\underline{I}(v)\} dv\right) du, \\
&\bar{F}_1(r) = \int_c^r \exp\{-\bar{I}(u) \left(\int_c^u \frac{h(v)}{\bar{\alpha}(v)} \exp\{\bar{I}(v)\} dv\right) du, \\
&\underline{F}_2(r) = \int_c^r \exp\{-\underline{I}(u)\} \left(\int_u^d \frac{h(v)}{\bar{\alpha}(v)} \exp\{\underline{I}(v)\} dv\right) du, \\
&\bar{F}_2(r) = \int_c^r \exp\{-\bar{I}(u)\} \left(\int_u^d \frac{h(v)}{\underline{\alpha}(v)} \exp\{\bar{I}(v)\} dv\right) du.
\end{aligned}
\tag{3.10}
$$

One then has the following inequalities (see BHATTACHARYA and RAMASUBRAMANIAN [3] for a proof in the general non-homogeneous case).

Proposition 3.1. *Let* $z \in R^k$, $r_0 > 0$, $0 < c < d < \infty$ *be given.*

(a) *Then for* $c < r = |x - z| < d$, *one has*

$$
\underline{\psi}(r)/\underline{\psi}(d) \le P_x(\eta'_d < \eta'_c) \le \bar{\psi}(r)/\bar{\psi}(d). \tag{3.11}
$$

(b) *Let* h *be a nonnegative continuous function on* $(0,\infty)$, *and let* $\eta = \eta'_c \wedge \eta'_d$. *Then with* $r = |x - z|$ *one has*

$$\underline{\max}\{2\bar{F}_1(d)(\underline{\psi}(r)/\underline{\psi}(d)) - 2\bar{F}_1(r), 2\underline{F}_2(r) - 2\underline{F}_2(d)(\bar{\psi}(r)/\bar{\psi}(d))\} \le E_x\int_0^n h(|X(s) - z|)ds$$
$$\le \underline{\min}\{2\underline{F}_1(d)(\bar{\psi}(r)/\bar{\psi}(d)) - 2\underline{F}_1(r), 2\bar{F}_2(r) - 2\bar{F}_2(d)(\underline{\psi}(r)/\underline{\psi}(d))\}. \tag{3.12}$$

The first part of the following theorem is obtained by letting $d \uparrow \infty$ in (3.11), while part (b) follows by letting $d \uparrow \infty$ in (3.12). The criteria for transience, recurrence, and positive recurrence extend (with minor improvements) those announced in an important paper by KHAS'MINSKII [9] for smooth L. The criterion for null recurrence presented below is different from that announced in [9]. The latter has never been verified and is probably in error. The first proofs of KHAS'MINSKII type criteria for transience and recurrence under certain growth conditions of the coefficients were given by FRIEDMAN [8], who arrived at the results independently. Proofs for cirteria for positive and null recurrence as they appear below were given by the author [1]. Extensions to the nonhomogeneous case have been obtained jointly with RAMASUBRAMANIAN [3]. Note that the functions $\bar{\psi}(r)/\bar{\psi}(d)$, $\underline{\psi}(r)/\underline{\psi}(d)$, $\underline{F}_i(r)$, $\bar{F}_i(r)$ $(i = 1,2)$ in (3.10) do not vary with r_0.

Theorem 3.2. (a) *The diffusion generated by* L *is recurrent if there exists* $r_0 > 0$ *and* $z \in R^k$ *such that*

$$\int_{r_0}^{\infty} \exp\{-\bar{I}(r)\}dr = \infty, \tag{3.13}$$

and is transient if for some $r_0 > 0$ *and* $z \in R^k$ *one has*

$$\int_{r_0}^{\infty} \exp\{-\underline{I}(r)\}dr < \infty. \tag{3.14}$$

(b) *In order that a diffusion be positive recurrent it is sufficient that it be recurrent and that there exists some* $r_0 > 0$ *and some* $z \in R^k$ *such that*

$$\int_{r_0}^{\infty} \frac{1}{\underline{\alpha}(r)}\exp\{\bar{I}(r)\}dr < \infty. \tag{3.15}$$

In order that a diffusion be null recurrent it is sufficient that it be recurrent and that for some $r_0 >$ *and some* $z \in R^k$

$$\lim_{N\to\infty} \frac{\int_{r_0}^{N} \exp\{-\bar{I}(r)\}(\int_{r_0}^{r} \frac{1}{\bar{\alpha}(u)}\exp\{\bar{I}(u)\}du)dr}{\int_{r_0}^{N} \exp\{-\underline{I}(r)\}dr} = \infty. \tag{3.16}$$

Remark 3.1. Note that the above criteria are exact if for some z the functions $A_z(x)$, $B(x) + C_z(x)$ are radial (with center at z) near infinity.

Remark 3.2. It was shown by KHAS'MINSKII [9] that for smooth L the recurrence of the diffusion generated by L is equivalent to the well posedness of the so called <u>exterior Dirichlet problem</u> for L in the exterior of bounded (nonempty) domains V with smooth boundary: for each continuous f on V the boundary value problem

$$Lh(x) = 0 \quad \text{for} \quad x \in R^k \setminus \bar{V}$$
$$h(x) \to f(a) \quad \text{as} \quad x \to a \in \partial V, \tag{3.17}$$

has a unique bounded solution h if and only if the diffusion generated by L is recurrent. A detailed proof for all V satisfying the exterior sphere property is given in the author's lecture notes [2]. An extension to the nonhomogeneous case is contained in the article with RAMASUBRAMANIAN [3].

Let us use (3.11) to study the growth of a positive recurrent diffusion. Let $M_n = \max\{|X(t)|:\eta_{2n} \le t \le \eta_{2n+2}\}$, $(n = 1, \ldots)$, where η_{2n} is defined as in (2.6). Let $c = r_0$, $z = 0$ in the definition of the functions $\underline{\psi}$, $\bar{\psi}$ in (3.10). Let g be an increasing positive function on $[0,\infty)$. Write $\beta = E_{\bar{q}}\eta_2$. By (3.11) one has, for an arbitrary initial distribution μ, $P_\mu(M_n > g(2n\beta))$ $\le \bar{\psi}(r_1)/\bar{\psi}(g(2n\beta))$; hence, by the Borel-Cantelli lemma $P_\mu(M_n > g(2n\beta)$ holds for infinitely many $n) = 0$ if $\sum\bar{\psi}(g(2n\beta))^{-1} < \infty$, i.e., if

$$\int_1^\infty \frac{1}{\bar{\psi}(g(u))} du < \infty. \tag{3.18}$$

The function g is said to be an <u>upper function</u> for the diffusion if $P_\mu(\sup\{t: |X(t)| > g(t)\} < \infty) = 1$ for all μ. Since $\eta_{2n}/n \to \beta$ a.s. (P_μ), it follows that g is an upper function if (3.18) holds. In the one-dimensional case this argument is due to MOTOO [13]. Since the excursions between successive η_{2n}'s are not independent, a modification of the reasoning is necessary to check if g is a lower function: g is a <u>lower function</u> if $P_\mu(\sup\{t:|X(t)| > g(t)\}$ $< \infty) = 0$. Since $P_\mu(M_n > g(\frac{1}{2}\beta n)|F_{\eta_{2n}^+}) \ge \underline{\psi}(r_1)/\underline{\psi}(g(\frac{1}{2}\beta n))$, a version of the second half of the Borel-Cantelli lemma (see NEVEU [14], p. 128) ensures that $P_\mu(M_n > g(\frac{1}{2}\beta n)$ for infinitely many $n) = 1$ if $\sum\underline{\psi}^{-1}(g(\frac{1}{2}\beta n)) = \infty$. Hence g is a lower function for the positive recurrent diffusion if

$$\int_1^\infty \frac{1}{\underline{\psi}(g(u))} du = \infty. \tag{3.19}$$

One of the most useful asymptotic results concerning stochastic processes from the point of view of applications is the central limit theorem. In view of the ground work that was done in Section 2 and in the first part of this section one may arrive at the following general result without difficulty. Details of the proof are to be found in the article with RAMASUBRAMANIAN [3].

Theorem 3.3. (a) Suppose that the diffusion generated by L is positive recurrent. Let f be a real Borel measurable function on R^k such that $\alpha = \int f(x)m(dx)$ is finite and

$$E_{\bar{q}}\left(\int_0^{\eta_2} |f(X(s)) - \alpha| ds\right)^2 < \infty. \tag{3.20}$$

Then, whatever be the initial distribution μ of the diffusion, the distribution of the stochastic process $Z_T(t) = T^{-1/2}(\int_0^{tT} f(X(s))ds - \alpha tT)$, $0 \leq t \leq 1$, converges weakly (as $T \to \infty$) to the Wiener measure W_{δ^2} with zero drift and variance parameter δ^2, where

$$\delta^2 = \sigma^2 / E_{\bar{q}}\eta_2,$$

$$\sigma^2 = E_{\bar{q}}\left(\int_0^{\eta_2} (f(X(s)) - \alpha)ds\right)^2 \tag{3.21}$$

$$+ 2\sum_{n=1}^{\infty} E_{\bar{q}}\left(\left(\int_0^{\eta_2} (f(X(s)) - \alpha)ds\right)\left(\int_{\eta_{2n}}^{\eta_{2n+2}} (f(X(s)) - \alpha)ds\right)\right).$$

(b) The hypotheses of part (a) are fulfilled if (i) the coefficients of L satisfy (3.13) and (3.15) for some $r_0 > 0$ and $z \in R^k$, and (ii) if the function f satisfies

$$\int_{r_0}^{\infty} \frac{g(v)}{\underline{\alpha}(v)} \exp\{\bar{I}(v)\}dv < \infty, \qquad \int_{r_0}^{\infty} \frac{g(v)h(v)}{\underline{\alpha}(v)} \exp\{\bar{I}(v)\}dv < \infty, \tag{3.22}$$

where $g(v) = \sup\{|f(x)| : |x - z| = v\}$ and

$$h(v) = \int_{r_0}^{v} \exp\{-\bar{I}(r)\}\left(\int_r^{\infty} \frac{g(u)}{\underline{\alpha}(u)} \exp\{\bar{I}(u)\}du\right)dr. \tag{3.23}$$

Remark 3.3. To prove part (a) one approximates $\int_0^T (f(X(s)) - \alpha)ds$ by $\sum_{i=0}^{N(T)} \int_{\eta_{2i}}^{\eta_{2i+2}} (f(X(s)) - \alpha)ds$. By good fortune $E_{\bar{q}} \int_0^{\eta_2} (f(X(s)) - \alpha)ds = 0$. This method runs into serious difficulties in the nonhomogeneous case, in view of the absence of a common centering for the integral and the sum.

Remark 3.4. One may make an argument similar to Remark 2.1 to assert that $\sigma^2 > 0$ (and, therefore, $\delta^2 > 0$) whenever f is not a constant a.s. (m). As is beautifully explained in DOBRUSHIN [6], this is in some sense the heart of the matter. It is quite unsatisfactory, as is often the case with results in the theory of stationary processes, if one cannot assert that the limiting Gaussian distribution is not degenerate.

Remark 3.5. It is not clear to the author if the hypotheses in part (a) or (b) guarantee the convergence of the integral $\int_0^{\infty}(\int f(x) \cdot T_s f(x)m(dx))ds$. In case of convergence, δ^2 is twice the integral. Note that the first inequality in (3.22) implies $\int|f(x)|m(dx) < \infty$, while the second ensures the convergence in (3.20). If $E_{\bar{q}}\tau^2_{\partial B(z:r_0)} < \infty$ (which will be the case if (3.22) holds with $g \equiv 1$), then the conclusion in part (a) holds for all bounded Borel f.

Remark 3.6. Relation (2.16) implies that a positive recurrent diffusion (with initial distribution m) is strong mixing in the sense of ROSENBLATT [15]. But they are rarely ϕ-mixing. Even the maximum correlation $\rho(t) = \sup\{[\int(T_t f(x))^2 m(dx)]^{1/2} : \int f(x)m(dx) = 0, \int f^2(x)m(dx) = 1\}$ does not go to zero (as $t \to \infty$), in general. For $\rho(t)$ is either identically 1 or decays exponentially fast. Even for a self adjoint operator L (acting on the appropriate subset of $L^2(R^k, dm)$) the latter possibility would imply that there is a gap between zero and the rest of the spectrum of L.

Remark 3.7. Let $f_1, f_2, \ldots, f_j$ all meet the conditions imposed on f in part (a). Then any linear combination of them will also meet the conditions. Thus $T^{-1/2}\int_0^T (f_i(X(s)) - \alpha_i)ds$ $(i = 1,2, \ldots, j)$ have an asymptotic j-dimensional Gaussian distribution. In case $E_{\bar{q}}\eta_2^2 < \infty$, Gaussian stochastic processes, indexed by appropriate subclasses of the Borel sigma-field of R^k arise in the limit.

Acknowledgement. The author wishes to thank Charles M. Newman for a helpful conversation.

*This research was supported by the United States National Science Foundation Grants MCS 79-03004 A01, CME 8004499.

References

1. Bhattacharya, R.N. (1978). Criteria for recurrence and existence of invariant measures for multidimensional diffusions. Ann. Probability 6, 541-553. Correction Note (1980) Ann. Probability 8
2. Bhattacharya, R.N. (1980): *Recurrence of Diffusions*. Lecture Notes. (MacMillan and the Indian Statistical Institute, to appear)
3. Bhattacharya, R.N., Ramasubramanian, S. (1980). Recurrence and ergodicity of diffusions (to appear)
4. Billingsley, P. (1968): *Convergence of Probability Measures* (Wiley, New York)
5. Chung, K.L. (1967): *Markov Chains with Stationary Transition Probabilities*, 2nd ed., Grundlehren der mathematischen Wissenschaften, Vol. 104 (Springer, Berlin, Heidelberg, New York)
6. Dobrushin, R.L. (1956): Central limit theorem for nonstationary Markov Chains, I, II. Theor. Probab. Appl. 1, 65-80, 329-383
7. Doob, J.L. (1953): *Stochastic Processes* (Wiley, New York)
8. Friedman, A. (1973): Wandering out to infinity of diffusion processes. Trans. Amer. Math. Soc. 184, 185-203
9. Khas'minskii, R.Z. (1960): Ergodic properties of recurrent diffusion processes and stabilization of the Cauchy problem for parabolic equations. Theory Probab. Appl. 5, 179-196

10. Mandl, P. (1968): *Analytical Treatment of One-Dimensional Markov Processes*, Grundlehren der mathematischen Wissenschaften, Vol. 151 (Springer, Berlin, Heidelberg, New York)

11. Maruyama, G., Tanaka, H. (1959): Ergodic property of N-dimensional homogeneous recurrent Markov processes. Mem. Fac. Sci. Kyushu Univ. Ser. A 13, 157-172

12. McKean, H.P., Jr. (1969): *Stochastic Integrals* (Academic Press, New York)

13. Motoo, M. (1959): Proof of the law of the iterated logarithm through diffusion equation. Ann. Inst. Statis. Math. 10, 21-28

14. Neveu, J. (1965): *Mathematical Foundations of the Calculus of Probability* (Holden-Day, San Francisco)

15. Rosenblatt, M. (1971): *Markov Processes. Structure and Asymptotic Behavior*, Grundlehren der mathematischen Wissenschaften, Vol. 184 (Springer, Berlin, Heidelberg, New York)

16. Stroock, D.W., Varadhan, S.R.S. (1979): *Multidimensional Diffusion Processes*, Grundlehren der mathematischen Wissenschaften, Vol. 233 (Springer, Berlin, Heidelberg, New York)

Qualitative Theory of Stochastic Non-Linear Systems

Ludwig Arnold*

Fachbereich Mathematik, Universität Bremen
D-2800 Bremen, Fed. Rep. of Germany

Summary

The aim of this survey paper is to sketch the problems and results of the qualitative theory of stochastic dynamical systems (SDS) and to give reference to papers where details can be found. Here an SDS is an ordinary differential equation $\dot{x} = f(x,\xi)$ with a random noise process ξ in the r.h.s. and random initial conditions $x(o) = x_o$. Qualitative theory studies the general nature of a solution in the entire time interval, e.g. recurrence and stability properties. Both the white noise and the real (i.e. non-white) noise case are covered. Emphasis is given to nonlinear systems (including multiplicative noise linear systems) and to stationary and Markovian noise. Standing reference is ARNOLD and KLIEMANN [7].

List of contents

---

*This paper was written during the author's stay at the Centre de Mathématiques Appliquées, Ecole Polytechnique, Palaiseau (France) in November 1980

1. Scope of the theory

1. Notion of an SDS

In this paper, a stochastic dynamical system (SDS) $\Sigma = (f, x_o, \xi)$ is just an ordinary differential equation in $\mathbb{R}^d$,

$$\dot{x}(t) = f(x(t), \xi(t)), \quad x(o) = x_o, \; t \in \mathbb{R}^+ \text{ or } \mathbb{R},$$

with random initial condition x_o and a (possibly generalized) stochastic noise process ξ with values in $\mathbb{R}^m$, entering into the (deterministic!) right-hand side f. The process ξ contains input noise as well as parameter noise and the influence of a random environment.

Classification of Σ according to f:

- f nonlinear,
- f linear (in state): $f(x,\xi) = A(\xi)x + b(\xi)$,
  A = multiplicative noise, b = additive noise,
- f linear in noise: $f(x,\xi) = f(x) + G(x)\xi$,
- f linear in state and noise: $f(x,\xi) = Ax + B\xi + \sum_{i=1}^{k} C_i x \xi_i$.

Classification of Σ according to noise ξ:

(i) ξ white noise = generalized Gaussian stationary process with constant spectral density on the whole real line.
Advantage: makes x a Markov process.
Drawbacks: needs special calculus (Ito calculus), in modeling often unrealistic (f has to be linear in ξ; ξ cannot be constrained to subsets of $\mathbb{R}^m$ as is often necessary due to the physical meaning of the disturbed parameters; ξ moves on a faster time scale than x).
The equation $\dot{x} = f(x) + G(x)\xi$ can be given a meaning by putting $\xi = \dot{W}$, W = Wiener process, and rewriting it as
$dx = f(x)dt + G(x)dW$, $x(o) = x_o$. This is short hand for the integral equation

$$x(t) = x_o + \int_o^t f(x(s))ds + \int_o^t G(x(s))dW(s).$$

The last integral is a stochastic integral in the sense of Ito (see e.g. ARNOLD [1]). If x_o is independent of W, the solution x (whose existence and uniqueness is always assumed in this paper without mentioning) is a strong Markov process, even a diffusion process whose infinitesimal generator on smooth functions is

$$L = \Sigma f_i \frac{\partial}{\partial x_i} + \frac{1}{2} \Sigma\Sigma \, (GG')_{ij} \frac{\partial^2}{\partial x_i \partial x_j} .$$

(ii) ξ real (i.e. non-white) noise = any ordinary stochastic process. Assume piecewise continuous sample paths.

Important cases:

- ξ Markov process, in particular diffusion: $d\xi = a(\xi)dt + b(\xi)dW$, this is 'coloured noise" = output of a nonlinear system with white input.
  Advantage: (x,ξ) Markov.
  Drawback: excludes smooth (e.g. differentiable) processes like periodic noise $\xi(t) = A\cos(\alpha t+\beta)$, A,α,β random.
- ξ (strictly) stationary: models non-evolutionary environment.

Here $\dot{x} = f(x,\xi)$ is to be interpreted samplewise as an ordinary differential equation for each noise trajectory (cf. BUNKE [13]). Existence and uniqueness is again taken for granted.
Note: x_o is allowed to anticipate ξ.

1.2 Qualitative theory

Qualitative theory studies the general nature of a solution on the entire time interval (asymptotic or long-term behavior) without solving the equation. The deterministic theory was founded at the end of the 19th century by H. Poincaré and A. M. Lyapunov (see NEMYTSKII and STEPANOV [30] or BHATIA and SZEGÖ [9]). It deals with problems like: invariant sets, critical and periodic points, limit sets, recursive concepts (recurrent/transient points, Poisson and Lagrange stability, minimal sets), stability theory (Lyapunov stability, attractors, regions of attraction), Lyapunov numbers/ Floquet theory, invariant measures/ergodic theory etc.

Qualitative theory of SDS was founded by KUSHNER [28] and KHASMINSKII [21] in the 1960ies. It deals with the same problems and concepts as deterministic theory. But now there are usually various randomized versions of each of the above concepts due to the various topologies for random variables.

Recurrence concepts (Kliemann [23]): Let $x(t,x_o)$ be the solution of $\dot{x} = f(x,\xi)$ (ξ white or real) starting in x_o (fixed). $M \subset \mathbb{R}^d$ is an invariant set if for each $x_o \in M$

$$P(x(t,x_o) \in M \text{ for all } t \in \mathbb{R}^+) = 1.$$

The points x_o in state space are classified according to recurrence properties of the solution starting there as follows:

x_o is weakly transient if there is an $\varepsilon > o$ such that $P(\text{there is a } t_o > o \text{ such that } |x(t,x_o)-x_o| \geqq \varepsilon \text{ for all } t \geqq t_o) = 1$.
x_o is strongly transient if there is an $\varepsilon > o$ and a fixed $t_o > o$ with

$$P(|x(t,x_o)-x_o| \geqq \varepsilon \text{ for all } t \geqq t_o) = 1 .$$

x_o is strongly recurrent if it is not weakly transient, i.e. if for each $\varepsilon > o$

$$P(|x(t_n,x_o)-x_o| < \varepsilon \text{ for } t_n \to \infty) > o.$$

x_o is weakly recurrent if it is not strongly transient, i.e. if for all $\varepsilon > o$ and $t_o > o$

$$P(|x(t,x_o)-x_o| < \varepsilon \text{ for some } t \geqq t_o) > o.$$

x_o is positively recurrent if it is weakly recurrent and for each $\varepsilon > o$

$$\limsup_{T\to\infty} \frac{1}{T} \int_o^T P(|x(t,x_o)-x_o|<\varepsilon)dt > o,$$

x_o is null-recurrent otherwise.

Stability concepts: Stability is a property of a solution rather than of a point in state space. Compare $x(\cdot,x_o)$ with $x \equiv o$ (say) on $\mathbb{R}^+$ if x_o is small.

Strong stability: The trivial solution is called strongly stable with probability 1 (w.p. 1), in probability (in pr.) or in p-th mean, if

$$\lim_{x_o \to o} \sup_{t \geq o} |x(t,x_o)| = o$$

w.p. 1, in pr. or in p-th mean, resp. It is strongly asymptotically stable w.p. 1 etc. if it is strongly stable w.p. 1 etc. and if w.p. 1 etc.

$$\lim_{T \to \infty} \sup_{t \geq T} |x(t,x_o)| = o \quad \text{for } |x_o| < \varepsilon .$$

Weak stability: The principle is "average first, take sup later".

Stability of the p-th moment:

$$\lim_{x_o \to o} \sup_{t \geq o} E|x(t,x_o)|^p = o .$$

Stability of the probability:

$$\lim_{x_o \to o} \sup_{t \geq o} P(|x(t,x_o)| > \varepsilon) = o, \text{ for all } \varepsilon > o.$$

The hierarchy of the modes of convergence is reflected by the hierarchy of the modes of stability. Of course, strong stability entails weak stability.

Growth of a solution is measured by its Lyapunov number

$$\lambda(x_o) := \limsup_{t \to \infty} \frac{1}{t} \log |x(t,x_o)| .$$

Stationary solutions: Stochastic analogue of a 'critical point' (i.e. constant solution). A solution $x^o = x^o(\cdot,x_o^o)$ on $\mathbb{R}$ is ξ-stationary if (x^o,ξ) is stationary. Wanted: x_o^o. A ξ-stationary solution is called weak if it is possibly defined on a new (bigger) probability space. It is called strong if it is defined on the original probability space and, moreover, x^o is ξ-measurable.

Invariant probability measures are always related to stationary solutions. Each stationary solution generates a shift-invariant probability in the canonical function space. There one can apply ergodic theory, and one can decompose each (stochastically continuous) process into ergodic components on minimal shift-invariant sets (ROZANOV [35]). However, while each invariant set in state space $\mathbb{R}^d$ generates a shift-invariant set in function space, it is not always the other way around. There is a one-to-one correspondence of the two kinds of invariant sets for Markov processes (ROSENBLATT [34]). If ξ is white, the Markov process x is stationary iff the initial distribution P_o is invariant with respect to its transition probability, i.e.

$$P_o(B) = \int_{\mathbb{R}^d} P(t,x,B)P_o(dx) \quad \text{for all } t \geq o.$$

2. Nonlinear f

2.1 White ξ

Qualitative theory of Ito equations of the form

$$dx = f(x)dt + G(x)dW, \quad x(o) = x_o,$$

has been intensively studied for the non-degenerate or elliptic case, i.e. under the condition

(ND) $\quad G(x)G(x)' > o \quad$ for all $\quad x \in \mathbb{R}^d$.

Recurrence theory (BHATTACHARYA [10] and [11], KHASMINSKII [21], FRIEDMAN [17]): A point $x_o \in \mathbb{R}^d$ is called recurrent, if for each $\varepsilon > o$

$$P(|x(t_n,x_o)-x_o| < \varepsilon \quad \text{for} \quad t_n \to \infty) = 1,$$

while $x_o \in \mathbb{R}^d$ is called transient if

$$P(|x(t,x_o)| \to \infty \quad \text{as} \quad t \to \infty) = 1.$$

We have the simple

Dichotomy: Let (ND) hold. If there is a recurrent point, then all points are recurrent and the diffusion is called recurrent. If there is no recurrent point, then all points are transient and the diffusion is called transient. Recurrence is equivalent to strong recurrence, and transience is equivalent to weak transience (see 1.2).

There are tests for checking transience and recurrence. If (ND) is not satisfied, the situation can become very complicated. For a particular case see 2.3. There is no dichotomy anymore, as the following example in $\mathbb{R}^2$ shows:

$$dx_1 = -x_1 dt,$$

$$dx_2 = f(x_2)dt + G(x_2)dW.$$

If x_2 is a recurrent diffusion in $\mathbb{R}^1$, (x_1,x_2) will be neither transient nor recurrent, but $\mathbb{R}^2 = (\{o\}\times\mathbb{R}^1) \cup (\mathbb{R}^1 \setminus \{o\}) \times \mathbb{R}^1$, where the first set is strongly recurrent and the second one strongly transient, see 2.3.
More information can be obtained in the degenerate case from support theorems (see KUNITA [27]).

Stationary solutions: under (ND) equivalent to positive recurrence or unique invariant probability P_o. The density of P_o for $d = 1$ reads

$$p_o(x) = c\, G(x)^{-2} \exp\left(\int_{x_o}^{x} 2f(y)G(y)^{-2}dy\right), \quad c \text{ norming constant.}$$

(GIKHMAN and SKOROKHOD [18]). For $d \geq 2$ there is an explicit construction by KHASMINSKII [21], including an ergodic theorem, and there are tests for positive or null-recurrence (BHATTACHARYA [10]). A general necessary and sufficient criterion for a transition probability $P(t,x,B)$ to admit an invariant probability is (KHASMINSKII [21])

$$\lim_{R\to\infty} \; \liminf_{T\to\infty} \frac{1}{T} \int_o^T P(t,x,|y|>R)dt = o \quad \text{for some} \quad x .$$

Stability theory: The deterministic Lyapunov theory can be genuinely carried over if a Lyapunov function is defined by $V \geqq o$, $LV \leqq 0$. Then $v(t) = V(x(t))$ is a supermartingale, and Doob's inequality gives

$$P(\sup_{t \geqq o} V(x(t,x_o)) > \varepsilon) \leqq \varepsilon^{-1} V(x_o)$$

from which we can immediately infer (for nice V) that

$$\lim_{x_o \to o} P(\sup_{t \geqq o} |x(t,x_o)| > \varepsilon) = o,$$

i.e. strong stability in probability. With these concepts, the stochastic theory looks very similar to the deterministic one, and the problem usually boils down to finding a Lyapunov function (KUSHNER [28], KHASMINSKII [21], ARNOLD [1]).

2.2 Real stationary ξ

There is no recurrence theory, but some Lyapunov stability theory for a general real noise ξ (BUNKE [13], KHASMINSKII [21]).

Stationary solutions: $\dot{x} = f(x,\xi)$ has a weak ξ-stationary solution iff it has a solution x for which

$$\lim_{R \to \infty} \frac{1}{T} \int_o^T P(|x(t)| > R)dt = o \quad \text{uniformly in } T > T_o \text{(or } T < -T_o\text{)}$$

(KHASMINSKII [21]).
This is satisfied, for example, if there is a bounded solution x. The necessity of enlarging the original probability space is shown by a deterministic equation $\dot{x} = f(x)$ (i.e. $\xi \equiv o$). If e.g. x is a periodic solution with period T, then $x(\cdot+\tau)$, τ uniformly distributed in $[o,T]$, is a weak ξ-stationary solution.
Weakly nonlinear equations are treated by BUNKE [13] via linearization.

2.3 Markovian ξ

Assume that ξ is a diffusion process in a connected set $Y \subset \mathbb{R}^m$ represented by the solution of the Ito equation

$$d\xi = a(\xi)dt + b(\xi)\,dW.$$

We add our differential equation $\dot{x} = f(x,\xi)$ in the form

$$dx = f(x,\xi)dt,$$

$x \in X \subset \mathbb{R}^d$, X connected. Then $z = (x,\xi)$ is again a diffusion process with state space $Z = X \times Y \subset \mathbb{R}^d \times \mathbb{R}^m$, but everywhere degenerate. Nevertheless, there is a complete

Recurrence theory (KLIEMANN [23]):

(i) A point $x_o \in X$ has a certain recurrence property (see 1.2) with respect to x (non-Markov) if all points $z_o = (x_o,y)$, $y \in Y$, have the same recurrence property for the Markov prozess $z = (x,\xi)$.

(ii) Control analysis: Associate the deterministic control system $\dot{x} = f(x,u)$, u continuous control with values in Y.
Then there is a decomposition of X into four disjoint sets, $X = A \cup B \cup C \cup D$, characterized by their control properties as follows:

$A \times Y$ = invariant ε-control set (invariant: no control takes you out; ε-control set: given two points and an $\varepsilon > o$, then there is a control u which takes the system starting in one of the points into the ε-neighborhood of the other point),

B x Y = ε-control set,

C x Y = points without bilateral control properties and without positive drift (i.e. $f(x,y) = o$ for some $y \in Y$),

D x Y = points without bilateral control properties and with positive drift (i.e. $f(x,y) \neq o$ everywhere).

For $d = 1$, m arbitrary, there is a simple algorithm yielding this decomposition.

(iii) Translation mechanism: To make sure that the noise is as rich as the controls we need the following nondegeneracy condition for ξ: For any continuous u with $u(o) = y$ and any $\delta > o$ we have

$$P(\max_{o \leq t \leq T} |\xi(t,y)-u(t)| < \delta) > o.$$

This is satisfied if ξ is a nondegenerate diffusion ($b(y)b(y)' > o$ for all $y \in Y$). Then we have

Invariant probabilities are (uniquely) sitting on sets of type A. Their existence is assured e.g. by compactness of the prospective support.

(iv) Example: The stability analysis of the undamped linear oscillator with random restoring force $\ddot{y} + \xi(t)y = o$ (see 3.3), ξ a diffusion in the interval $Y \subset \mathbb{R}^1$, makes it necessary to do recurrence theory for the $(y,\dot{y})$ - system projected onto the unit circle S^1 and described by the nonlinear equation for the angle φ,

$$\dot{\varphi} = f(\varphi,\xi) = -\sin^2\varphi - \xi \cos^2\varphi, \quad \varphi \in [0,2\pi), \; \xi \in Y \subset \mathbb{R}^1.$$

The result of the control analysis for the case where Y is an interval of the negative half-line is shown in figure 1.

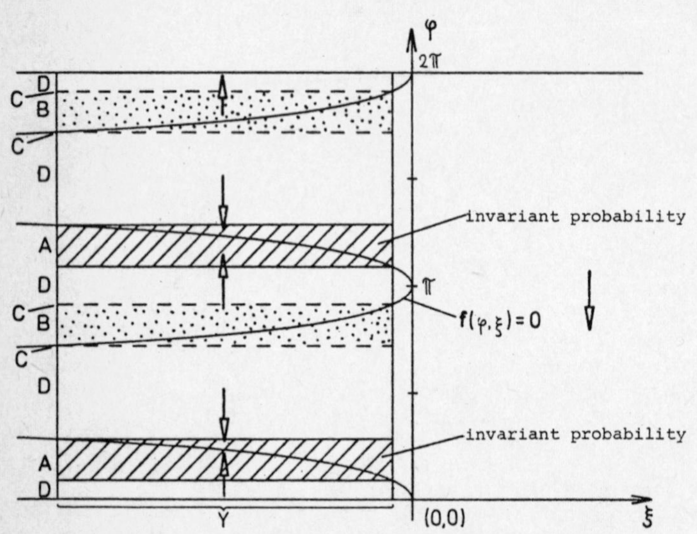

Figure 1. Control-theoretic decomposition of the state space $[0,2\pi) \times Y$ for the angle of the noisy undamped linear oscillator

Dichotomic noise: If in $\dot{x} = f(x) + G(x)\xi$ ξ is a two-state continuous time Markov process (random telegraph signal or dichotomic noise) then there is an explicit formula for the invariant probability of x, see HORSTHEMKE [19].

3. Linear f

3.1 White ξ

In this case, the SDS looks like this:

$$\dot{x} = \underbrace{Ax}_{\text{systematic part}} + \underbrace{B\xi}_{\text{additive noise}} + \underbrace{\sum_{i=1}^{k} C_i x \xi_i}_{\text{multiplicative noise part}},$$

where $A, B, C_1, \ldots, C_k$ are fixed matrices and ξ, $\xi_1, \ldots, \xi_k$ are independent vector and scalar white noise processes, resp.

3.1.1 Additive noise

Here $\dot{x} = Ax + B\xi$, add a linear read-out map $y = Cx$, $y \in \mathbb{R}^p$. For additive noise, recurrence and invariant probabilities are the most important problems. Stability reduces to looking at the undisturbed system $\dot{x} = Ax$. There is an algebraic recurrence theory (cf. BROCKETT [12], ERIKSON [16], ZAKAI and SNYDERS [41]) and there are necessary and sufficient algebraic criteria for existence and uniqueness of an invariant probability for y. For example (SNYDERS [38]), y has an invariant probability iff every vector x that is both unstable and controllable is unobservable, in symbols iff

$$(\ker \phi^{+}(A) \oplus \ker \phi^{o}(A)) \cap \langle A|B \rangle \subset N(C,A),$$

where

$$\langle A|B \rangle = \sum_{i=1}^{d} \text{image } (A^{i-1}B) = \text{controllable subspace of } \mathbb{R}^d,$$

$$N(C,A) = \bigcap_{i=1}^{d} \ker (CA^{i-1}) = \text{unobservable subspace of } \mathbb{R}^d,$$

ϕ^{+}, ϕ^{o} = factors in the minimal polynomial of A having roots with positive or zero real parts, resp.

3.1.2 Multiplicative noise

Here $\dot{x} = Ax + \sum_{i=1}^{k} C_i x \xi_i = (A + \sum_{i=1}^{k} \xi_i C_i)x$ is a mathematical model for white parameter excitation in $\dot{x} = Ax$. For multiplicative noise, stability of $x \equiv o$ and growth of solutions are the most important problems. KHASMINSKII ([20], [21]) had the idea of projecting x onto $S^{d-1} = \{y \in R^d : |y| = 1\}$,

$$x = |x|s, \quad s = x/|x| \in S^{d-1}, \quad |x| \in \mathbb{R}^{+},$$

and looking at $|x|$ and s separately. Happily enough, s is decoupled from $|x|$ (thus Markov) and satisfies a nonlinear Ito equation

$$ds = f(s)dt + \sum_{i=1}^{k} G_i(s)\xi_i,$$

while $|x|$ depends only on s and on noise in the form

$$d \log|x| = Q(s)dt + \sum_{i=1}^{k} P_i(s)\xi_i .$$

If s is ergodic (i.e. has a unique invariant probability P_o on S^{d-1}) it follows from the last equation that for every $x_o \in \mathbb{R}^d$

$$\lim_{t\to\infty} \frac{1}{t} \log |x(t,x_o)| = \int_{S^{d-1}} Q(s)dP_o = \lambda \quad \text{w.p. } 1,$$

in other words: the exakt (i.e. lim sup replaced by lim) Lyapunov number of any solution is w.p. 1 equal to the deterministic constant λ . After that, stability analysis is now simple: $\lambda < o$ means strong exponential asymptotic stability w.p. 1, $\lambda > o$ means exponential blow-up, while for $\lambda = o$ the central limit theorem tells us that the distribution of $|x|$ will spread.

Unfortunately, in relevant applications s is not ergodic, so we can have several stationary solutions on S^{d-1} and thus several (up to d) different Lyapunov numbers. However, the recurrence analysis of s on S^{d-1} by means of control arguments shows that the case

$$x(t,x_o) \sim \exp(\lambda t)s^o(t) \quad (t\to\infty),$$

λ fixed, s^o stationary (unique up to reflection at o) on S^{d-1}, is generic. For particular cases see KLIEMANN ([23], appendix on stochastic stability).

3.1.3 Additive and multiplicative noise

For recurrence theory of $\dot{x} = Ax + B\xi + \Sigma C_i x\xi_i$ see BROCKETT [12].

3.2 Real stationary ξ

3.2.1 Additive noise

For $\dot{x} = Ax + \xi$ with ξ stationary we have much more possibilities for stationary solutions than in the white noise case because (i) x_o can anticipate ξ and (ii) ξ can be adapted to the structure of the undisturbed system $\dot{x} = Ax$, see ARNOLD and WIHSTUTZ [8] for a complete description.

Examples: Without loss of generality, we can assume that A is in Jordan canonical form. We restrict our considerations to invariant subspaces.

(i) Re $\lambda(A) < o$: There is a unique strong ξ-stationary solution iff

$$x_o^o = \lim_{T\to-\infty} \text{in pr.} \int_T^o \exp(-At)\xi(t)dt =: (\text{pr}) \int_{-\infty}^o \exp(-At)\xi(t)dt$$

exists. In this case, x_o^o is the starting variable, so the stationary solution reads

$$x^o(t) = (\text{pr}) \int_{-\infty}^t \exp(A(t-s))\xi(s)ds.$$

Sufficient conditions are $E|\xi(t)| < \infty$ (BUNKE [13]) or $\xi(t) \equiv \xi_o$.

(ii) Re $\lambda(A) > o$: There is a unique strong ξ-stationary solution iff

$$x_o^o = \lim_{T\to\infty} \text{in pr.} -\int_o^T \exp(-At)\xi(t)dt =: -(\text{pr})\int_o^\infty \exp(-At)\xi(t)dt$$

exists. In this case, x_o^o is the (anticipating!) starting variable, and the stationary solution is

$$x^o(t) = -(\mathrm{pr})\int_t^\infty \exp(A(t-s))\xi(s)ds.$$

(iii) $\lambda(A) = o$: $\dot{x} = \xi$ (scalar): $x^o(t) = x_o^o + \int_o^t \xi(s)ds$ is a strong ξ-stationary solution iff $z(t) = \int_o^t \xi(s)ds$ fulfills

$$\lim_{T\to\infty} \frac{1}{T}\int_o^T \exp(iuz(t))dt = L(u) \text{ w.p. 1, for each } u \in \mathbb{R},$$
(or $T\to-\infty$)

where $L(u)$ is a (random) characteristic function (OREY [31]). The strong solution is unique up to adding an invariant random variable. $\dot{x} = \xi$ has a unique wide-sense stationary solution for a wide-sense stationary ξ iff the spectral distribution F of ξ satisfies

$$\int_{-\infty}^{\infty} \lambda^{-2} dF(\lambda) < \infty$$

(ARNOLD, HORSTHEMKE and STUCKI [3]).
(iv) $\lambda(A) = \pm ai$ $(a>o)$: can be handled similarly to (iii), see the same references.
(v) Re $\lambda(A) = o$, but A unstable: There is always a class of ξ's for which there are strong stationary solutions.

3.2.2 Multiplicative noise

To let notation look not too unusual we write $A(t)$ instead of $\xi(t)$, so we deal with the equation $\dot{x} = A(t)x$, $A(t)$ a stationary and ergodic (say) matrix-valued stochastic process. Again growth and stability, i.e. the long-term behavior of the fundamental matrix $\phi(t)$, are of interest.

Multiplicative ergodic theorem of OSELEDEC [32] (for a more recent account see RUELLE [36], a new proof is given by RAGHUNATHAN [33]): There are r fixed numbers $(1\leq r\leq d)$ $\lambda_1 < \lambda_2 < \ldots < \lambda_r$ and (random) subspaces $E_1, E_2, \ldots, E_r$ of $\mathbb{R}^d$, where dim $E_i = d_i$ is fixed, $\sum_{i=1}^r d_i = d$, such that

$$\mathbb{R}^d = E_1 \oplus \ldots \oplus E_r, \quad \phi(t,\omega)E_i(\omega) = E_i(\Theta_t\omega),$$

$$\Theta_t = \text{shift associated with } A(t),$$

and for the (exact) Lyapunov numbers

$$\lim_{t\to\infty} \frac{1}{t} \log |x(t,x_o)| = \lambda_i \quad \text{w.p. 1, uniformly for } x_o \in E_i.$$

This determines the radial behavior of x. For studying the angular behavior, we project x again onto S^{d-1} obtaining for $s = x/|x|$ and $|x|$ the equations

$$\dot{s} = f(s,A) = (A-q(s,A))s, \quad q = \tfrac{1}{2}s'(A+A')s, \quad |x| = |x_o|\exp\int_o^t q(s,A)du,$$

and we look for A-stationary projected solutions $s_i^o \in S^{d-1} \cap E_i =: S_i$.

Floquet representation (WIHSTUTZ [40]): There is a fundamental matrix $\phi(t)$ of $\dot{x} = A(t)x$ of the form

$$\phi(t) = S(t)\exp(Rt + o(t)) \quad \text{for} \quad t \to \infty,$$

where $R = \text{diag}(\lambda_1,\ldots,\lambda_r)$ (λ_i repeated d_i times) and the columns of $S(t)$ consist of (roughly speaking) A-stationary solutions of $\dot{s} = f(s,A)$ in $S_1,\ldots,S_r$. There is at least one (and in general only one) stationary $s_i^o \in S_i$. In case there are less than d_i stationary solutions in S_i we fill-up $S(t)$ by other linearly independent solutions in S_i. If we transform x by $x = S(t)y$ we obtain a decoupled asymptotically constant coefficient linear system

$$\dot{y} = (R + o(1))y.$$

3.3 Markovian ξ

We continue the discussion in 3.2.2 of the multiplicative noise case $\dot{x} = A(t)x$ where now $A(t)$ is a Markovian stationary ergodic matrix-valued process. As

$$|x(t)| = |x_o|\exp(t\lambda(t)), \quad \lambda(t) = \frac{1}{t}\int_o^t q(s(u), A(u))du,$$

the recurrence behavior of the Markov process $z = (s,A)$, $\dot{s} = f(s,A)$, with state space $S^{d-1} \times \mathbb{R}^{d\times d}$ determines stability and growth of x. But this is exactly the situation of 2.3 where (x,ξ) is now (s,A). The control analysis yields that in generic cases there is (up to reflection at O) exactly one invariant ε-control set on S^{d-1}, one unique stationary solution s^o living there, such that for any x_o w. p. 1

$$\lim_{t\to\infty} \lambda(t) = Eq(s_o^o,A(o)) = \lambda \ (=\lambda_r \text{ of } 3.2.2)$$

(KLIEMANN [22], [23]). So all solutions have one and the same deterministic exact Lyapunov number λ, and any solution looks asymptotically like

$$x(t,x_o) \sim \exp(\lambda t)s^o(t) \quad (t\to\infty).$$

Now stability analysis is again very simple: We only have to check whether $\lambda > 0$ or $\lambda > 0$ resulting in exact stability diagrams in parameter space (KLIEMANN and RÜMELIN [24]).
The cases $d = 2$ and $d = 3$ have been completely investigated ($d = 2$: KLIEMANN [22], $d = 3$: SOMMER [39]).

Examples: (i) Stabilization by noise (ARNOLD [4]):
The system

$$\dot{x} = \begin{pmatrix} a & \xi \\ -\xi & -b \end{pmatrix} x, \quad a,b>o,\ a-b<o,\ \xi \text{ Ornstein-Uhlenbeck process with variance } \sigma^2,$$

has the stability diagram shown in figure 2 thus exhibiting the property of being stabilizable by "turning on" noise.
(ii) The damped linear oscillator with random restoring force

$$\ddot{y} + 2\beta\dot{y} + (1+\sigma\xi(t))y = 0, \quad \xi \text{ Ornstein-Uhlenbeck process with variance } 1,$$

has with respect to the parameters β and σ^2 the stability diagram shown in figure 3.
Note that for $\beta > 1$ stability is first being improved by putting noise on the system. Only beyond a certain critical intensity noise acts as a destabilizing effect.

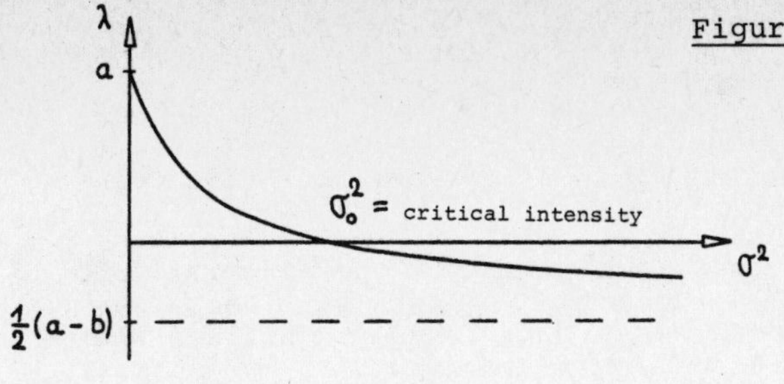

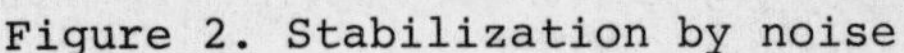
Figure 2. Stabilization by noise

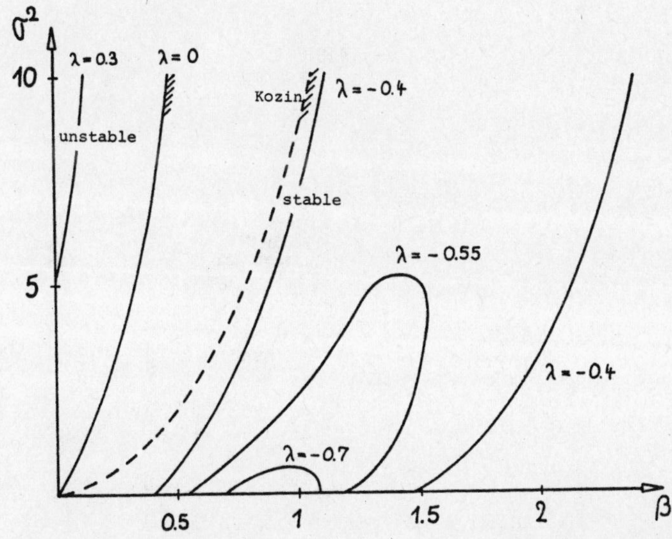

Figure 3. Stability diagram of the damped linear oscillator. Level curves of the Lyapunov number $\lambda(\beta,\sigma^2)$ are drawn. The diagram shows the improvement of the result of KOZIN [26]

4. Linearization

If we linearize a nonlinear SDS $\dot{x} = f(x,\xi)$ around its trivial solution (say) we obtain the multiplicative noise linear SDS

$$\dot{y} = A(\xi(t))y, \qquad A(\xi) = \frac{\partial f}{\partial x}\bigg|_{x=0} .$$

For white ξ KHASMINSKII [21] showed that the findings of the stability analysis of the linear system essentially carry over to the nonlinear one. For a real stationary ξ, the stable manifold theorem of RUELLE ([36], [37]) tells us the same for the multiplicative ergodic theorem (see 3.2.2): There are manifolds M_i tangent to E_i of dimension d_i with Lyapunov number (locally) λ_i, $i=1,\dots,r$.

5. Generalizations

One can consider SDS as stochastic flows on manifolds (ELWORTHY [15], RUELLE [36]), SDS in infinite dimensions (CHOW [14], RUELLE [37]), SDS with more general noise processes, e.g. semimartingales (METIVIER and PELLAUMAIL [29]), etc.

6. Applications

Qualitative theory of SDS applies whenever a physical system is subject to the action of something which can be modeled as a random process. This is particularly the case if one includes fluctuations (internal noise) or considers changes of the environment (external noise), see e.g. ARNOLD and LEFEVER [5].

Qualitative theory is the mathematical base for the calculation of long-term objects like invariant measures, growth rates etc. by numerical or Monte-Carlo procedures. It tells us whether and where those objects (uniquely) exist and how they depend on parameters. Then one can go on and investigate noise-induced phase transitions or bifurcations (ARNOLD, HORSTHEMKE and LEFEVER [2], HORSTHEMKE [19]), stabilization and destabilization by noise (ARNOLD [4]) and other noise-induced changes of qualitative behavior (KLIEMANN [25]).

References

1. Arnold, L.: *Stochastic differential equations: Theory and applications* (Wiley, New York 1974)
2. Arnold, L., Horsthemke, W., Lefever, R.: White and coloured external noise and transition phenomena in nonlinear systems. Z. Phys. B29, 367-373 (1978)
3. Arnold, L., Horsthemke, W., Stucki, J.: The influence of external real and white noise on the Lotka-Volterra model. Biometrical J. 21, 451-471 (1979)
4. Arnold, L.: A new example of an unstable system being stabilized by random parameter noise. Inform. Communication of Math. Chem. 7, 133-140 (1979)
5. Arnold, L., Lefever, R. (eds.): *Stochastic nonlinear systems in Physics, Chemistry and Biology*, Springer Series in Synergetics, Vol. 8 (Springer, Berlin, Heidelberg, New York 1981)
6.
7. Arnold, L., Kliemann, W.: "Qualitative theory of stochastic systems", in *Probabilistic Analysis and Related Topics*, ed. by A. Bharucha-Reid, Vol. 3 (Academic Press, New York 1981)
8. Arnold, L., Wihstutz, V.: Stationary solutions of linear systems with stationary additive noise. Preprint Universität Bremen 1981
9. Bhatia, N.P., Szegö, G.P.: *Stability Theory of Dynamical Systems*, Grundlehren der mathematischen Wissenschaften, Vol. 161 (Springer, Berlin, Heidelberg, New York 1970)
10. Bhattacharya, R.N.: Criteria for recurrence and existence of invariant measures for multidimensional diffusions. Ann. Prob. 6, 541-553 (1978)
11. Bhattacharya, R.N.: *Asymptotic Behavior of Several-Dimensional Diffusions*, Springer Series in Synergetics, Vol. 8 (Springer, Berlin, Heidelberg, New York 1981)
12. Brockett, R.W.: Parametrically stochastic linear differential equations. Mathematical Programming Study 5, 8-21 (1976)
13. Bunke, H.: Gewöhnliche Differentialgleichungen mit zufälligen Parametern (Akademie-Verlag, Berlin 1972)
14. Chow, P.-L.: "Stochastic partial differential equations in turbulence related problems", in *Probabilistic Analysis and Related Topics*, ed. by A. Bharucha-Reid, Vol. 1 (Academic Press, New York 1978)
15. Elworthy, D.: Stochastic differential equations on manifolds. Lecture notes, Dept. of Mathematics, University of Warwick, 1978
16. Erikson, R.V.: Constant coefficient linear differential equations driven by white noise. Ann. Math. Stat. 2, 820-823 (1971)
17. Friedman, A.: *Stochastic Differential Equations and Applications*, Vol. I,II (Academic Press, New York 1975, 1976)
18. Gihman, I.I.,Skorohod, A.V.: *Stochastic Differential Equations*, Ergebnisse der Mathematik und ihrer Grenzgebiete, Vol. 72 (Springer, Berlin, Heidelberg, New York 1972)

19. Horsthemke, W.: *Noise Induced Transitions*, Springer Series in Synergetics, Vol. 8 (Springer, Berlin, Heidelberg, New York 1981)
20. Khasminskii, R.Z.: Necessary and sufficient conditions for the asymptotic stability of linear stochastic systems. Theory Prob. Appl. 12, 144-147 (1967)
21. Khasminskii, R.Z.: *Stability of Systems of Differential Equations With Random Disturbances of Their Parameters* (Nauka, Moscow 1969) (in Russian)
22. Kliemann, W.: "Some exact results on stability and growth of linear parameter excited stoachstic systems", in *Stochastic Control and Stochastic Differential Systems*, ed. by M. Kohlmann, W. Vogel, Lecture Notes in Control and Information Sciences, Vol. 16 (Springer, Berlin, Heidelberg, New York 1979) pp 456-471
23. Kliemann, W.: Qualitative Theorie nichtlinearer stochastischer Systeme. PhD Thesis (Universität Bremen 1980)
24. Kliemann, W., Rümelin, W.: On the growth of linear systems parametrically disturbed by a diffusion process. Report Nr. 27 Forschungsschwerpunkt Dynamische Systeme (Universität Bremen 1980)
25. Kliemann, W.: "Qualitative theory of stochastic dynamical systems — applications to life sciences", in *Stochastic Methods in Life Sciences*, ed. by M. Ianelli, G. Koch, Lecture Notes in Biomathematics (Springer, Berlin, Heidelberg, New York 1981)
26. Kozin, F.: Stability of the linear stochastic system", in *Stability of Stochastic Dynamical Systems*, ed. by R.F. Curtain, Lecture Notes in Mathematics, Vol. 294 (Springer, Berlin, Heidelberg, New York 1972) pp 186-229
27. Kunita, H.: Supports of diffusion processes and controllability problems, in *Proc. Intern. Symp. Stoch. Diff. Equs*, ed. by K. Ito , Kyoto 1986 (Wiley, New York 1978) pp. 163-185
28. Kushner, H.: *Stochastic Stability and Control* (Academic Press, New York 1967)
29. Metivier, M., Pellaumail, J.: *Stochastic Integration* (Academic Press, New York 1980)
30. Nemytskii, V.V., Stepanov, V.V.: *Qualitative Theory of Differential Equations* (Princeton 1960)
31. Orey, S.: Stationary solutions for linear systems with additive noise. Preprint, University of Minnesota 1980
32. Oseledec, V.I.: A multiplicative ergodic theorem. Lyapunov characteristic numbers for dynamical systems. Trans. Moscow Math. Soc. 19, 197-231 (1968)
33. Raghunathan, M.S.: A proof of Oseledec's multiplicative ergodic theorem. Israel J. Math. 32, 356-362 (1979)
34. Rosenblatt, M.: *Markov Processes, Structure and Asymptotic Behavior*, Grundlehren der mathematischen Wissenschaften, Vol. 184 (Springer, Berlin, Heidelberg, New York 1971)
35. Rozanov, Y.A.: *Stationary Random Processes* (Holden Day, San Francisco 1967)
36. Ruelle, D.: Ergodic theory of differentiable dynamical systems. Publ. Math. IHES 50, 275-306 (1979)
37. Ruelle, D.: Characteristic exponents and invariant manifolds in Hilbert space. Preprint IHES/P/80/11, March 1980
38. Snyders, J.: Stationary probability distributions for linear time-invariant systems. SIAM J. Control and Optim. 15, 428-437 (1977)
39. Sommer, U.: Die Wachstumszahlen dreidimensionaler linearer parametererregter Systeme. To appear in ZAMM 61 (4/5) (1981)
40. Wihstutz, V.: Ergodic theory of linear parameter-excited systems. Preprint, Universität Bremen 1980
41. Zakai, M., Snyders, J.: Stationary probability measures for linear differential equations driven by white noise. J. Diff. Equ. 1, 27-33 (1970)

Part V

External Fluctuations and Noise Induced Transitions

Noise Induced Transitions

Werner Horsthemke

Service de Chimie Physique II, Université Libre de Bruxelles,
B-1050 Bruxelles, Belgium

1. Introduction

Coupled to a fluctuating environment, nonlinear open systems can display types of behavior that are impossible under corresponding deterministic external constraints. Indeed, when the characteristics of the external fluctuations (e.g. variance, correlation time) cross certain thresholds, transition phenomena can take place without any change at all in the average state of the environment. This new class of nonequilibrium transitions is called noise induced transitions. They are characterized by the fact that the system no longer adjusts its macroscopic behavior to the average value of the external constraints. In this paper I will address mainly the more theoretical aspects of these noise induced transitions. Their importance for natural systems is discussed in the paper by LEFEVER.

Let me begin by specifying the kind of system and environment I intend to deal with in particular here: I will consider only systems that are spatially homogeneous. This is a satisfactory approximation in applications if either the transport is fast compared to the "reaction" kinetics or if the system is artificially kept homogeneous, e.g. by stirring. Furthermore it is desirable to avoid complications from other noise sources in the system. Therefore I will deal only with macroscopically large systems and assume that the thermodynamic limit, system size $V \to \infty$, has been taken. Hence it is safe to neglect the effect of internal fluctuations which scale with an inverse power of V. In order to obtain as far as possible exact analytical results, I will restrict myself to the class of systems, the state of which can be satisfactorily described by one intensive variable, i.e. a kinetic equation of the type:

$$\dot{x} = f(x,\lambda). \tag{1}$$

λ denotes an external parameter which depends on the state of the environment and $x \in [b_1, b_2]$, often $b_1 = 0$, $b_2 = \infty$. In most cases the kinetic equation is linear in the external parameter λ and can thus be written:

$$\dot{x} = h(x) + \lambda\, g(x). \tag{2}$$

In most applications, at least over the time spans one is interested in, the environment is constant on the average. Hence I will assume in the following that the external parameter is given by a stationary random process λ_t. For convenience it is decomposed in its systematic part λ and the fluctuating part ζ_t, i.e.

$$\lambda_t = \lambda + \zeta_t \quad \text{with} \quad E\{\zeta_t\} = 0 \tag{3}$$

Here I will restrict myself to situations in which the external parameter is the cumulative effect of a multitude of small, similar, weakly coupled contributions. In the light of the Central Limit Theorem one can then conclude that ζ_t should be Gaussian. Furthermore I will assume that the environment is without aftereffect, i.e. ζ_t is Markovian. As far as applications are concerned, this property holds on a sufficiently coarse-grained time scale. Invoking now DOOB's theorem [1], which states that the only nonsingular Gaussian Markov process is the Ornstein-Uhlenbeck process, we conclude that the fluctuations of the external parameter are given by the stochastic differential equation (SDE):

$$d\zeta_t = -\gamma\zeta_t dt + \sigma dW_t , \qquad \zeta_0 \; N(0,\sigma^2/2\gamma). \tag{4}$$

The Ornstein-Uhlenbeck noise has an exponentially decreasing correlation function, a form that is indeed widely found in applications:

$$C(\tau) = E\{\zeta_t\zeta_{t+\tau}\} = (\sigma^2/2\gamma)\ \exp(\ -\gamma|\tau|). \tag{5}$$

The correlation time is given by

$$\tau_{cor} = \gamma^{-1} \tag{6}$$

and the spectrum $S(\nu)$, related to $C(\tau)$ by a Fourier transform, is the Lorentzian:

$$S(\nu) = (\sigma^2/2\ \pi)(\nu^2+\ \gamma^2)^{-1}\ . \tag{7}$$

In applications it is often the case that the environmental fluctuations are rapid compared to the typical macroscopic time of the system:

$$\tau_{cor} \ll \tau_{macro}\ . \tag{8}$$

It is then tempting to neglect the correlations in the external noise completely and pass to the idealisation $\tau_{cor} = 0$. However, some circumspection has to be exerted in taking this limit. If all other characteristics of the noise ζ_t are kept constant, then for $\tau_{cor} \to 0$, i.e. $\gamma \to \infty$,

$$S(\nu) \to 0\ . \tag{9}$$

Hence to avoid this noiseless limit, the variance has to be increased concomitantly with the decrease in the correlation time, i.e.

$$\gamma\to\infty\ ,\ \sigma\to\infty \quad \text{such that}\quad \sigma^2/\gamma^2 = \bar{\sigma}^2 = \text{const.} \tag{10}$$

This limit is known as the white noise limit, since now

$$S(\nu) \to \bar{\sigma}^2/2\pi \quad \text{for all } \nu, \tag{11}$$

i.e. a flat (white) spectrum. The process

$$\bar{\sigma}\xi_t = \lim \zeta_t \tag{12}$$

is known as Gaussian white noise:

$$E\{\xi_t\} = 0,\ E\{\xi_t\xi_{t+\tau}\} = \delta(\tau). \tag{13}$$

It is easily proven that the integrated O.-U. noise

$$V_t = \int_0^t \zeta_s ds \tag{14}$$

converges in the limit (10) towards the Wiener process $\bar{\sigma}W_t$, i.e.

Brownian motion in position space. Hence we have that in some sense

$$\int_0^t \xi_s ds = W_t \quad \text{or} \quad \xi_t = \dot{W}_t. \tag{15}$$

2. White Noise Induced Transitions

I will first investigate the influence of extremely rapid external fluctuations on the macroscopic behavior of nonlinear open systems. To this aim I suppose that the above presented white noise idealisation is adequate and describe therefore, using (15), the system by the SDE

$$dx_t = [h(x_t) + \lambda g(x_t)]dt + \bar{\sigma} g(x_t) dW_t \tag{16}$$

(In the following $\bar{\sigma}=\sigma$.) Since (16) is obtained as the white noise limit of a differential equation with O.-U. noise, it has to be interpreted in the sense of Stratonovic, according to WONG and ZAKAI's theorem [2] and its extension by BLANKENSHIP and PAPANICOLAOU [3]. (For the problems I consider here, qualitatively the same results are obtained if (16) were interpreted as an Ito equation.) The solution of (16) is a diffusion process and its transition probability density is the fundamental solution of the Fokker-Planck equation (FPE):

$$\partial_t p(x,t) = -\partial_x [f(x,\lambda) + \tfrac{1}{2} g'(x) g(x)] p(x,t) + \tfrac{1}{2}\sigma^2 \partial_{xx} g^2(x) p(x,t) \tag{17}$$

I will mainly be interested in the stationary behavior of the system. If the boundaries b_1 and b_2 are natural, the stationary solution of the Fokker-Planck equation is given by

$$p_s(x) = N\, g(x)^{-1} \exp\{ \frac{2}{\sigma^2} \int^x \frac{f(z,\lambda)}{g^2(z)} dz \} \tag{18}$$

provided it is normalisable on $[b_1, b_2]$. In the following I will not retain the complete information contained in $p_s(x)$, but will consider only its extrema x_m. There are three main reasons for this procedure: i) The number and location of the extrema are the most distinguishing features of $p_s(x)$ and contain essential information on the stationary behavior of the system. ii) The extrema are so to speak the continuation of the deterministic steady states. Indeed, for $\sigma^2 \downarrow 0$ x_m tends to x_s, the zeroes of the rhs of (1). iii) If $p_s(x)$ is normalisable, then the process x_t is ergodic. Hence we can interpret $p_s(x)$ as a measure for that part of the time, that an arbitrary sample path spends in an infinitesimal vicinity of x. This motivates the usual identification of the extrema of $p_s(x)$, which are preferentially seen in an experiment, with the macroscopic steady states: the maxima, where the process spends relatively much time, as the stable steady states and the minima, which the process leaves rather quickly as unstable ones.

The extrema of $p_s(x)$ are easily determined from the relation

$$[h(x_m) + \lambda g(x_m)] - (\sigma^2/2) g'(x_m) g(x_m) = 0. \tag{19}$$

It contains two terms: the one in brackets equal to zero corresponds to the equation for the deterministic steady states x_s. The second one describes the influence of the external noise. It vanishes identically for additive noise, i.e. $g(x) \equiv \text{const}$, and hence $x_m = x_s$ for <u>all</u> intensities σ^2 of the external noise. Since the influence of the environmental fluctuations does not depend on the state of the system in this case, additive noise has only the expected disorganising effect, namely it smears

$p_S(x)$ out around the deterministically stable steady states. The situation is however quite different for mutliplicative noise, i.e. $g(x) \neq$ const. Here the effect of the environemental fluctuations does depend on the state of the system. We expect the following behavior: If σ^2 is sufficiently small, the roots of (19) do not differ in any essential way from the deterministic steady states. If however the intensity σ^2 of the noise increases and if $g(x)$ is nonlinear in a suitable way, the extrema of $p_s(x)$ can be essentially different, in number and location, from the deterministic steady states. That is, if the intensity σ^2 crosses certain threshold values, the shape of $p_s(x)$ can change drastically. The external noise can thus deeply modify the macroscopic behavior of the system, namely induce new nonequilibrium transitions. In addition to the disorganising effect, which it shares with additive noise, multiplicative noise can create new "potential wells", i.e. "stabilise" macroscopic states without any counterpart under deterministic environmental conditions. If e.g. $h(x)$ and $g(x)$ are polynomials,

$$h(x) = \Sigma^{n} a_\nu x^\nu \quad , \; g(x) = \Sigma^{m} b_\nu x^\nu \tag{20}$$

there are two possible types of noise induced transitions: i) If $2m-1 \leq \max(n,m)$, then the number of extrema does not exceed the number deterministic steady states. However their location can be strongly altered by the noise as shows the Verhulst model [4]:

$$\dot{x} = \lambda x - x^2 . \tag{21}$$

In a constant environment this system has one transition point at $\lambda=0$, where the trivial steady state $x_s=0$, stable for $\lambda<0$, becomes unstable and a new branch of steady states $x_s=\lambda$, which is stable for $\lambda>0$, emerges. In a white noise environment we have

$$(S) \quad dx_t = (\lambda x_t - x_t^2)\, dt + \sigma x_t \, dW_t \tag{22}$$

and (19) reads in this case

$$(\lambda - \sigma^2/2)x_m - x_m^2 = 0. \tag{23}$$

In addition to the deterministic transition point $\lambda=0$, the Verhulst model possesses in white noise surroundings a second transition point at $\lambda=\sigma^2/2$, which is induced by the external noise: For $\lambda<0$ $p_s(x) = \delta(x)$; for $0<\lambda<\sigma^2/2$ a genuine probability density exists, but $x=0$ is still the most probable value. Only for $\lambda>\sigma^2/2$ has $p_s(x)$ a maximum near the nontrivial deterministic steady state. As far as the extrema are concerned, the deterministic bifurcation diagram of the Verhulst model has been shifted by $\sigma^2/2$. ii) If $2m-1 > \max(n,m)$, then the number of extrema may exceed the number of deterministic steady states. In this case types of behavior become possible for the system which are forbidden under deterministic environmental conditions. An example for these striking phenomena is presented in LEFEVER's paper. Noise induced transitions occur of course also for nonpolynomial phenomenological equations and two such systems which are of importance in applications are studied in [5,6].

3. Colored Noise Induced Transitions

In the preceding section it was established that white external noise can drive the system far away from thermodynamic equilibrium. Indeed, when the intensity of the environmental fluctuations increases beyond certain threshold values, new types of nonequilibrium transitions are observed. These transitions take place without any change in the average state of the environment and their major characteristic is that the system no longer adjusts its macroscopic behavior to the average value of the external constraints. These theoretical results are however obtained by adopting the white noise idealisation. While this is convenient from a mathematical point of view, - a system coupled to a white noise environment is markovian -, real fluctuating environments have a nonvanishing correlation time, i.e. are nonwhite, as stated in the introduction. Hence the validity of the white noise idealisation has to be investigated. This question has two aspects: i) It is necessary to establish that noise induced transitions are no artefacts of the white noise idealisation but that qualitatively the same transition phenomena occur for external noises with short but nonzero correlation time. ii) A quantitative evaluation of the effect of correlations on noise induced transitions is desirable.

3.1 Ornstein-Uhlenbeck Noise

Let us come back to the situation described in 1, namely a system coupled to an environment whose fluctuations can be modelled by an Ornstein-Uhlenbeck process. A convenient way to study (2) and (4) in the neighborhood of white noise, i.e. γ large but finite, is to write these equations in the scaled form [3,8]:

$$dx_t^\varepsilon = f(x_t^\varepsilon,\lambda)dt + \varepsilon^{-1}\zeta_t g(x_t^\varepsilon)dt \tag{24}$$

$$d\zeta_t = \varepsilon^{-2}\zeta_t dt + \varepsilon^{-1}\ \sigma dW_t \tag{25}$$

where

$$E\{\zeta_t\}=0,\ E\{\zeta_t\zeta_{t+\tau}\}=(\sigma^2/2)\exp(-|\tau|/\varepsilon^2),\ \tau_{cor}=\gamma^{-1}=\varepsilon^2. \tag{26}$$

With this scaling the white noise limit (10) corresponds to $\varepsilon\to 0$. The main difficulty in any approach using nonwhite noise is that the temporal evolution of the system is non-Markovian. However in the present situation we can still work in the framework of the Markovian theory by considering the twodimensional process $(x_t^\varepsilon,\zeta_t)$. The pair process, made up of the state variable of the system and the external fluctuating parameter, is always Markovian, if the external noise itself is described by a Markov process. Ergodic markovian noise is commonly called colored noise, the most important representative of which is the Ornstein-Uhlenbeck noise considered here. It follows from (24) and (25) that $(x_t^\varepsilon,\zeta_t)$ is a diffusion process. It is thus governed by the FPE:

$$\partial_t p^\varepsilon(x,z,t) = (\varepsilon^{-2}F_1 + \varepsilon^{-1}F_2 + F_3)p^\varepsilon(x,z,t) \tag{27}$$

with

$$F_1 = \partial_z z + (\sigma^2/2)\partial_{zz} \tag{28}$$

$$F_2 = -z\partial_x g(x) \tag{29}$$

$$F_3 = -\partial_x f(x,\lambda). \tag{30}$$

In general the stationary probability density of (27) cannot be calculated exactly. Hence approximation schemes have to used to determine the stationary behavior of the system. There are too basic possibilities: i) Determine an approximate evolution operator with nice properties, e.g. of the Fokker-Planck type, for the one-time probability density $p^\varepsilon(x,t)$ of the system as done by SANCHO and SAN MIGUEL [7], cf. their paper in this volume. ii) Solve (27) in the form of a systematic perturbation expansion in ε. We have recently developed such a perturbation scheme, which is presented in detail in [8]. The main result is that up to the first significant order in ε the stationary probability density $p_s^\varepsilon(x)$ for the state variable is given by

$$p_s^\varepsilon(x) = p_s(x)\{1+\varepsilon^2[C-f'(x)+(f(x)g'(x))/g(x)-f^2(x)/(\sigma^2 g^2(x))]\}$$
$$= p_s(x)\{1+\varepsilon^2(C-u(x))\} \tag{31}$$

where

$$C = -\sigma^{-2}\int_{b_2}^{b_1}(f^2(x)/g^2(x))p_s(x)dx \tag{32}$$

and $p_s(x)$ is the stationary probability density in the white noise limit as given by (18). Hence we have for the extrema of $p_s^\varepsilon(x)$:

$$(f(x_m)-(\sigma^2/2)g'(x_m)g(x_m))$$
$$+\varepsilon^2\{[f(x_m)-(\sigma^2/2)g'(x_m)g(x_m)][C-u(x_m)]-(\sigma^2/2)\,g^2(x_m)u'(x_m)\} = 0. \tag{33}$$

This answers the questions raised at the beginning of this section: Noise induced transitions are not artefacts due to the white noise idealisation. The roots of (33) are essentially given by the first term, i.e. the white noise results are robust and apply qualitatively to colored noises with short correlations. This is true provided $C<\infty$, which holds in all systems where noise induced transitions have been described so far.

3.2 Dichotomous Markov Noise and Phase Diagrams

If we want to study the influence of colored noise on nonlinear systems for arbitrary values of the correlation time, then obviously we have to go beyond the situation covered by the above perturbation scheme. As already mentioned, in general this impossible for O.-U.-noise, the type of noise most frequently encountered in natural systems. That is why we will now turn our attention to the special class of colored noises which allow for an exact solution of the steady state problem for arbitrary values of the noise characteristics. To my knowledge this class consists essentially of the dichotomous Markov noise, also known as the random telegraf signal [9,10]. Its state space consists of only two levels and for the sake of clarity I will only consider here the symmetric dichotomous Markov noise:

$$I_t \in \{-\Delta, +\Delta\}. \tag{34}$$

The transition probability of this markovian process obeys the following master equation:

$$\frac{d}{dt}\begin{pmatrix} P_{\Delta j}(t) \\ P_{-\Delta j}(t)\end{pmatrix} = -\frac{\gamma}{2}\begin{pmatrix} 1 & -1 \\ -1 & 1\end{pmatrix}\begin{pmatrix} P_{\Delta j}(t) \\ P_{-\Delta j}(t)\end{pmatrix} \tag{35}$$

where

$$P_{ij}(t) = \mathrm{Prob}(I_t=i\,|\,I_0=j)\ ,\quad i,j=\pm\Delta.$$

If I_t is started with equal probability in the two states, then it is a stationary process with

$$E\{I_t\} = 0 \tag{36}$$

and the correlation function:

$$E\{I_t I_{t+\tau}\} = \Delta^2\exp(-\gamma|\tau|). \tag{37}$$

Thus the dichotomous Markov noise, though being rather artificial in the sense that it takes only two values, has the appealing feature to possess the wide-spread form of an exponentially decreasing correlation function. It is intuitively obvious that, as the O.-U. noise, the dichotomous noise converges in the limit

$$\Delta\to\infty,\ \gamma\to\infty: \qquad \Delta^2/\gamma = \text{const} = \sigma^2/2 \tag{38}$$

towards Gaussian white noise: $I_t \to \sigma\xi_t$. The temporal evolution of a nonlinear system like (2) subjected to dichotomous Markov noise is described by the equation

$$\dot{x}_t = h(x_t) + \lambda g(x_t) + I_t g(x_t). \tag{39}$$

Again x_t is a non-Markovian process, but the pair process (x_t, I_t) is Markovian. It is intuitively clear that the evolution equation for the joint probability consists of two parts, one describing the evolution of x_t for a fixed value of the dichotomous noise, $I_t = \Delta$ or $I_t = -\Delta$, and one governing the evolution of the dichotomous Markov noise, which is of course given by the master equation (35):

$$\begin{aligned} \partial_t p(x,\Delta,t) &= -\partial_x[h(x)+(\lambda+\Delta)g(x)]\,p(x,\Delta,t) - (\gamma/2)q(x,t) \\ \partial_t p(x,-\Delta,t) &= -\partial_x[h(x)+(\lambda-\Delta)g(x)]\,p(x,-\Delta,t) + (\gamma/2)q(x,t) \end{aligned} \tag{40}$$

where $q(x,t) = p(x,\Delta,t) - p(x,-\Delta,t)$. The evolution equation for the probability density $p(x,t)$ of the state variable x alone can be derived from (40) and reads

$$\begin{aligned} \partial_t p(x,t) &= -\partial_x f(x,\lambda)\,p(x,t) \\ &+\Delta^2 \partial_x g(x) \int_{-\infty}^{t} dt'\,\exp\{-[\gamma+\partial_x f(x,\lambda)](t-t')\}\partial_x g(x)\,p(x,t'). \end{aligned} \tag{41}$$

Its non-Markovian character is clearly displayed by the memory kernel. The stationary solution of (41) is given by

$$p_s(x) = N \frac{g(x)}{\Delta^2 g^2(x) - f^2(x,\lambda)} \exp\{\gamma \int^{x} \frac{f(y,\lambda)}{\Delta^2 g^2(y) - f^2(y,\lambda)}\,dy\} \tag{42}$$

and vanishes identically outside the interval $U = [x_s(\lambda-\Delta), x_s(\lambda+\Delta)]$ where

$$h(x_s(\lambda\pm\Delta)) + (\lambda\pm\Delta)g(x_s(\lambda\pm\Delta)) = 0.$$

The extrema of $p_s(x)$ are given by the equation:

$$\begin{aligned} &[h(x_m)+\lambda g(x_m)] - (\Delta^2/2)g'(x_m)g(x_m) + \\ &(2/\gamma)[h(x_m)+\lambda g(x_m)][h'(x_m)+\lambda g'(x_m)] - (1/\gamma)[h(x_m)+g(x_m)]^2 \frac{g'(x_m)}{g(x_m)} = 0 \end{aligned} \tag{43}$$

for $x_m \in U$. In the white noise limit (38) only the first two terms survive, yielding (19). This confirms again explicitly the robustness of the results obtained in the white noise idealisation. The last two terms are corrections due to a nonvanishing correlation time, i.e. finite γ. (43) holds for arbitrary γ and clearly there can be additional modifications of the macroscopic behavior of the system if the correlation time increases, i.e. γ decreases, as reflected by these last two terms.

To illustrate this let us consider again the Verhulst model (21) and restrict ourselves to the case that in a nonfluctuating environment genuine growth is possible, i.e. $\lambda>0$. The equation for the extrema reads

$$\lambda x_m - x_m^2 - (\Delta^2/\gamma)x_m + (2/\gamma)x_m(\lambda-x_m)(\lambda-2x_m) - (1/\gamma)x_m(\lambda-x_m)^2 = 0 \qquad (44)$$

and has the roots

$$x_0 = 0 \qquad (45)$$

$$x_{1,2} = (1/6)(4\lambda+\gamma\pm\sqrt{(2\lambda-\gamma)^2 + 12\Delta^2}).$$

An analysis of this expression together with the behavior of $p_s(x)$ near the boundaries of the support U permits us to construct a phase diagram of the steady state behavior of the Verhulst model in the (Δ,γ)-plane, as displayed in Fig. 1.

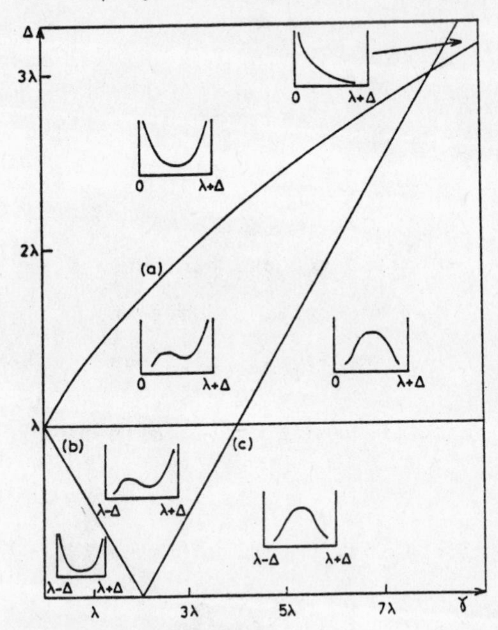

Fig.1 Phase diagram for the steady state behavior of the Verhulst system as a function of the amplitude Δ and the inverse of the correlation time γ of the external dichotomous Markov noise for fixed positive λ. The shape of $p_S(x)$ is sketched for the different regions. Curve (a) is given by $\lambda(1+\lambda/\gamma)=\sigma^2/2$

This phase diagram shows that either by increasing the amplitude Δ of the external noise or by changing its correlation time $1/\gamma$, the system undergoes a series of transitions. The neighborhood of white noise corresponds of course to the upper right corner of the phase diagram. Here curve (a) is given by $\lambda=\sigma^2/2 - O(1/\gamma)$. As expected, the results of the white noise analysis are explicitly recovered for short correlations in the external noise.

4. Perspectives

4.1 Dynamics of Noise Induced Transitions

So far I have only dealt with the stationary behavior of systems coupled to a fluctuating environment. The temporal behavior and especially the dynamics of noise induced transitions is a very arduous problem, even in the white noise idealisation. In general the time dependent solution of the FPE (17) cannot be determined exactly. For the Verhulst model this problem has been tackled by studying the eigenvalues of the FPE [11,12]. The knowledge of the eigenvalues is however not sufficient to determine if the dynamics of the macroscopic states, i.e. the extrema of the probability density, display critical slowing down at certain noise induced transition points. Consider for instance HONGLER's model [13]:

$$(S)\ dx_t = -\frac{1}{2\sqrt{2}}\tanh(2\sqrt{2}x_t)dt + \frac{\sigma}{4}\frac{1}{\cosh(2\sqrt{2}x_t)}dW_t . \tag{46}$$

This process can be transformed to the Ornstein-Uhlenbeck process by the change of variables $z=\sinh(2\sqrt{2}x)$ and hence the time depent probability $p(x,t)$ is given by

$$p(x,t) = \frac{2\sqrt{2}\cosh(2\sqrt{2}x)}{\sqrt{\pi(\sigma^2/2)(1-e^{-2t})}} \quad \exp\{-\frac{1}{2}\frac{\sinh^2(2\sqrt{2}x)}{(\sigma^2/4)(1-e^{-2t})}\} \tag{47}$$

if $p(x,0)=\delta(x)$ which I will consider here for simplicity. The extrema of $p(x,t)$ are given by

$$x_1(t) = 0 \quad \text{for all } t\geq 0, \tag{48}$$

$$x_{2,3}(t) = \pm(2\sqrt{2})^{-1}\ \text{arccosh}(\sigma\sqrt{1-e^{-2t}}/2) \quad \text{for all } t\geq t_c = -0.5\ \ln(1-4/\sigma^2). \tag{49}$$

Thus the stationary probability density has one peak, centered at $x=0$, for $\sigma^2<4$ and becomes bimodal for $\sigma^2>4$. Critical behavior occurs at $\sigma^2=4$. As far as the dynamics of the extrema is concerned, we have the following picture: If for $\sigma^2>4$ we start with a δ-peak on $x=0$, i.e. the unstable state, then this peak will split in two at $t=t_c$. As Hongler already remarked, the time it takes the system to leave the unstable state tends to infinity if σ^2 approaches the critical value 4. However the eigenvalues of the FPE corresponding to (46) are obviously those of the Ornstein-Uhlenbeck process, i.e.

$$\mu_n = 0,\ 1,\ 2,\ \ldots \tag{50}$$

and are <u>not at all</u> influenced by the intensity of the white noise.

4.2 Spatial Inhomogeneities and Nucleation

The second major open problem in the theory of noise induced transitions is the inclusion of spatial inhomogeneities. The need for this is well examplified by the experimental results, obtained for a three-dimensional chemical system, namely the Briggs-Rauscher reaction [14]. The influence of light intensity fluctuations on this photosensitive reaction, realized in a continuous stirred tank reactor, was investigated and the main experimental result is given in Fig. 2.

There the optical density of the reacting mixture at 460 nm, a measure for the concentration of I_2, is plotted versus the (mean) intensity of the incident light. The solid lines correspond to a nonfluctuating intensity, the broken lines to a noisy light intensity, which is Gaussian distributed with a relative variance of 13% and has an exponentially decreasing correlation function with $\tau_{cor}\simeq 0.05$ sec, about one twentieth of the typical macroscopic time. Fig. 2 depicts a region of bistability between an oscillating state A of low concentration in I_2 (the vertical bars represent the amplitude) and a stationary state B of high con-

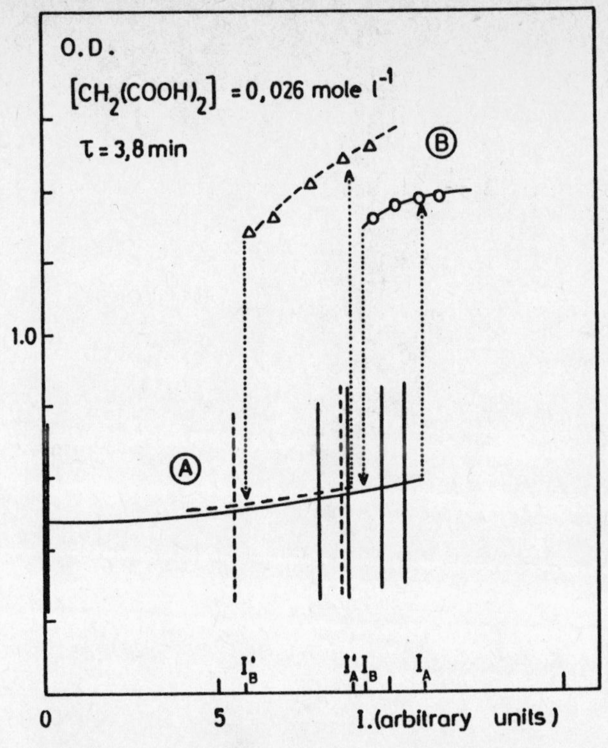

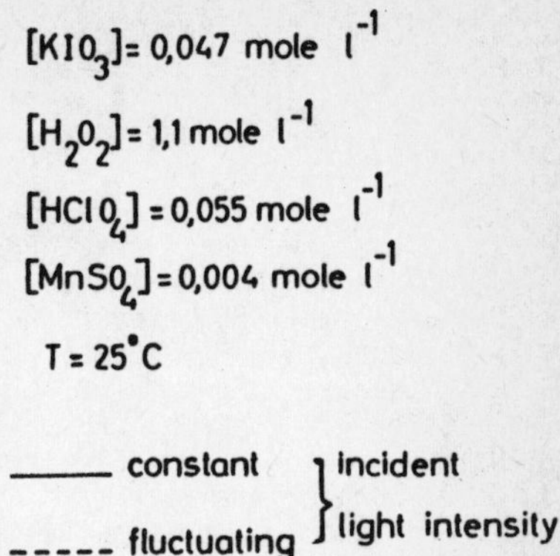

Fig. 2 Plot of the optical density (at 460 nm) of the Briggs-Rauscher reaction versus the (mean) incident light intensity. The concentration of the reactants, fed into the reactor, the temperature and the renewal time τ of the reactor (volume 30 cm^3) are indicated

tration in I_2 for a nonrandom light intensity. Using the fluctuating light source leads to a broadening of the region of bistability and to such a shift to lower (mean) intensities that it is disjunct with that of the nonfluctuating case. This implies that for intensities between I'_A and I_B the system undergoes a transition if, keeping the mean intensity constant, the noise is switched on or off. Note that in contradistinction to zerodimensional one-variable systems, as e.g. the Verhulst model and the electrical circuits studied experimentally by KABASHIMA et al [15], the macroscopic states, i.e. the extrema of the probability density, are very well defined. Furthermore during the time span of the experiment, about one hour for every value of the mean intensity, no spontaneous transition between the two states in the region of bistability was observed. The most plausible explanation for this phenomenon is that also for noise induced bistability a nucleation mechanism occurs: Due to the thorough stirring any "droplets", being in the other state, are broken up in the reacting mixture before they can grow to the finite critical size to tip the system over into the other state. Thus the stirring stabilises the two states and leads to narrow peaks in the probability density. From a theoretical point of view these results show that a quantitative description of noise induced phenomena in non-zerodimensional systems makes the inclusion of spatial dependence in the equations necessary, even if the system remains homogeneous macroscopically. Hence it is desirable to develop a theory of reaction-diffusion systems coupled to a fluctuating environment.

Acknowledgement: This work was supported by the Instituts Internationaux de Physique et de Chimie, fondés par E. Solvay, and by the Belgian Government, Actions de Recherche Concertées, convention n° 76/81II3.

References

1. J. L. Doob, Ann. Math. 43, 351 (1942)
2. E. Wong, M. Zakai, Ann. Math. Stat. 36, 1560 (1965)
3. G. Blankenship, G. C. Papanicolaou, SIAM J. Appl. Math. 34, 437 (1978)
4. W. Horsthemke, M. Malek-Mansour, Z. Physik B 24, 307 (1976)
5. W. Horsthemke, R. Lefever, Phys. Lett. 64A, 19 (1977)
   R. Lefever, W. Horsthemke, Bull. Math. Biol. 41, 469 (1979)
6. R. Lefever, W. Horsthemke, Proc. Natl. Acad. Sci. U.S.A. 76, 2490 (1979)
7. J. M. Sancho, M. San Miguel, Z. Physik B 36, 357 (1980)
8. W. Horsthemke, R. Lefever, Z. Physik B (in press)
9. K. Kitahara, W. Horsthemke, R. Lefever, Phys. Lett. 70A, 377 (1979)
10. K. Kitahara, W. Horsthemke, R. Lefever, Y. Inaba, Progr. Theor. Phys. (in press)
11. A. Schenzle, H. Brand, Phys. Rev. A20, 1628 (1979)
12. M. Suzuki, K. Kaneko, F. Sasgawa, preprint 1980
13. M. O. Hongler, Helv. Phys. Acta 52, 280 (1979)
14. P. De Kepper, W. Horsthemke, C. R. Acad. Sc. Paris C 287, 251 (1978)
15. S. Kabashima, T. Kawakubo, Phys. Lett. 70A, 375 (1979)
    S. Kabashima, S. Kogure, T. Kawakubo, T. Okada, J. Appl. Phys. 50, 6296 (1979)

Noise Induced Transitions in Biological Systems

René Lefever

Chimie Physique II, Université Libre de Bruxelles,
B-1050 Bruxelles, Belgium

1. Introduction

As a result of their complexity, natural environments exhibit a great degree of variability which biological systems often perceive as a kind of external noise. This situation is particularly evident for biological populations, be it populations of animals or populations of simple cells. The cells of a tissue for example are subjected to the influence of numerous environmental factors: concentration of oxygen and nutrients in the tissue, hormones, chemical agents, temperature, etc... Most of these factors exhibit temporal variations over a broad spectrum of frequencies due to periodicities of seasonal, circadian or autonomous metabolic origin. Futhermore these factors also undergo the erratic noisy type of fluctuations which to a greater or lesser degree always affect the behavior of complicated dynamical systems. The combined action of these various sources of variability makes that cellular properties such as the capability of the cells to replicate are always to some extent uncertain and stochastic.

This fact and similar facts in ecological systems have been acknowledged by many authors [1-5] which therefore have stressed the necessity to incorporate explicitly the influence of environmental fluctuations in the phenomenological description of biological systems. Clearly indeed these fluctuations may be responsible for large scale effects. To give a simple example, let me just briefly mention the mechanism of sex determination in turtles. For several species of these reptiles it has unambiguously been demonstrated that there is no strict genotypic control of sex determination [6]. This process quite sensitively depends on the temperature at which the eggs are incubated. A small increase in this temperature can switch the populations of newborn turtles from completely male to completely female.

In view of the many observations of this type which have been made, it is not astonishing that there exists a vast literature dealing with problems in population growth, epidemiology, genetics, with strategies for optimal harvesting, etc..., in which the question of the influence of environmental randomness has been raised (for a general introduction see [5,7]). Outside population biology also, this question has been considered in connection with the behavior of neurons [8,9], with the development of the nervous system [10], with the stability of some enzymatic pathways [11,12,13].

Obviously, these studies constitute a general context to which the noise induced transitions discussed in the paper of HORSTHEMKE in this volume belong. I would like to pursue the analysis of these phenomena futher here in the case of systems of biological interest. In the white noise limit, two main classes of noise induced transitions can be distinguished: (i) Transitions which correspond to shifts in the location of critical points or of bifurcation points which already are displayed by varying the systemic parameters under deterministic conditions.

(ii) Transitions which stabilize macroscopic states without that there exists a counterpart behavior in the deterministic case. These transitions have been called *peak splitting transitions* to distinguish them from the former ones [14,15]. The Verhulst model $\dot{x} = \lambda x - x^2$ furnishes in population dynamics, when λ fluctuates, the prototypic example of noise induced transitions of type (i). In contrast the models analyzed below, namely a model for genic selection and the Hodgkin-Huxley model for sodium and potassium activation in nerve membranes, both exhibit the second class of transitions. In view of their unexpected character, these transitions certainly are the more interesting to analyze at least from a fundamental point of view.

2. Peak splitting transition in a model for genic selection

I consider the kinetic equation [16]

$$\dot{x} = \frac{1}{2} - x + \beta x(1 - x) \tag{1}$$

which describes the changes in gene frequency in a haploid population at a particular locus on the chromosome for which there are only two possible alleles in the population, say A and a. The frequencies of these alleles in the population are respectively x and 1 - x; β is a selection coefficient which depends on the state of the environment: positive, resp. negative, values of β favor the allele with frequency x, resp. 1 - x. The term 1/2 - x describes the change in frequency due to random mutations in the population. It is assumed that the total population a + A = 1 is constant, i.e. there is no immigration on the territory. The stationary state value of the frequency x under deterministic environmental conditions is given by the relation

$$\beta = (x_s - 1/2)/[x_s(1 - x_s)]. \tag{2}$$

Obviously the population frequency is restricted to the interval [0,1] for values of β which may range between $-\infty$ and $+\infty$. Futhermore x_s is a single valued function of β over this entire interval and it can easily be shown that for any β, x_s is an absolute attractor, i.e. the stationary state belonging to the *switching curve* x_s (labelled 0 in Fig.1) are asymptotically globally stable for any biologically acceptable initial condition.

Let us now discuss the stationary behavior of this system when the external parameter β fluctuates. I first consider that these fluctuations are rapid compared to the time scale of evolution of x, i.e.

$$\tau_{cor\beta} << \tau_{macro} \simeq 1/\omega_x \tag{3}$$

and thus I can assume that the white noise idealisation is adequate. Accordingly the probability density of x is the solution of the Fokker-Planck equation associated to the stochastic differential equation which replaces (1). Interpreting this equation in the Stratonovic sense yields for the extrema x_m of the stationary probability density the simple polynomial equation

$$\frac{1}{2} - x_m - \beta x_m(1 - x_m) - \frac{\sigma^2}{2} x_m(1 - x_m)(1 - 2x_m) = 0. \tag{4}$$

The term multiplied by σ^2 reflects the multiplicative character of the noise in this system. Its presence increases the degree of the polynomial by one as compared to the deterministic case. If we now suppose that on the average the environment favors neither one of the alleles, i.e.

$$E\{\beta_t\} = \beta = 0, \tag{5}$$

then the stationary values of the extrema are

$$x_{m_1} = 1/2 \tag{6}$$

and

$$x_{m\pm} = \frac{1}{2}[1 \pm (1 - \frac{4}{\sigma^2})^{\frac{1}{2}}]. \tag{7}$$

Accordingly, for $\sigma^2 < 4$ the probability density $p_s(x)$ has a unique extremum which is a maximum and corresponds to the deterministic stationary state. At $\sigma^2 = \sigma_c^2 = 4$, x_{m_1} is a triple root and for $\sigma^2 > \sigma_c^2 = 4$, x_{m_1} becomes a minimum of $p_s(x)$; the original peak corresponding to a stable stationary state under deterministic conditions has split into two peaks having maxima at $x_{m\pm}$. These peaks tend to 0, resp. 1, i.e. the asymptotes of $x_s(\beta)$ as $\sigma \to \infty$. For $\beta \neq 0$, the transition is a hard one. For $\beta > 0$ ($\beta < 0$), the peak corresponding to the extrema moves towards 1 (0) when σ increases. When σ crosses some critical value σ_c' (larger than σ_c) a second peak appears at a finite distance from the original one, near the other boundary of the phase space and corresponds to the asymptotes for $\beta \to -\infty$ ($\beta \to +\infty$). These facts show that as to the extrema of $p_s(x)$, in the (β,σ^2)-half plane we have a cusp catastrophe with critical point at (0,4).

The fact that this critical point occurs necessarily with a finite value of σ, i.e. $\sigma_c^2 = 4$, clearly shows that it corresponds to a transition which has no deterministic equivalent. It thus is a well distinct transition from a noise induced shift, like the one observed in the Verhulst model. In that case, by definition the effect persists even in the limit $\sigma^2 \to 0$: $\lambda = \sigma^2/2 \to 0$.

On the other hand, it is also interesting to strip for a moment this model from its biological context and to reconstruct it into a meaningful chemical form in order to be able to look at its behavior from a more thermodynamical point of view. One possible realisation of (1) is the chemical reaction scheme

$$A + X + Y \underset{k_2}{\overset{k_1}{\rightleftharpoons}} 2Y + A' \tag{8}$$

$$B + X + Y \underset{k_4}{\overset{k_3}{\rightleftharpoons}} 2X + B', \tag{9}$$

A,B,A',B' are concentrations coupled to the environment; X,Y are the state variables of the system. Obviously these reactions conserve the total number of X and Y particles in the course of time

$$X(t) + Y(t) = N = \text{const.} \tag{10}$$

Using this relation and defining the dimensionless parameters

$$\alpha = \frac{k_2A'}{k_2A' + k_4B'} = \frac{1}{2} \qquad \beta = \frac{k_3B + k_4B' - k_1A - k_2A'}{k_2A' + k_4B'} \tag{11}$$

the time evolution of the concentration X/N can straightforwardly be written in the form of (1). If k_2A', k_4B' are fixed and such that $\alpha = 1/2$, it is still possible for β to vary on the interval $(-\infty,+\infty)$ as

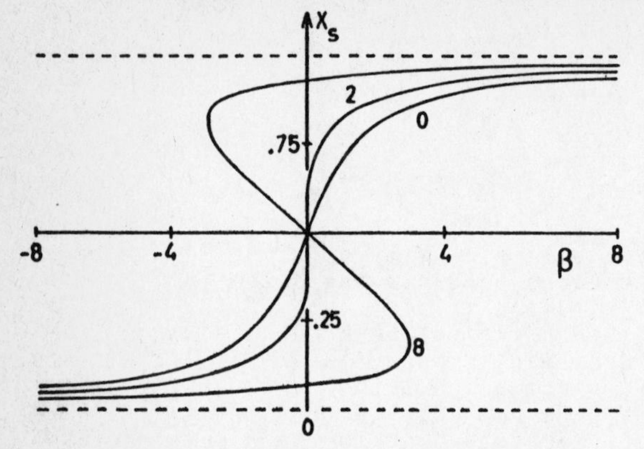

Fig.1 Extrema of the probability density $p_s(x)$ given by (4) as a function of and for three values of $\sigma^2/2$. The deterministic stationary states x_s correspond to the curve labelled 0.

a result of fluctuations in the values of A and B. The remarkable feature which merits to be underlined is that this remains possible *even* if one imposes that the average values $\langle A\rangle, \langle B\rangle, \langle A'\rangle, \langle B'\rangle$ be such that

$$\frac{\langle A'\rangle\langle B'\rangle}{\langle A\rangle\ \langle B\rangle} = \frac{k_1k_3}{k_2k_4}\ . \tag{12}$$

Under deterministic environmental conditions, (12) excludes the possibility of any instability or critical phenomenon as it means that the open chemical system (8-9) is in exact equilibrium with its environment, i.e. there are no exchanges of matter between it and the environment. This examplifies in a striking way that predictions concerning the behavior of non linear systems based on the average properties of the environment may become totally irrelevant when the latter is fluctuating, and this even if the fluctuations are completely structureless, i.e. δ-correlated.

3. Influence of noise correlations on noise induced transitions

Having established that the model for genic selection (1) exhibits an unexpected non equilibrium transition when the intensity of the external noise increases, it seems worthwhile to gain more insight into this phenomenon by analyzing the influence of the correlation time of the environmental fluctuations. It is futhermore highly desirable to incorporate the effect of the noise correlations because this permits to improve significantly the modelisation of real situations. As already indicated in the paper of HORSTHEMKE, once one accepts to model environmental noise by a Markovian colored noise process, the natural choice is the so called Ornstein-Uhlenbeck process ζ_t. Since the genetic model does not belong to the special class of systems for which the probability density of the pair process (x_t,ζ_t) can be evaluated exactly, it is only by approximate methods that futher information can be obtained. Using the perturbation expansion in the band width parameter $\varepsilon^2 = \tau_{cor}$ [17], or the method due to SANCHO and SAN MIGUEL [18], it is possible to cover the immediate neighborhood of the gaussian white noise. To the first order in the correlation time it is then found that the stationary probability density has an extrema which is a triple root for

$$\sigma^2 = 4 - 4\tau_{cor}\ . \tag{13}$$

Qualitatively the white noise result is recovered. Quantitatively one sees that when the rapidity with which the noise fluctuates slows down, the critical point occurs for a smaller value of σ^2.

Working in the opposite extreme case $\tau_{cor} >> \tau_{macro}$, an approximation of $p_s(x)$ is the transformation of the Ornstein-Uhlenbeck process via the switching curve

$$p_o(x) = p_s[\beta(x_s)] \left| \frac{d\beta(x_s)}{dx} \right| \text{, with } p_s(\cdot) = p_s \text{ of the O-U process.}$$

This approximation can therefore be called the switching curve approximation. It was first introduced in [16], and can be justified rigourously by the perturbation scheme reported in the paper of HORSTHEMKE, but of course using a different scaling. Intuitively it amounts to the adiabatic elimination of x_t. This is possible because the correlation time of the noise is much larger than the typical relaxation time of the system. Thus the process will always stay close to the curve of the deterministic stationary states. The behavior of the extrema of $p_o(x)$ is easy to determine. It shows that there is a lower limit to the decrease of the critical variance due to the effect of correlations. One finds for this limit: $\sigma_c = 4/3$.

These results suggest that the peak splitting transition takes place at all values of the correlation time, provided the variance of the noise exceeds some *finite* minimal value.

4. (Δ,γ)-phase diagram of model (1) subjected to dichotomous noise

To push futher the analysis of the influence of noise correlations, the only way which remains open in order to obtain explicit results is the special class of noise whose state space is restricted to jumps between two well-defined levels, namely the Markovian dichotomous noise. Assuming thus that $\beta_t = I_t$ where I_t has a state space which consists of only two levels, say $\pm\Delta$ and a mean value and correlation function given by

$$E\{I_t\} = 0, \; E\{I_t I_{t+\tau}\} = \Delta^2 e^{-\gamma\tau} \tag{14}$$

with $\tau_{cor} = 1/\gamma$, the behavior of the stationary probability of x can be studied in complete detail as a function of γ and Δ [19]. This probability density is defined on an interval, called the support U of $p_s(x)$, the boundaries of which are the stationary state solutions of the deterministic phenomenological equation (1) corresponding to the values $\beta\pm\Delta$ of the noise (see Fig.2). Obviously only for $|\Delta| \to \infty$, does this support correspond to the interval [0,1] defined by the asymptotes of x_s.

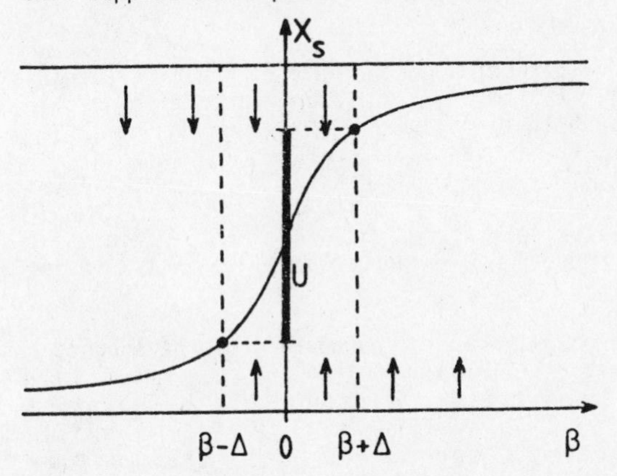

Fig.2 State space of the random process (x_t,β_t). The arrows indicate the direction of evolution of x_t. Evidently the whole probability mass will be inside the support U for $t \to \infty$.

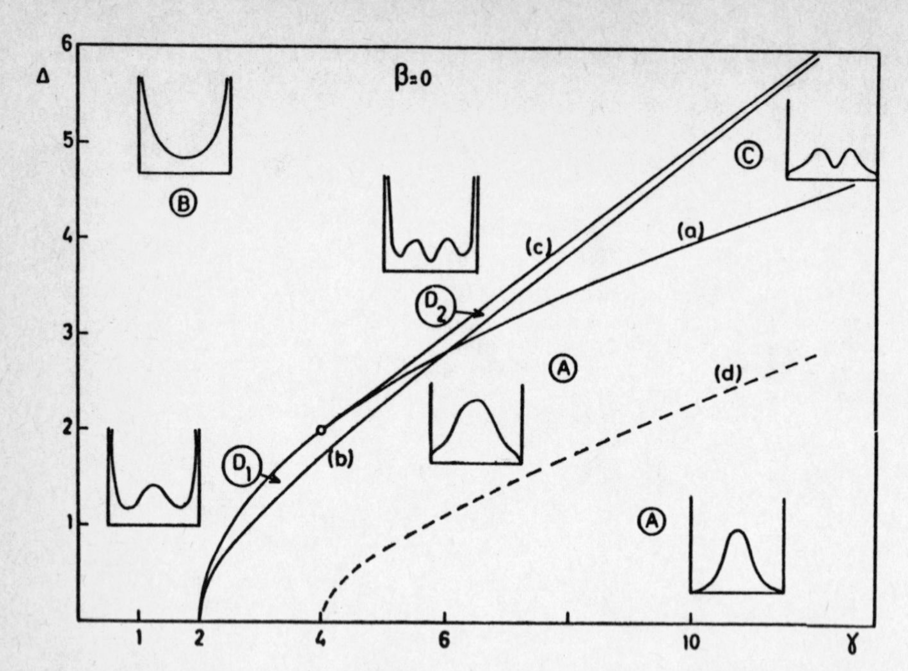

Fig.3 (Δ,γ)-phase diagram of model (1). Curves a, b, c and d correspond to the relations: (a) $\Delta^2 = 2\gamma - 4$, (b) $\gamma = 2\sqrt{\Delta^2+1}$, (c) $\gamma = 2\Delta$, (d) $\gamma = 4\sqrt{\Delta^2+1}$.

The behavior of $p_s(x)$ in terms of the amplitude Δ and correlation time γ^{-1} of the noise is displayed in Fig.3. This figure summarizes the domains in which $p_s(x)$ either diverges or vanishes at the boundaries of the support U and also sketches the number and the nature of the extrema found in these domains (see [19] for a detailed mathematical discussion). At most $p_s(x)$ may have five extrema. One of them

$$x_{m_0} = 1/2 \tag{15}$$

corresponds to the deterministic stationary state and obviously belongs to the support U for all Δ and γ. The other roots given by

$$x_{m_{1,2,3,4}} = \frac{1}{2\Delta}\{\Delta \pm [\Delta^2 - \gamma \pm (\gamma^2 - 4\Delta^2)^{1/2}]^{1/2}\} \tag{16}$$

may exist and are physically acceptable only if $\gamma \geq 2\Delta$.

In the white noise case, one has found that for $\sigma^2 = \sigma_c^2 = 4$ x_{m_0} is a triple root of the equation of the extrema. The same situation occurs here on the line (denoted a in Fig.3) $\Delta^2 = 2\gamma - 4$ or equivalently in term of $\sigma^2 \equiv 2(\Delta^2/\gamma)$ (The white noise limit is recovered for $\Delta^2 \to \infty$, $\gamma \to \infty$ such that $\Delta^2/\gamma = \sigma^2/2$)

$$\sigma_c \equiv 2(\Delta^2/\gamma)_c = 4 - \frac{8}{\gamma} = 4 - 8\tau_{cor}. \tag{17}$$

For short correlation times, (17) qualitatively agrees with (13) obtained in the case of Ornstein-Uhlenbeck noise. There is however now an upper limit $\tau_{cor} = 1/2$ beyond which any critical behavior is impossible.

The investigation of the behavior of $p_s(x)$ near the boundaries of the support yields in Fig.3, the lines b, c, d. The point $\gamma = 4$, $\Delta = 2$ is particularly remarkable because x_m is then a five fold root of the equation of the extrema. In its neighbourhood three domains B, D , D coexist in which $p_s(x)$ displays quite different features. For small changes in Δ and γ abrupt transitions from one domain to another are possible.

The results suggest that two distinct effects, simultaneously at work, determine the form of the stationary probability density of x.
-*first*, a *peak damping* effect which the multiplicative noise considered here has in common with additive noise. This only reflects the expected intuitive effect that the randomness of the environment tends to disorganize the system.
-*second*, the *peak splitting* effect which is specific of multiplicative noise and corresponds to a local change in the shape of $p_s(x)$ taking place at $x = 1/2$.

In regions A and B the peak damping effect dominates. In A however since the correlation time is short, the system still adjusts to the average state of the environment. In B on the contrary, the correlation time is large and the system has the time to relax to the neighbourhood of the deterministic stationary states corresponding to $\beta \pm \Delta$. In region C the peak splitting effect dominates. Clearly the bimodal probability density is not a compromise between the behavior in regions A and B. Indeed the two maxima never move outwards sufficiently far to reach the boundaries of the support and to induce a divergence of $p_s(x)$ there. $p_s(x)$ changes already from non divergent to divergent behavior when the maxima are still at a finite distance from the boundaries. This transition is observed in D_2 which thus may be viewed as a compromise between C and B. On the other hand, D_1 results from a compromise between A and B.

The main outcomes of the analysis can thus be summarized as follows: (i) The peak splitting transition which in the white noise limit takes place at $\sigma^2 = 4$ is recovered for (Δ,γ) values belonging to domains C and D_2. It occurs for $\gamma > 4$, i.e. the noise must fluctuate on a time scale which is at least four times faster than the one of the deterministic system. (ii) For $\gamma < 4$, the peak damping effect alone is observed. For large values of Δ this leads to divergences of $p_s(x)$ at the boundaries of U. (iii) Increasing the correlation time of the noise decreases the critical variance at which the peak splitting takes place. The effect is more pronounced for dichotomous noise than for Ornstein-Uhlenbeck noise. This is understandable since dichotomous noise stays all the time near one or the other of the extreme states $\beta \pm \Delta$ while Ornstein-Uhlenbeck noise stays mainly around the average state β. It is also interesting to remark that the influence of increasing the correlation time of the noise does not always decrease the value of σ_c. HONGLER's model is a counterexample, see Table 1. In this model the switching curve has no built in asymptotes to which it tends for values of the fluctuating parameter approaching $\pm\infty$. This feature combined with the fact that for the Ornstein-Uhlenbeck noise the average state of β_t

Table 1 Influence of τ_{cor} on the critical variance σ_c^2

Model	O-U noise	Dichotomous noise
model (1)	$\sigma_c^2 = 4 - 4\tau_{cor}$	$\sigma_c^2 = 4 - 8\tau_{cor}$
Hongler's model [20]	$\sigma_c^2 = 4 + 4\tau_{cor}$	$\sigma_c^2 = 4 - 8\tau_{cor}$

is the most probable state clearly makes the transition more difficult to induce when the correlations increase. With dichotomous noise, HONGLER's model presents qualitatively the same response as the genetic model. This is a consequence of the structure of dichotomous noise which imposes a finite support to $p_s(x)$ in any case and futhermore favors completely the extreme values of the fluctuating parameter.

5. Noise induced transitions in electrically excitable membranes

Over recent years the problem of fluctuations in electrically excitable membranes has been the subject of intensive studies [21-24]. The attention however has been focused exclusively on fluctuations which arise from the inherently probabilistic nature of the molecular processes internal to the membrane. The question of the influence of large externally driven constraints has not been dealt with up to now. The existence of noise induced transitions however and the fact that in the study of this phenomenon information can be gathered concerning the kinetic properties of the systems considered furnish a motivation for investigating the influence of external noise on membrane systems. I shall address this question within the framework of the Hodgkin-Huxley model for sodium and potassium activation in nerve.

Under resting conditions the nerve membrane sustains a potential difference of about -70mV. During the action potential the permeability of the nerve membrane with respect to Na and K changes while the potential exhibits a spike with maximum near 55mV; simultaneously separate channels in the membrane for the Na and K ions open and currents carried by these two ions flow through the membrane. The conductance of Na and K currents obeys to the kinetic equation (sodium inactivation is not considered here)

$$g = \nu\alpha_\nu(v)[g^{(\nu-1)/\nu} - g] - \nu\beta_\nu(v)g, \quad \nu=4 \text{ for K}, \ \nu=3 \text{ for Na}. \tag{18}$$

This equation is empirical: The coefficients

$$\alpha_4 = (v + 10)[\exp(v/10 + 1) - 1]^{-1}/100, \quad \beta_4 = \exp(v/80)/8$$
$$\alpha_3 = (v + 25)[\exp(v/10 + 5/2) - 1]^{-1}/10, \quad \beta_3 = 4\exp(v/18) \tag{19}$$

are chosen to obtain the best fit of the experimentally observed steady state currents under constant voltage clamp conditions. The Na and K kinetics mainly differ by the exponent ν in (18) and by their time scale Qualitatively however they are very similar and in both cases yield steady state conductances

$$g_s = [1/(1 + \beta/\alpha)]^\nu \tag{20}$$

which are monotonously increasing functions of v having asymptotes at 0 and 1 for v tending to $-\infty$ and $+\infty$.

In the following I shall discuss the behavior of these conductances when the potential

$$v_t = v + I_t \tag{21}$$

fluctuates due to the addition of the dichotomous Markov process I_t. The average value of the potential v is chosen so that $g_s(v) = 0.5$. Fig.4 reports the results for the K system. For a given value of Δ, the probability density diverges at the upper (lower) boundary of the support U, if γ is chosen to the left of γ_+ (γ_-). This accounts for the existence of regions A and D in which $p_s(g)$ has only one extremum. If Δ is larger than 42.5 mV and γ increases from zero and crosses the γ_--line, $p_s[g_s(v-\Delta)]$ vanishes and $p_s(g)$ acquires an additional extremum which of course is a maximum. When it crosses the γ_+-line the disappearance of the divergence at the upper boundary of U leads to the disappearance of the minimum. Below 42.5 mV, $p_s(g)$ has a saddle point for $\gamma = \gamma_s$ at which the minimum and maximum of B coalesce. Inside γ_s, i.e. in C, $p_s(g)$ is monotonously increasing from the lower to the upper boundary of U. The influence of γ on the location of the extrema is given

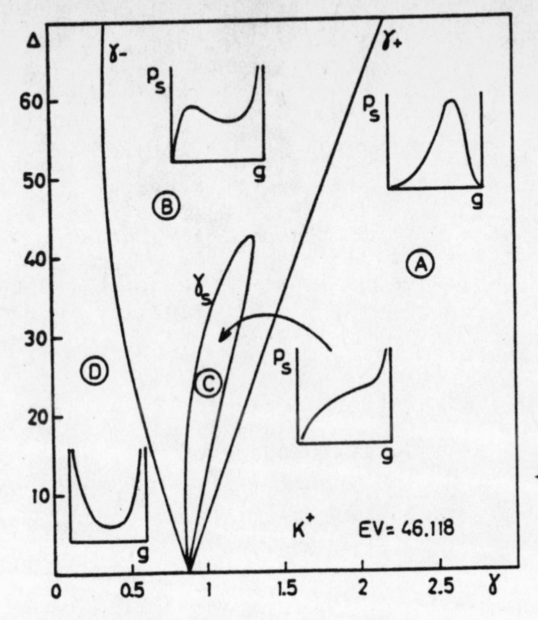

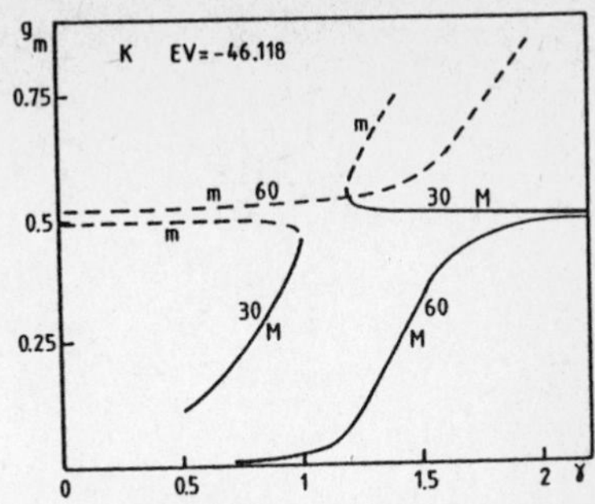

Fig.5 Location of the extrema of the K system as a function of γ for $\Delta=30$ and $\Delta=60$. Broken line: minima; full line: maxima.

Fig.4 (Δ,γ)-phase diagram of the K system. The average potential $v=-46.118$mV corresponds to $g_4=0.5$ under deterministic voltage clamp conditions.

in Fig.5. Below $\Delta \simeq 42.5$ mV, provided γ is not taken in C, there always is an extremum in the neighbourhood of $g = 1/2$. This extremum is a minimum to the left of C and a maximum to the right of C. The location of the second extremum in B increases rapidely with γ from values which lie near the lower boundary of U to values lying near the upper boundary. Above $\Delta = 42.5$ mV, the minimum which starts out near $g = 1/2$ for small γ, moves rapidely towards the upper boundary which it attains at $\gamma = \gamma_+$. The maximum which appears at γ_- approaches $g = 1/2$ when $\gamma \to \infty$.

Fig.6 gives the results for the sodium system. The characteristic time scale of the Na channel being faster than that of the K channel as can be seen from the coefficients α_ν and β_ν, the region of interest is shifted to higher values of γ. The difference in the power ν of the auxilary variable for Na and K leads to an essential difference in their phase diagram. For the Na system the γ_+ and γ_- curves cross each other for a value of the depolarisation which remains physiologically accep-

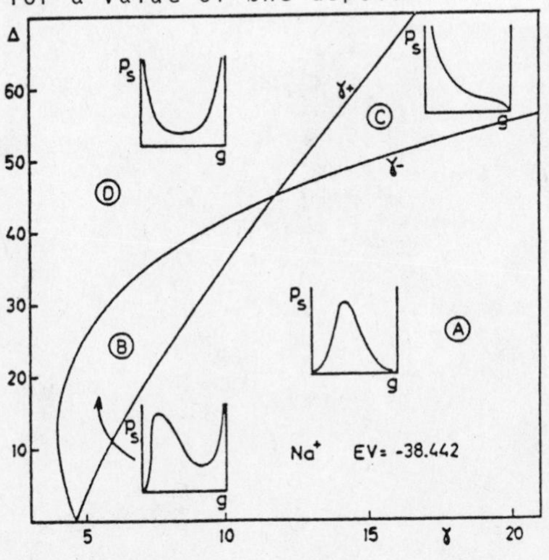

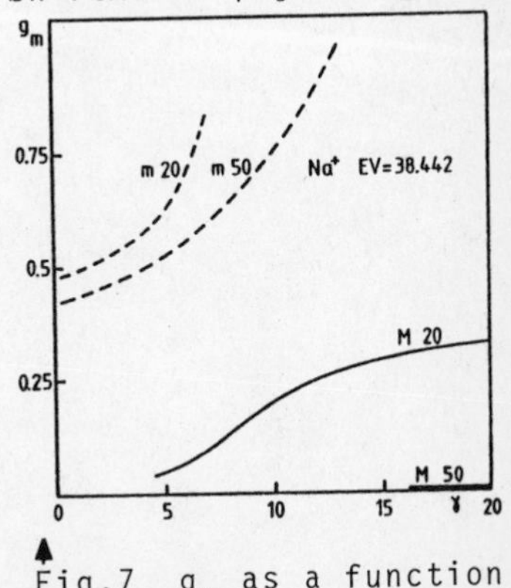

Fig.7 g_m as a function of γ.

Fig.6 (Δ,γ)-phase diagram of the Na system.

table. Consequently: (i) In contrast with the K system, the region C where $p_s(g)$ is a monotonous fuction is restricted to the values of Δ larger than 45 mV. (ii) The divergence occurs at the lower boundary of U, whereas for K it occurs at the upper boundary. (iii) The two extrema region B now occupies a finite domain in the (Δ,γ)-plane. Though the behavior of $p_s(g)$ in C is qualitatively identical in both systems it is important to underline that the mechanism of transition to this region is completely different in the Na and K systems: in the K system one has a soft transition corresponding to a local event taking place inside the support, namely the coalescence of two extrema into a saddle oint. This phenomenon does not affect the behavior of $p_s(g)$ near the boundaries of the support. This aspect ties it in with the noise induced transitions of the peak splitting type. In contradistinction, in the Na system region C appears via a hard transition caused by the abrupt change of behavior at one of the support boundaries.

The marked difference between the K and Na channels is futher seen in Fig.7. Whatever the value of Δ , for small, there is always a minimum near g = 1/2. It moves towards the upper boundary of the support which it attains for $\gamma = \gamma_+$. For $\gamma \geq \gamma_-$ the location of the additional extremum of $p_s(g)$ (maximum) which appears at the lower boundary increases slowly towards g = 1/2. The two curves overlap only if $\Delta < 45$ mV.

References

1. R. Levins, Proc. Nat. Acad. Sci. USA 62, 1061 (1969)
2. R. C. Lewontin, D. Cohen, Proc. Nat. Acad. Sci. USA 62, 1956 (1969)
3. R. M. May, Stability and Complexity in Model Ecosystems, Princeton University Press, Princeton (1973)
4. C. E. Smith, H. C. Tuckwell: In *Mathematical Problems in Biology*, ed. by P. v.d. Driessche, Lecture Notes in Biomathematics, Vol. 2, Springer, Berlin, Heidelberg, New York 1974
5. N. S. Goel, N. Richter-dyn, Stochastic Models in Biology, Academic Press, New York (1974)
6. J. J. Bull, R. C. Vogt, Science 206, 1186 (1980)
7. D. Ludwig: *Stochastic Population Theories*, Lecture Notes in Biomathematics, Vol. 3, Springer, Berlin, Heidelberg, New York 1974
8. J. R. Clay, N. S. Goel, J. Theor. Biol. 39, 633 (1979)
9. R. M. Capocelli, L. M. Ricciardi, Kibernetika 8, 214 (1971)
10. J. P. Changeux, A. Danchin, Nature 264, 705 (1976)
11. H. S. Hahn, A. Nitzan, P. Ortoleva, J. Ross, Proc. Nat. Acad. Sci. USA 71, 4067 (1974)
12. A. Boiteux, A. Goldbeter, B. Hess, Proc. Nat. Acad. Sci. USA 72, 3829 (1975)
13. J. de la Rubia, M. G. Velarde, Phys. Lett. 69A, 304 (1978)
14. W. Horsthemke, R. Lefever, Biophys. J. (to appear)
15. W. Horsthemke (preprint)
16. L. Arnold, W. Horsthemke, R. Lefever, Z. Physik B29, 367 (1978)
17. W. Horsthemke, R. Lefever, Z. Physik (in press)
18. J. M. Sancho, M. San Miguel, Z. Physik B36, 357 (1980)
19. K. Kitahara, W. Horsthemke, R. Lefever, Y. Inaba, Prog. Theor. Phys. (in press)
20. M. O. Hongler, Helv. Phys. Acta 52, 280 (1979)
21. H. Lecar, R. Nossal, Biophys. J. 11, 1048 and 1068 (1971)
22. A. A. Verveen, L. J. De Felice, Prog. Biophys. Mol. Biol. 28, 191 (1974)
23. C. F. Stevens, Nature 270, 391 (1977)
24. B. Neumcke, W. Schwarz, R. Stämpfli, Biophys. J. 31, 325 (1980)

Multiplicative Ornstein Uhlenbeck Noise in Nonequilibrium Phenomena

M. San Miguel*
Physics Department, Temple University
Philadelphia, PA 19122, USA

J.M. Sancho
Departamento de Fisica Teorica, Universidad de Barcelona
Diagonal 647, Barcelona 28, Spain

1. Introduction

Stochastic differential equations of the Langevin type for a finite set of variables are a common tool to study a variety of physical, chemical and biological systems. The more recent interest in this type of equation is mainly due to its success in describing nonequilibrium situations of open systems. When dealing with these equations it is often assumed that the fluctuating term does not depend on the state of the system ("additive noise") and, invoking a difference in time scale, that the white noise idealization is appropriate. Nevertheless remarkable novel features of these equations appear when removing these two constraints. We are here precisely concerned with this last situation. That is, we consider stochastic differential equations of the form

$$\dot{q}_\mu(t) = v_\mu[q(t)] + g_{\mu\nu}[q(t)]\xi_\nu(t) \quad \mu = 1, \ldots, N; \quad \nu = 1, \ldots, M \tag{1.1}$$

where $\xi(t)$ is not a white noise but has a finite correlation time ("colored noise"). The q dependence of $g_{\mu\nu}$ gives its "multiplicative" character to the noise term. It is our purpose here to elucidate some phenomena appearing in nonequilibrium systems described by (1.1) with special emphasis on the effect of considering a finite correlation time as compared to the white noise case. The interest in this problem is not only purely mathematical. In fact, there are at least two important sources of these equations for a realistic description of a system. The first one is the elimination of fast variables from the equations of motion. A careful adiabatic elimination procedure [1] from a set of additive white noise Langevin equations leads in general to colored multiplicative noise. Projection methods applied to a set of phenomenological deterministic equations [2] also lead to (1.1). In these cases the fluctuations are originated by the system itself and should be considered as internal fluctuations. A simple and interesting example of this situation is provided by the study of nonequilibrium Brownian motion and the Smoluchowski equation. This is

* Permanent address: Departamento di Fisica Teorica, Universidad de Barcelona, Diagonal 647, Barcelona 28, Spain

discussed in Sect.3. A second source, to which this paper is mainly devoted, is the phenomenological modelling of a fluctuating environment or of a superimposed external fluctuation. This is mathematically achieved by letting a parameter in a phenomenological equation of motion become a random variable with prescribed statistics [3]. An example of this is a spin in a fluctuating magnetic field as considered when studying magnetic resonance absorption phenomena [4]. The meaning of this external noise concept is particularly clear in recent experiments in which externally generated noise is applied to the system. These systems include illuminated chemical reactions [5], electrical circuits [6,7], liquids crystals [8] and Rayleigh-Benard systems [9]. Noise forced systems have also been considered in computer simulations [10]. In these situations the noise is controlled in the laboratory independently of the system, and the noise characteristics, in particular the correlation time, are at our disposal. Therefore the white noise approximation cannot describe all the richness of possible situations.

A main mathematical complication associated to (1.1) is the loss of the Markovian character of the q-process present in the white noise case [11], and also the associated loss of the ordinary well known Fokker Planck description. This can be overcome by enlarging the space of variables, for example considering $\xi_\nu(t)$ as a second set of variables satisfying equations of motion driven by white noise. This does not simplify matters in practice. For example, for a single variable q one ends with a two variables problem for which the stationary solution is not in general known. In Sect.2 we present special exact models and different approximations in which it is possible to use a Fokker Planck equation (FPE) for the probability density of these nonMarkovian processes. The possibility of a Fokker-Planck treatment is, of course, of great practical convenience. On physical grounds we assume in the following a Gaussian, stationary, Markov noise. This restricts us [12] to the Ornstein-Uhlenbeck noise characterized by zero mean value and correlation

$$\gamma(t - t') = \langle\xi(t)\xi(t')\rangle = (D/\tau)\exp\{-|t - t'|/\tau\} \quad . \tag{1.2}$$

The Ornstein-Uhlenbeck noise is characterized by two independent parameters, the correlation time τ and what we call the noise intensity D which characterizes the white noise limit

$$\gamma(t - t') \xrightarrow[\tau\to 0]{D \text{ fixed}} 2D\delta(t - t') \quad . \tag{1.3}$$

We present in Sect.2 two approximation schemes which correspond respectively to choosing τ or D as smallness parameters. The noise spectrum is characterized by the Fourier transform S(w) of $\gamma(t - t')$. S(w) is sketched, for small τ, in Fig.1., where the meaning of the two parameters D and τ becomes clear.

The theoretical analysis of some of the above external noise situations predicts [3,13] a kind of nonequilibrium phase transition in which the external parameters governing the transition are the noise parameters. Sect.4 is devoted to the analysis

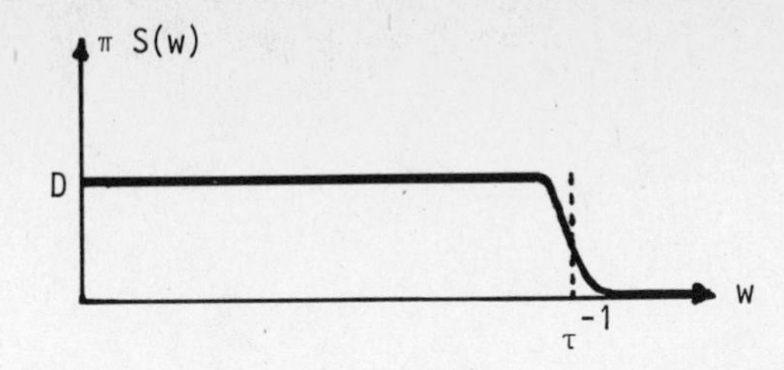

Fig.1

of these so-called noise induced phase transitions [3] by means of the Fokker-Planck formalism described in Sect.2. The transition point is defined as the point in parameter space in which the most probable state changes value. Although obvious differences exist with respect to ordinary phase transitions, the nature of these nonequilibrium transitions is not yet completely understood in such matters as the possible existence of diverging modes and slowing down phenomena [14,15]. Despite this fact good experimental evidence has been reported [5-8]. As argued before the consideration of nonwhite noise corresponds to actual experimental situations. Indeed, experimental results are usually given in terms of a noise intensity defined as the integral of S(w), that is D/τ (the area in Fig.1). This is a diverging quantity in the strict white noise limit often used to analyze such results. Therefore one is forced to consider the effect of a nonvanishing correlation time to fully understand the experimental situation. Another reason to consider colored noise is to check if the white noise results are mathematical artifacts or if they remain valid for a small correlation time. This is important because the experimental situation is quite often close to the white noise case. Our approximate scheme for small τ gives results which support the consistency of the white noise results with the ones for small τ and it also provides a systematic way to take into account corrections to the white noise limit. We can then study the modification arising in the transition law and also mean values as a function of τ (Sect.4.3). In particular, we report on changes in the white noise threshold value of the parameters (Sect.4.2), and on the possibility of new transitions absent for $\tau = 0$ (Sect.4.3 and [13]). The alternative approximation for small D, in whith the effect of τ is exactly taken into account, provides a method to study the effect of larger correlation times, and also a consistency check of the results for small τ. In particular the modification of threshold values is considered (Sects.4.2,3).

The Ornstein Uhlenbeck noise has also been studied [16] in the limit of large τ [13], but this approximation gives no insight into the validity of white noise. Other colored noise that has been considered is the two step Markov process [17]. Although it is then possible to obtain exact solutions for the stationary solution, this noise is of a very special nature, and no exact dynamical model has been reported. On the other hand, a few models whose dynamical evolution is exactly known exist when considering the Ornstein Uhlenbeck noise [13]. The model of [18] is of this class. Nevertheless, it is interesting to note that our results using Ornstein-

Uhlenbeck noise are in general in qualitative agreement with the ones obtained using the two step Markov process.

2. General Formalism

We start for simplicity with the case in which $N = M = 1$ in (1.1). Considering the stochastic Liouville equation [19] for the probability density associated to (1.1) for each realization of $\xi(t)$ and averaging over the realizations of $\xi(t)$, we have

$$\partial_t P(q,t) = - \partial_q v(q) P(q,t) - \partial_q g(q) \langle \xi(t) \delta(q(t) - q) \rangle \tag{2.1}$$

where $P(q,t)$ is the probability density at time t. To handle the average left in (2.1) we use the Gaussian property of $\xi(t)$, as characterized by NOVIKOV [20], by considering $\delta[q(t) - q]$ as a functional of $\xi(t)$. We then arrive at [13]

$$\partial_t P(q,t) = - \partial_q v(q) P(q,t) + \partial_q g(q) \partial_q \int_0^t dt' \gamma(t - t') \langle [\delta q(t)/\delta\xi(t')] \delta(q(t) - q) \rangle \quad . \tag{2.2}$$

The key quantity appearing in (2.2) is the response function $\delta q(t)/\delta\xi(t')$. Different approximations to this function lead to different approximated FPE for the basic and formally exact equation (2.2).

2.1 Exact Models

If $\delta q(t)/\delta\xi(t')$ can be computed exactly and does not depend explicitly on $\xi(t)$, (2.2) becomes an exact FPE, local in time, with time dependent diffusion coefficient $D(t)$. This is the case if we have additive noise (g = cte) and linear drift ($v = Aq + B$). Furthermore, there exists a class of equations which are of linear additive form in the variable $Q = \int dq\, g^{-1}(q)$. This class is characterized by the condition

$$g(q) \partial_q\, g^{-1}(q) v(q) = A = \text{constant} \tag{2.3}$$

and then $B = (v/g) - AQ$. In these circumstances, $\delta Q(t)/\delta\xi(t') = \exp\{A(t - t')\}$, and transforming back to the q-variables, the exact FPE for this class of models becomes [13]

$$\partial_t P(q,t) = - \partial_q (AQ + B) g(q) P(q,t) + D(t) \partial_q\, g(q) \partial_q\, g(q) P(q,t) \tag{2.4}$$

$$D(t) = [D/(1 - A\tau)][1 - \exp(-(\tau^{-1} - A)t)] \quad . \tag{2.5}$$

The time dependent solution of (2.4) is easily obtained from the solution in Q space and it only has no trivial steady state for $A < 0$. As can be seen from (2.4), the whole effect of the colored noise, as compared to the white noise limit, is a renormalization of the diffusion coefficienct so that D is replaced by D(t). (An explicit example is given in [13]). The class of models with N variables for which the transformation to linear additive form is possible is discussed in [21,22].

2.2 Approximation for Small τ

When $\delta q(t)/\delta\xi(t')$ can not be calculated exactly we need a smallness parameter to establish a systematic approximation. The two independent noise parameters available are D and τ. (These are here regarded as dimensionless parameters, see Sect. 4.3.) We consider now the common experimental situation of a noise with approximately flat and very wide spectrum ($\tau \ll 1$) so that $\gamma(t - t')$ decays very fast. Then, the main contribution to the integral in (2.2) comes from $\delta q(t)/\delta\xi(t')$ evaluated in the vicinity of $t = t'$

$$\frac{\delta q(t)}{\delta\xi(t')} = \left.\frac{\delta q(t)}{\delta\xi(t')}\right|_{t'\to t} + \left.\frac{d}{dt'}\frac{\delta q(t)}{\delta\xi(t')}\right|_{t'\to t}(t - t') + \dots$$

$$= g(q(t)) + [v\partial_q g - g\partial_q v]\Big|_{q(t)}(t - t') + \dots \quad . \tag{2.6}$$

This expansion is powers of $(t - t')$ implies, when substituted under the integral sign in (2.2), an expansion in powers of τ. (For the evaluation of the response function and its derivatives at equal times see [13].) Substituting (2.6) in (2.2) and neglecting transients by extending the integrals to ∞ we arrive at the following FPE:

$$\partial_t P(q,t) = -\partial_q [v(q) + D(\partial_q\, g(q))h(q)]P(q,t) + D\partial_q^2\, g(q)h(q)P(q,t) + O(\tau^2) \tag{2.7}$$

$$h(q) = g(q) - \tau[\{\partial_q\, g(q)\}v(q) - \{\partial_q\, v(q)\}g(q)] \quad . \tag{2.8}$$

The first term of $h(q)$ corresponds to the white noise limit in Stratonovich's sense [11] and the second term is a first order correction. The approximation amounts to substitute $g(q)$ by $h(q)$ in the ordinary FPE for white noise. The second order term in (2.7) involves in general $\xi(t)$. The average in (2.2) introduces then a third order derivative which breaks the FP form in order τ^2. Some insight in the approximation is gained by applying it to the exact models (2.3). The approximated equation is exactly of the form (2.4) with $D(t)$ replaced by $D(1 + \tau A)$. Recalling (2.4), it is clearly seen that this is obtained by neglecting transients and expanding to first order in τA. Artificial boundaries at $1 + \tau g\partial_q(v/g) = 0$ are present in (2.7). This point fixes the values of q and of the parameters for which the approximation becomes meaningless since the first order correction term is no longer smaller than the white noise term. For the exact models this point is given by $|\tau A| \sim 1$, where the expansion breaks down. At least for these models our systematic expansion in powers of τ is convergent for $|\tau A| < 1$. Despite these mathematical problems associated with an approximation which is not uniform for all values of q, (2.7) is an equation easy to handle that gives reasonable answers in specific examples as shown below.

The above results can be extended to the case of $N \neq 1$ variables and $M \neq 1$ stochastic forces as follows [22]. We assume uncorrelated noises of different correlation times τ_μ and we include the noise intensities in $g_{\mu\nu}(q)$. Proceeding as above one obtains

$$\partial_t P(q,t) = - \partial_{q_\mu} v_\mu(q) P(q,t) + D\partial_{q_\mu} g_{\mu\nu}(q)\partial_{q_\beta}[g_{\beta\nu}(q) - \tau_\nu M_{\beta\nu}(q)]P(q,t) \quad (2.9)$$

$$- D\tau_\nu \partial_{q_\mu} g_{\mu\nu}(q)\partial_{q_\beta} K_{\beta\gamma\nu}(q)\langle\xi_\gamma(t)\delta^n(q(t) - q)\rangle + O(\tau^2) \quad (2.9)$$

$$M_{\beta\nu} = v_\rho \partial_{q_\rho} g_{\beta\nu} - (\partial_{q_\rho} v_\beta) g_{\rho\nu}$$

$$K_{\beta\gamma\nu} = g_{\rho\gamma}\partial_{q_\rho} g_{\beta\nu} - (\partial_{q_\rho} g_{\beta\gamma}) g_{\rho\nu} \quad . \quad (2.10)$$

The third term on the right hand side of (2.9) has no analog counterpart in (2.7), and it involves an average that when handled as in (2.1) will introduce third order derivatives. Therefore no FPE valid in first order in τ exists in general. Important exceptions are the cases in which $K_{\beta\gamma\nu}$ vanishes. These include the case $N = M = 1$ and the case of additive noises. It also includes, for $N = M$, the case in which (1.1) can be transformed to additive noise equations by means of a nonlinear change of variables. In general the $K_{\beta\gamma\nu}$ term gives to leading order a contribution

$$D^2 \tau_\nu \partial_{q_\mu} g_{\mu\nu} \partial_{q_\beta} K_{\beta\gamma\nu} \partial_{q_\delta} g_{\delta\gamma} P(q,t) \quad (2.11)$$

which is of order $D\tau^2$ instead of $D\tau$. If the correlation time is the same for all the stochastic forces ($\tau_\nu = \tau \; \forall\nu$), still a FPE exists to order τ: when the derivatives in (2.11) are shifted to the left, the term with third order derivatives cancels. Finally, it is worth mentioning that there are special cases in which a FPE in higher order in τ exists [22]. An interesting application of (2.10) is the consideration of the presence of weak additive white noise in a problem with multiplicative colored noise for a single variable. (See Section 4.3)

2.3 Approximation for Small D

As pointed out before, higher order terms in the expansion (2.6) lead to terms which are in general not of the FP form. Nevertheless it exists a subseries in that expansion, characterized by coefficients $D\tau^n$, which is of the FP form (i.e., it contains only second order derivatives). The remaining terms have coefficients nonlinear in D. Therefore a summation of the FP terms represents a partial resummation of that expansion, based on considering D as a smallness parameter and, in this limit, it takes exactly into account the effect of $\tau \neq 0$. A formal expression for the sum of this subseries can be obtained and it corresponds to a different approximation of the one given in (2.6) for the response function $\delta q(t)/\delta\xi(t')$ in (2.2): Since we are interested in small D, the appropriate approximation is to evaluate $\delta q(t)/\delta\xi(t')$ to lowest order in $\xi(t)$. The response function can be written as the operator expression [23]

$$\delta q(t)/\delta\xi(t') = \theta(t - t')g(q(t'))[\partial q(t'), q(t)] \tag{2.12}$$

where the square bracket stands for a commutator and the appropriate evolution operator is

$$U(t_2,t_1) = T \exp \int_{t_1}^{t_2} ds[v(q(s)) + g(q(s))\xi(s)]\partial q(s) \ . \tag{2.13}$$

Here T is the time ordering operator. A "linear response" approximation is obtained by neglecting the second term in the exponential of (2.13). We then obtain

$$\delta q(t)/\delta\xi(t') = \theta(t - t')\exp\{\partial_q \, v(t - t')\}\partial_q \, g(q) \, \exp\{-\partial_q \, v(t' - t)\}\Big|_{q(t)} \cdot \tag{2.14}$$

Substituting in (2.2) and neglecting transients we have

$$\partial_t P(q,t) = - \partial_q \, vP(q,t) + \partial_q \, g \int_0^\infty ds\gamma(s) \, \exp\{-\partial_q \, vs\}\partial_q \, g \, \exp\{\partial_q \, vs\} \ . \tag{2.15}$$

This is also the result obtained by VAN KAMPEN [19] using a cumulant expansion. A commutator expansion of the operator expression under the integral sign in (2.15) gives rise to the FP terms of order $D\tau^n$ mentioned above. Of course to order $D\tau$ we recover (2.7). In Sect.4 we present examples in which this resummation can be explicitly carried out.

3. Corrections to the Smoluchowski Equation [24]

The description in position space q of a Brownian particle under the influence of some potential is a good example of how multiplicative colored noise appears after the elimination of a fast variable, in this case the momentum p. The starting Langevin equations in phase space are

$$\dot{q}(t) = p(t) \quad ; \quad \dot{p}(t) = - \lambda p(t) - \partial_q\phi(q(t)) + \xi_w(t) \tag{3.1}$$

where λ is the damping coefficient, ϕ the potential and $\xi_w(t)$ a white noise of intensity $\lambda k_B T$. A formal integration of the second equation leads to a nonMarkovian equation for q(t) driven by an Ornstein-Uhlenbeck process of correlation time λ^{-1}. Expanding the memory kernel in powers of λ^{-1} one obtains an equation of the form (1.1):

$$\dot{q}(t) = - \lambda^{-1}[\partial_q\phi(q(t))][1 + \lambda^{-2}\partial_q^2\phi(q(t))] + [1 + \lambda^{-2}\partial_q^2\phi(q(t))]\xi(t) + \ldots \tag{3.2}$$

Equation (2.7) applied to (3.2) results in a corrected Smoluchowski equation appropriate to find nonequilibrium steady state solutions. The leading term corresponds to the Smoluchowski equation, which can then be interpreted as a white noise approximation in position space. For a harmonic potential ϕ this system provides an interesting example of the exact model of Sect.2.1.

4. Ornstein-Uhlenbeck Noise and Nonequilibrium Transitions

4.1 Generic Results for Small τ

In this Section we are concerned with the implications of considering a nonvanishing correlation time of the noise in the study of nonequilibrium phase transitions driven by the noise parameters. Some general conclusions can be drawn from (2.7). These are obtained in connection with the maximum of the stationary distribution which is identified with the macroscopic state: For natural boundary conditions the stationary solution of (2.7) is

$$P_{st}(q) = \{N/h(q)\}\exp\{\int dq\, v(q)/D\, g(q)h(q)\} \qquad (4.1)$$

whose extrema are the solutions of

$$v(q) - D\, g(q)\, \partial_q g(q) + D\tau g(q)[v(q)\partial_q^2\, g(q) - g(q)\partial_q^2 v(q)] = 0 \quad . \qquad (4.2)$$

The first term corresponds to the deterministic stationary points and the first two terms give the white noise limit extrema. The existence of the third term will in general modify quantitatively the white noise results, but more important is that it can also lead to new qualitative features. In particular, it can introduce in (4.2) higher orders of q, and so new extrema and new transitions in which τ acts as the control parameter. In contrast with the white noise limit, even for additive noise the extrema of $P_{st}(q)$ will not in general coincide with the deterministic stationary states. These general features of the colored noise that follow from (4.2) are in agreement with the results of a different calculation using a two step Markov process for the noise [17]. With respect to the exact models of Sect.2.1 we already pointed out that the whole effect of τ in the steady state properties is a renormalization of the diffusion coefficient so that D is replaced by $D/(1 - A\tau)$ in the FPE. Therefore no new transitions absent for $\tau = 0$ will appear, but the threshold value D_{th} will be shifted to a renormalized value $D_{th}/(1 - A\tau)$.

4.2 A Genetic Model

This model [16] is defined by the following stochastic equation of motion in some dimensionless units

$$\dot{q} = (1/2) - q + q(1 - q)\xi(t) \quad . \qquad (4.3)$$

Such an equation is obtained by letting an external parameter fluctuate with mean value zero. In the small τ limit, the stationary distribution is given by (4.1) and it goes to zero in the boundaries of the interval of definition. The extrema obtained from (4.2) are

$$q_1 = 0 \quad ; \qquad q_{2,3} = \frac{1}{2}\,[1 \pm (1 - 2/\{D(1 + \tau)\})^{1/2}] \quad . \qquad (4.4)$$

This shows a transition from a distribution with a single peak at q_1 to a bimodal distribution with peaks at $q_{2,3}$. The transition occurs at $D_{th} = 2/(1 + \tau) \simeq 2(1 - \tau)$, which goes for $\tau \to 0$ into the white noise value $D_{th} = 2$. The separation of the peaks increases with increasing values of D.

Let us now consider the small D limit. The operator under the integral sign in (2.15) can be expressed in this case as

$$\int_0^\infty ds\gamma(s)\exp\{-\partial_q vs\}\partial_q g \exp\{\partial_q vs\}$$

$$= \partial_q\left[(-\frac{1}{2} + q(1-q))\sum_{n=0}^{\infty} \tau^{2n+1} + q(1-q)\sum_{n=1}^{\infty} \tau^{2n}\right]$$

$$= \partial_q[1/(1-\tau^2)]q(1-q)[q(1-q)(1+\tau) - \tau/2] \tag{4.5}$$

where the sum of the series is valid for $\tau < 1$. This also gives us an upper bound for the validity of the small τ approximation. Substituting (4.5) in (2.15) we obtain a FPE, valid to order D, which only differs from the one obtained from (2.7) for the small τ approximation by the replacement of D by $D/(1-\tau^2)$. For small D the symmetric peaks of the stationary distribution are then located at

$$q_{2,3} = \frac{1}{2}[1 \pm \{1 - 2(1-\tau)/D\}^{1/2}] \quad . \tag{4.6}$$

Therefore, the threshold value is now $D_{th} = 2(1-\tau)$ which is the same obtained to first order in τ. For this value to be consistent with the approximation we must require $\tau \leq 1$. For comparison, we summarize in Table 1 the values obtained for D_{th} in different situations.

Table 1

White noise [16]	$D_{th} = 2$
Small τ approximation [13]	$D_{th} = 2(1-\tau)$
Small D approximation	$D_{th} = 2(1-\tau)$
Large τ approximation [16]	$D_{th} = 2/3$
Two step Markov noise [17]	$D_{th} = 2(1-2\tau)$

The common qualitative feature of these results is that the threshold value of D decreases with increasing τ. Other aspects of the transition in this model have been studied experimentally in [7]. In particular, evidence of the transition has been presented.

4.3 Stratonovich's Model

We now consider the model [25] defined by the following stochastic equation for a variable x(t')

$$d_{t'}x(t') = (\alpha - \gamma)x(t') - \beta x^3(t') + x(t')\bar{\xi}(t') \tag{4.7}$$

where $(\alpha - \gamma)$ and β are positive parameters characteristic of the system and $\bar{\xi}(t')$ has intensity D' and correlation time τ'. From a mathematical point of view this model has the advantage that all the terms in (4.7) have the same parity. This is not the case for the related Verhulst equation [13,17]. External noise situations for which the model (4.7) is of physical relevance are, for example, the analysis of superimposed noise in a parametric oscillator [6] and the electrohydrodynamic transition in liquid crystals [8,26]. To proceed with our approximation schemes we rescale variables: $q = [\beta/(\alpha - \gamma)]^{1/2}x$, $t = (\alpha - \gamma)t'$. Equation (4.7) becomes then

$$d_t\, q(t) = q(t) - q^3(t) + q(t)\xi(t) \tag{4.8}$$

where $\xi(t)$ has intensity $D = D'/(\alpha - \gamma)$ and correlation time $\tau = \tau'(\alpha - \gamma)$. The dimensionless parameters D and τ are the only ones left in the model.

In the small τ approximation the stationary distribution (4.1) is

$$P_{st}(q) = N\, q^{D^{-1}-1}(1 - 2\tau q^2)^{(1/4D\tau)-(2D)^{-1}-1} \tag{4.9}$$

where N is the normalization constant. The maxima are located at

$$q_1 = 0 \quad \text{for} \quad D > 1 \quad ; \quad q_2 = [(1 - D)/(1 - 6\tau D)]^{1/2} \quad \text{for} \quad D < 1 \quad . \tag{4.10}$$

In the experimental setup of [6], x is the amplitude of an oscillatory current, and α a pumping current. An oscillatory to nonoscillatory transition corresponding to (4.10) is observed. The noise $\bar{\xi}(t')$ is Gaussian with a flat spectrum between 0.01 and 10^5 Hz. Given the smallness of the correlation time τ' a white noise approximation was used to analyze the results [6]. However, we stress that the experimental noise intensity corresponds to D/τ. We are now in a position to study the effect of the two parameters D and τ characterizing the noise: i) From (4.10) we predict that the transition occurs at $D_{th} = 1$ ($D'_{th} = \alpha - \gamma$). This result is also the white noise prediction and remains unchanged when varying the spectrum of the noise while keeping its intensity D' fixed. ii) According to our result written in terms of the original variables, the most probable value of the oscillatory current follows the mean field law

$$x_2 = A(D'_{th} - D')^{1/2} \quad . \tag{4.11}$$

This law reported in [6] is again not modified by changing the noise spectrum. iii) The amplitude in (4.11) is given by $A = \beta^{-1/2}(1 - 6\tau'D')^{-1/2}$. In the white noise limit $A = \beta^{-1/2}$ is independent of the noise characteristics while here it depends on the dimensionless number $D'\tau'$. Therefore, for a fixed intensity of the noise D' the amplitude of the oscillatory current will increase when narrowing the spectrum of $\bar{\xi}(t')$ (increasing τ'). In fact to first order in τ

$$[x_2^2 - x_2^2(w)]/x_2^2(w) \simeq 6\tau'D' \tag{4.12}$$

where $x_2(w)$ represents the white noise result. Furthermore, keeping τ' fixed, the constant of proportionality A depends on the noise intensity differently to what happens in the white noise limit. It is finally interesting to note the analogy of these results to the ones of [27] where a nonequilibrium phase transition for a system composed of Brownian charged particles in a periodic potential and in the presence of an electric field is analyzed. As discussed in Sect.3 the Smoluchowski approximation can be understood as a white noise limit. In this limit a mean field law like (4.11) is obtained. The effect of the corrections to the Smoluchowski equation is again a modification of the amplitude A.

The stationary distribution (4.10) allows us also to study the dependence on τ of the different mean values. This information is useful since it can be more accurately checked in numerical calculations than the most probable value. We quote only here the results for the first two moments

$$\begin{aligned} \langle q \rangle &= (2\tau)^{-1/2}\Gamma[(2D)^{-1} + 1/2]\ \Gamma[(4D\tau)^{-1}]\ \Gamma^{-1}[(2D)^{-1}]\ \Gamma^{-1}[(4D\tau)^{-1} + 1/2] \\ &= \langle q \rangle_W (4D\tau)^{-1}\Gamma[(4D\tau)^{-1}]\ \Gamma^{-1}[(4D\tau)^{-1} + 1/2] \end{aligned} \tag{4.13}$$

$$\langle q^2 \rangle = 1 \tag{4.14}$$

where $\langle q \rangle_W$ is the white noise result. The second moment is thus independent of the noise characteristics. For the first moment, using the Stirling approximation for Γ, we have to leading order that $\langle q \rangle = \langle q \rangle_W(1 + D\tau/2)$ and therefore

$$(\langle q \rangle - \langle q \rangle_W)/\langle q \rangle_W \simeq D\tau/2 \quad . \tag{4.15}$$

The domain of validity of the result (4.9) can be analyzed as follows: The artificial boundary mentioned in Sect.2.2 appears here at $q^2 = 1/2\tau$, which goes to ∞ in the white noise limit. This puts a restriction on the values of q for which the approximation is meaningful. An upper bound for the value of τ can be obtained by requiring that the term of order τ^2 in the expansion leading to (2.7), must be smaller than the first order term. This justifies neglecting higher order terms. In this model such a condition reduces to $\tau < 1/2$ which is also the condition for (4.9) to be normalizable.

The result (4.9) gives rise to a phase diagram with the same qualitative features (transition lines, regions in parameter space and shape of $P_{st}(q)$ in these regions) as the one for the Verhulst equation. We refer to that for the interpretation of results [13]. Transition lines occur here for $D = 1$ and $D = (1/4\tau) - 1/2$. The first one corresponds to the white noise result and does not depend on the noise spectrum. The white noise transition is then seen to be consistent with small τ results. The results (4.12) and (4.15) refer to this transition. The second line corresponds to possible transitions not present in the white noise limit. As it was also the case for the Verhulst equation, the phase diagram is qualitatively similar to the one obtained by considering in this model a two step Markov process for the noise [7]. Even numerical estimates are of the same order of magnitude.

We can also consider in this model the small D approximation. The operator entering in (2.15) explicitly written in (4.5) becomes for this model

$$\partial_q\left[q + (2q^3)\sum_{n=1}^{\infty} (-1)^n 2^{n-1}\tau^n\right] = \partial_q\, q[1 - 2\tau q^2/(1 + 2\tau)] \tag{4.16}$$

where the sum is valid for $\tau < 1/2$. This upper bound for τ agrees with the one discussed above. Substituting (4.16) in (2.15) we obtain an equation of the form (2.7) where here $h(q) = q[1 - 2\tau g^2/(1 + 2\tau)]$. In the small τ approximation we had $h(q) = q(1 - 2\tau q^2)$. Therefore, the results for small D can be obtained from those for small τ by formally replacing τ by $\tau/(1 + 2\tau)$. As a consequence, the white noise transition line remains at $D = 1$ and the second transition line is shifted to $D = 1/4\tau$. Since in this last line D has the value 1/2 for the limiting value of $\tau = 1/2$, our result predicts no transition when $D < 1/2$ and $\tau < 1/2$.

Finally and as an application of (2.9) we wish to consider the effect of a small additive noise which will always be present. This noise will model internal fluctuations of the system and may be taken to be white. So we add now a Gaussian white noise $\xi_2(t)$ to (4.8) of intensity D_2. This corresponds to the general situation (1.1) with $N = 1$, $M = 2$, $g_{11} = \sqrt{D}\, q$, $g_{12} = \sqrt{D_2}$, $\tau_1 = \tau$ and $\tau_2 = 0$. Being interested in the effect of external fluctuations we shall work in the limit in which $\varepsilon = D_2/D \ll 1$. Of the $M_{\beta\nu}$ and $K_{\beta\gamma\nu}$ quantities in (2.10), only $M_{12} = 2\sqrt{D_2}q^3$ and $K_{112} = (D_1D_2)^{1/2}$ contribute here. The quantity K_{112} gives rise to a term with a third order derivative of order $\tau\varepsilon D^2$. Since we are considering both τ and ε as smallness parameters this term can be neglected and we obtain

$$\partial_t P(q,t) = -\partial_q[(1 + D)q - (1 + 2\tau D)q^3]P(q,t) + \partial_q^2 D[\varepsilon + q^2(1 - 2\tau q^2)]P(q,t) + O(\varepsilon\tau,\tau^2) \tag{4.17}$$

It should be noted that the recipe of adding the two diffusion coefficients in the FPE is strictly correct when both noises are white [28], but here it is only approximately valid because we neglect a crossed term of order $\varepsilon\tau$. The equation for the extrema of the stationary distribution of (4.17) turns out to be the same as the one in the limit $D_2 = 0$. Therefore (4.10) and the subsequent discussion remains unchanged when also considering small additive fluctuations. One can also see from (4.17) that the transition lines remain unchanged. The main effect introduced by $\xi_2(t)$ is rounding off divergences at the boundaries. Indeed, a lengthy analysis shows that in agreement with white noise results [28], $P_{st}(q)$ does not become strictly infinity or zero at the boundaries but a quantity of order ε^{-1} or ε respectively.

5. Perspectives

Although the improvement of the existing theoretical descriptions of the effect of noise in nonequilibrium transitions, and also the consideration of other systems seems desirable, the main features of time independent properties of the steady state seem by now reasonably well understood. More intriguing aspects appear to be the time dependent properties of the transition. These include the decay of fluctuations in the steady state near threshold and the nonlinear relaxation to the steady state. Some information exists on the eigenvalue spectrum for the dynamical evolution [28,14,15] but only limited analytical results are available for relaxation near threshold [14,29]. With this motivation and in order to get a better understanding of the phenomena we have undertaken [30] a numerical analysis based on noise simulation which is also helpful to check the predictions on time independent properties. Preliminary results confirm in particular the τ-dependence of threshold and mean values discussed here.

References

1. H. Haken: Synergetics, An Introduction, Springer Series in Synergetics, Vol.1, 2ed. (Springer, Berlin, Heidelberg, New York 1978)
2. H. Mori, T. Morita, K.T. Mashiyama: Progr. Theor. Phys. *63*, 1865 (1980); *64*, (1980)
3. W. Horthemke: "Nonequilibrium Transitions Induced by External White and Colored Noise", in *Dynamics of Synergetic Systems*, ed. by H. Haken, Springer Series in Synergetics, Vol.6 (Springer, Berlin, Heidelberg, New York 1980) p.67 and references therein
4. R. Kubo: in "Fluctuations, Relaxation and Resonance in Magnetic Systems", ed. by D. Ter Haar (Oliver and Boyd, Edinburgh 1962)
5. P. De Kepper, W. Horsthemke: C.R. Acad. Sci. Paris Ser. C*287*, 251 (1978)
6. S. Kabashima, S. Kogure, T. Kawakubo, T. Okada: J. Appl. Phys. *50*, 6296 (1979)
7. S. Kabashima, T. Kawakubo: "Experiments on Phase Transitions Due to the External Fluctuation", in *Systems Far from Equilibrium,* Proceedings, Sitges Conf. on Statistical Mechanics, Sitges, Spain, June 1980, ed. by L. Garrido, Lecture Notes in Physics, Vol.132 (Springer, Berlin, Heidelberg, New York 1980) p.395
8. S. Kai, T. Kai, M. Takata, K. Hirakawa: J. Phys. Soc. Jpn. *47*, 1379 (1979)
9. J.P. Gollub, J.F. Steinman: Phys. Rev. Lett. *45*, 551 (1980)
10. J.P. Crutchfield, B.A. Huberman: Phys. Lett. *77*A, 407 (1980)
11. L. Arnold: "Stochastic Differential Equations" (Wiley, New York 1974)
12. J.L. Doob: Ann. Math. *43*, 351 (1942)
13. J.M. Sancho, M. San Miguel: Z. Phys. B*36*, 357 (1980)
14. M. Suzuki, K. Kaneko, F. Sasagawa: "Phase Transitions and Slowing Down in Nonequilibrium Stochastic Processes", Preprint (1980)
15. K. Kitahara, K. Ishii: "Relaxation of Systems Under the Influence of Two Level Markovian Noise", contributed paper to Statphys 14 (Edmonton 1980)
16. L. Arnold, W. Horsthemke, R. Lefever: Z. Phys. B*29*, 367 (1978)
17. K. Kitahara, W. Horsthemke, R. Lefever: Phys. Lett. *70*A, 377 (1979)
    K. Kitahara, W. Horsthemke, R. Lefever, Y. Inaba: Progr. Theor. Phys. (1980)
18. M.O. Hongler: Helv. Phys. Acta *52*, 280 (1979)
19. N.G. Van Kampen: Phys. Rep. *24*C, 171 (1976)
20. E.A. Novikov: Sov. Phys. JETP *20*, 1290 (1965)
21. M. San Miguel: Z. Phys. B*33*, 307 (1979)

22. M. San Miguel, J.M. Sancho: Phys. Lett. *76*A, 97 (1980)
23. L. Garrido, M. San Miguel: Progr. Theor. Phys. *59*, 40 (1978)
24. M. San Miguel, J.M. Sancho: J. Stat. Phys. *22*, 605 (1980)
25. R.L. Stratonovich: "Topics in the Theory of Random Noise", Vol.2 (Gordon and Breach, New York 1967)
26. A. Schenzle, H. Brand: J. Phys. Soc. Jpn. *48*, 1382 (1980)
27. T. Schneider, E.P. Stoll, R. Morf: Phys. Rev. B*18*, 1417 (1978)
28. A. Schenzle, H. Brand: Phys. Rev. A*20*, 1628 (1979)
29. Y. Hamada: "Dynamics of the Noise Induced Phase Transition of the Verhulst model". Preprint (1980)
30. J.M. Sancho, M. San Miguel, S. Katz, J.D. Gunton: To be published

Part VI

Stochastic Behavior in Model Systems

Weak Turbulence in Deterministic Systems

J.-P. Eckmann
Département de Physique Théorique, Université de Genève

We consider "discrete dynamical systems", i.e. iterations of maps $F:\mathbb{R}^m \to \mathbb{R}^m$:

$$x_{n+1} = F(x_n).$$

These maps are assumed to be "dissipative", i.e. they contract volumes. Note that this does not mean that they must also contract lengths. The regions in phase space $(\mathbb{R}^m)$ in which almost all orbits land are called "attractors". After many iterations (large n) only attractors seem to be relevant. One would thus like to classify the possible attractors and the dynamics on them. This difficult task has only been solved for very special classes of dynamical systems. In the last few years, another, more modest, strategy has emerged. It consists in describing what one may call SCENARIOS. Instead of F, one considers F_μ, depending on a parameter μ and F_o is assumed to have a very simple attractor. As the parameter μ is varied, one tries to find successions of controllable changes of attractors, leading eventually to weakly aperiodic regimes. We present 3 such scenarios

I. The Ruelle-Takens-Newhouse scenario : 3 successive Hopf bifurcations (3 successive crossings of pairs of complex conjugated eigenvalues of DF_μ across the unit circle) lead under mild conditions to a strange attractor (generically).

- General reference : D. Ruelle, Les attracteurs étranges. La Recherche, Vol. 11, 132 (1980).
- Original reference : D. Ruelle, F. Takens. Commun. Math. Phys. 20, 167 (1971).

II. The Manneville-Pomeau scenario : One saddle node bifurcation (two real eigenvalues near +1) is associated with (type I) intermittency.

- General reference : Y. Pomeau, P. Manneville, Different ways to turbulence in dissipative dynamical systems. Physica 1D, 219-226 (1980).
- Original reference : Y. Pomeau, P. Manneville. Comm. Math. Phys. 74, 189 (1980).

III. The Feigenbaum-Collet-Eckmann-Koch-Lanford scenario : Two successive pitchfork bifurcations (two successive crossings of one eigenvalue at -1) lead often to infinite cascades of further such bifurcations with universal scaling between bifurcation points. In the limit, an aperiodic regime is reached.

- General reference : P. Collet, J.-P. Eckmann. Iterated Maps on the Interval as Dynamical Systems. Progress in Physics, Vol. 1, Birkhäuser, Boston (1980).
- Original reference : M. Feigenbaum. J. Stat. Phys. 19, 25-52 (1978).

Stochastic Problems in Population Genetics: Applications of Itô's Stochastic Integrals

Takeo Maruyama

National Institute of Genetics, Mishima, Shizuokaken 411, Japan

1. Introduction

Stochastic models have a rather long history in population genetics. The model often refered to as the FISHER-WRIGHT type has been studied extensively by S. WRIGHT, M. KIMURA and others. Thise theoretical works have influenced experimental population genetics, but their impact on the topical problem of molecular evolution and protein polymorphism can be symbolized by the birth of the neutral hypothesis advanced by [1, 2, 3]. Most of the early stochastic models dealt with one locus at which two alleles are present, where one dimensional diffusion theory can be successfully applied. In these early studies, much of the effort was spent in searching for analytic solutions of problems. Particularly when the subject can be formulated in one dimensional theory, it can be almost always solved analytically [4]. Although they may appear to be mathematically simple, the stochastic models and their analyses have been playing an important and indispensable role in population genetics and molecular evolutionary theory.

On the other hand, except for some special cases, however important biologically, analytic solution to multiple allele problems and linked locus problems have been very difficult to obtain. However, experimental population genetics now strongly demands solutions to such stochastic models so that further biological insight can be made. In fact, many problems which now appear to be biologically realistic seem to be hopelessly difficult for analytic solutions. Therefore it is natural that a number of numerical methods have been developed. There are two basically different methods. One is to describe the process in terms of a classical diffusion equation and solve it numerically. The other is to carry out simulations on the models. Since a diffusion equation or an equation derived from it gives the expectation of a quantity and the numerical solutions are more stable, it is usually more desirable to do so than the simulations. However, perhaps with few exceptions, the numerical integration of a diffusion equation is limited to processes with no more than two or three space variables.

Recently we developed a new numerical method which enables us to handle various stochastic models in very general form, [5, 6, 7]. Because of the availability of modern high-speed computers, the method proved to be useful, permitting us to parameterize the problem and seek concrete answers. Here I outline the method and present some of the findings on various problems, which are biologically important.

2. Mathematical Method

Let x_i be random variables representing gene frequencies or gamete frequencies.

In most of the cases dealt with in population genetics the random variables form a stochastic process which can be regarded as a diffusion process. Then the set of random variables x_i satisfies a system of stochastic differential equations,

$$dx_i = \sum_j a_{ij}(x_1,x_2,\ldots)dB_j + b_i(x_1,x_2,\ldots)dt \qquad (1)$$

where a_{ij}'s and b_i's are functions of x_i's and B_j's are independent Brownian motions. The coefficients a_{ij} are elements of the non-negative definite square root of the covariance matrix whose ith row and jth column element is given by

$$\sigma_{ij} \equiv \lim_{\Delta t \to 0} \frac{1}{\Delta t} E\{\Delta x_i \Delta x_j\}$$

where E{ } indicates the expectation. In practice, it is usually easy to determine the covariance matrix $[\sigma_{ij}]$. And then $[a_{ij}]$ is obtained by taking square root of the matrix $[\sigma_{ij}]$. The other coefficient b's represent the infinitesimal mean change in x_i. The differential of the Brownian motion represents the random change which occurs in an infinitely small time interval.

Since (1) requires operations involving infinitesimally small time intervals, it is not possible to simulate it by computer. As with differential equations of deterministic processes, (1) can be approximated by difference equations,

$$\Delta x_i = \sum_j a_{ij}(x_1, x_2, \ldots)\Delta B_j + b_i(x_1, x_2, \ldots)\Delta t \tag{2}$$

where ΔB_j indicates a white noise with variance equal to the time increment Δt. Let t_k be discrete time and $x_i(t_k)$ be the random variable x_i at time t_k. Then, we can rewrite (2) as

$$x_i(t_{k+1}) = x_i(t_k) + \sum_j a_{ij}(x_1(t_k), x_2(t_k), \ldots) \{B_j(t_{k+1}) - B_j(t_k)\}$$

$$+ b_i(x_1(t_k), x_2(t_k), \ldots) \{t_{k+1} - t_k\} . \tag{3}$$

This equation certainly gives the values of x_i at these discrete times t_k but not between these specified times. Now let

$$x_i(t) = x_i(t_k) \tag{4}$$

for $t_{k+1} > t \geqq t_k$. With this definition, the path constructed by (2) or equivalently (3) is defined for all values of time, and each path forms a step function having discontinuity at t_k's. If we are dealing with differential equations of a deterministic process, it is almost obvious that solutions given by a difference equation converge to those of the original equations, and there are many different improved methods available, such as Runge-Kutta approximation. However with a differential equation involving differentials of random processes this is not simple and in fact requires highly technical mathematics to show the convergence of $x_i(t)$ given by (3) and (4) to the solution of the original equation (1), [8]. The reason is that the second order moment $\Sigma\{B_j(t_{k+1}) - B_j(t_k)\}^2$ does not vanish as the time increment decreases to zero, while a corresponding quantity in differential equations for deterministic processes can be made arbitrarily small by letting Δt small, [9].

In some models of population genetics, we assume that the possible number of variables, which are usually either the gene frequencies or the gamete frequencies, is fixed. In many recent models dealing with molecular evolution, the number of variables is not constant, but changes in time. This is because the number of possible allelic states of a gene or gamete has been shown to be extremely large, but those represented in a population at a time is rather small. Furthermore in reality, if once a particular variable appeares and then disappears from the population it will never come back again to the same population.

Therefore it is necessary to incorporate newly arising variables into the system (1), while the vanishing variable must be removed from the system. It is easy to eliminate a vanishing variable. We rename the variables by omitting the one which we want to remove. For adding a new variable we do as follows: During the time interval of $\Delta t_k = t_{k+1} - t_k$, the new variable

$$x_{n+1}(t_{k+1}) = \varepsilon \tag{5}$$

is introduced with a probability which can be specified by the model. For instance, in the case of a gene mutation at one locus in a population of N diploid organisms with mutation rate v, the probability is $2Nv\Delta t/\varepsilon$, provided that this is much less than unity. I tested the accuracy of the method using the FISHER-WRIGHT model in which every mutant is new to the population. We denote by N and v the population size and mutation rate. Two selection schemes have been used. One is that all mutants are selectively neutral, and the other is an overdominance model in which every heterozygote has relative fitness $1 + s$, while it is equal to unity for every homozygote. The accuracy test consists of comparing the average level $(\bar{h})$ of heterozygosity, defined as the mean of $1 - \Sigma x_i^2$, and the allele frequency spectrum, $\Phi(x)$, which is the average number of alleles having frequency x. For the neutral model, both of these quantity can be obtained exactly by using formulae of [10, 11], for the overdominant model, the $\bar{h}$ can be calculated numerically if Ns is not large.

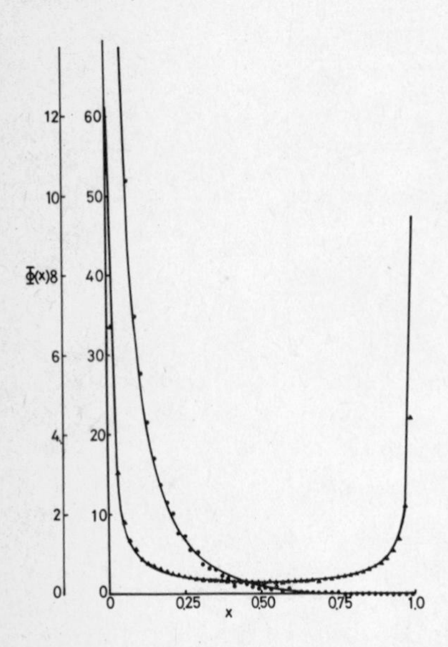

Fig.1 Theoretical allele frequency distributions $\Phi(x) = 4Nv\, x^{-1}(1 - x)^{4Nv-1}$ and the distributions obtained by computer simulation for neutral alleles. Symbols ▲ and ● refer to the cases of $4Nv = 0.08$ and $4Nv = 4$, respectively. The left and right scales of the ordinate refer to the distributions for $4Nv = 0.08$ and $4Nv = 4$, respectively.

Comparisons of the allele frequency spectrums obtained by the simulation and by the exact formulae are presented for two cases, $4Nv = 0.08$ and $4Nv = 2$. The values of $\bar{h}$ obtained by (3) and (5) are compared with the corresponding exact values. Agreement between the simulation results and the exact expectations are good both

Table 1 Mean heterozygosities for testing the accuracy of the computer simulation used. The theoretical values for neutral alleles were obtained by $\bar{h} = 4Nv/(1 + 4Nv)$ and those for overdominant alleles were obtained by numerical integration of formula given in [10].

$4Nv$	$2Ns$	Mean heterozygosity	
		Theoretical	Simulation
0.02	0	0.0196	0.0195
0.1	0	0.0909	0.0897
0.2	0	0.1667	0.1658
0.5	0	0.3333	0.3316
0.004	2	0.0080	0.0079
0.012	6	0.0970	0.0964
0.026	13	0.4548	0.4501
0.060	30	0.6265	0.6360

for the gene frequency spectrum, Fig.1, and the average level of the heterozygosity, Table 1.

3. Overdominance Model

Overdominant selection is a powerful mechanism for maintaining genetic variability, and therefore a number of authors have assumed that it is the major factor responsible for the observed variation in populations. However, the model has been extensively studied only for the case of an infinite population, while all natural populations are finite in size and thus stochastic factors play some role in determining the fate of mutants. A thorough study of the model under the assumptions of stochastic processes is needed, and here I intend to explore some of its characteristic features, which are biologically important, (see [7]).

The model under consideration is specified by the following diffusion covariance and drift terms:

$$\sigma_{ij} = x_i(t)(\delta_{ij} - x_j(t))$$

and

$$b_i = \{-V + S(f(t) - x_i(t))/(1 - sf(t)\}$$

where $\delta_{ij} = 1$ if $i = j$, $\delta_{ij} = 0$ otherwise, $V = 2Nv$, $S = 2Ns$ and $f(t) = \Sigma\, x_i(t)^2$.

Mean heterozygosity: The mean heterozygosity $(\bar{h})$ is shown as a function of Nv and Ns in Fig.2. It is certain that the average heterozygosity under overdominant selection is always higher than that for neutral mutations when $4Nv$ remains the same. It is biologically very important to note that in the presence of overdominant selection $\bar{h}$ increases more rapidly particularly when s/v is large. This implies that the average heterozygosity is more sensitive to the variation of population sizes than in the case of neutral mutations. The change in $\bar{h}$ is particularly sensi-

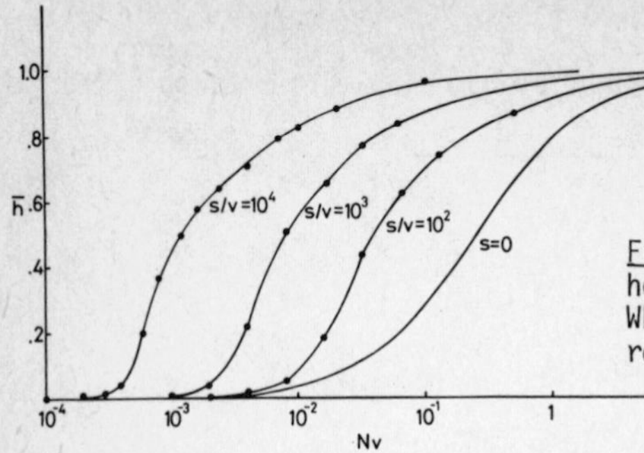

Fig.2 Relationships between Nv and mean heterozygosity $(\overline{h})$ for overdominant alleles. When s and v remain constant, each curve represents the relationship between Nv and $\overline{h}$.

tive when the value of $\overline{h}$ changes from about 0.05 to 0.6, which is indeed the range that includes practically all the experimentally cases.

Variance of heterozygosity: The relationship between the mean $(\overline{h})$ and variance (σ_h^2) of heterozygosity is important, and has been studied by [12] for the neutral model. However, it has not been possible to obtain the relationships for other models. The present method certainly enables us to carry out the calculations for any model that can be formulated in a form of (1). I have done that for the overdominance model. The results are presented in Fig.3. The relationship depends on the ratio of s/v. As the value of s/v becomes large, graphs of the relationships tend to a parabolic curve with its maximum at $\overline{h} = 1/4$. Interestingly, the graphs assume their minima at $\overline{h} = 1/2, 2/3, 3/4$ and probably at all the other $\overline{h} = (n - 1)/n$ for $n = 5, 6, \ldots$.

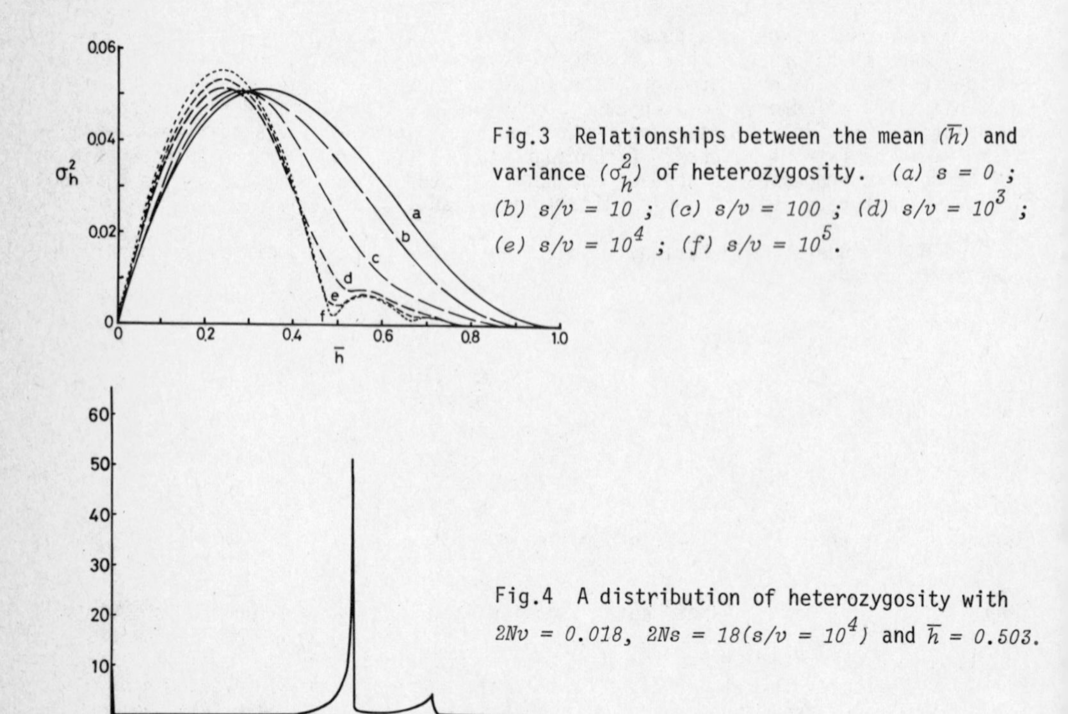

Fig.3 Relationships between the mean $(\overline{h})$ and variance (σ_h^2) of heterozygosity. (a) $s = 0$; (b) $s/v = 10$; (c) $s/v = 100$; (d) $s/v = 10^3$; (e) $s/v = 10^4$; (f) $s/v = 10^5$.

Fig.4 A distribution of heterozygosity with $2Nv = 0.018$, $2Ns = 18 (s/v = 10^4)$ and $\overline{h} = 0.503$.

Distribution of heterozygosity: Beside the mean and variance of heterozygosity. the present method enables us to obtain its distribution. This has been studied by [12] who conjectured that the distribution for neutral mutations will have peaks at $h = 0, 1/2, 2/3, \ldots$. In the overdominance model, these peaks are much more pronounced, and particularly when the value of s/v is large, almost all the mass is localized at the peaks. Therefore if the average heterozygosity is 0.5 or less, as observed in nature, a population stays mostly at $h = 0$ or $h = 0.5$, but practicall none at other values of h. For instance if $\bar{h} = 0.20$, which is a typical value observed in nature a population is at $h = 0.5$ for 40 percent of time and at $h = 0$ for the rest. A typical case of the distribution is given in Fig.4.

4. Linkage Disequilibrium

During the past several years, linkage data, which are mostly for molecular polymorphisms, are becoming available in many species and there have been a number of attempts to use linkage disequilibrium theory in testing rival hypotheses, [13, 14]. Also some of the known linked loci in man have been extensively studied and turn out to have a rather important bearing in human welfare, (e.g., the HLA system studied by [15]). Since techniques of genotype identication and detection of linkage are rapidly advancing, we expect, in the near future, that more data subject to linkage analysis will be produced and its theory will become important.

The first mathematical theory of linkage for finite populations was advanced by [16] and followed by [17], but the theory advanced by these studies has a limitation in applications because it deals only with mean valves, while other quantities of the distribution are also necessary. The present method is readily applicable and enables us to obtain almost any quantity, for the method provides sample paths of the process. I will demonstrate the method in obtaining some of the quantities associated with linkage problems, which has not been possible previously.

We consider a population consisting of N diploid individuals ($2N$ gametes), and two linked loci where the recombination value between them is c. We assume that, every mutant is new to the population and the mutation rate is v at one locus and u at the other. Let A_i's be alleles at one locus and B_i's at the other, and let $x_{ij}(t)$ be the frequency of gametes carrying A_i and B_j, where t denotes the time measured in units of $2N$ generations. Let $p_i(t) = \sum_j x_{ij}(t)$ and $q_j(t) = \sum_i x_{ij}(t)$. Then (1) for this process is

$$dx_{ij}(t) = \sum_{kl} a_{kl}^{ij}\, dB_{kl} + [-(U + V)x_{ij}(t) + C\{p_i(t)q_j(t) - x_{ij}(t)\}dt \tag{6}$$

where $U = 2Nu$, $V = 2Nv$, $C = 2Nc$, B_{kl}'s are independent Brownian motions and a_{kl}^{ij}'s are elements of the non-negative definite square root of the covariance matrix

$$[x_{ij}(t)\{\delta_{ik}\delta_{jl} - x_{kl}(t)\}]$$

with $\delta_{ij} = 1$ if $i = j$ and $\delta_{ij} = 0$ otherwise. Eq.(6) can be easily changed into a system of difference equations, and then a term accounting for newly arisen mutants can be added.

The following three quantities were studied.

$$D \equiv \sum_{ij} (p_i q_i - x_{ij})^2 \tag{7}$$

$$H_A H_B \equiv \{\sum_{i \neq j} p_i p_j\}\{\sum_{i \neq j} q_i q_j\} \tag{8}$$

and

$$R \equiv D/H_A H_B \tag{9}$$

where the time variable t is omitted. These quantities are biologically important, and usually experimentally measurable.

Assuming $V = U$, the means and variances of these quantities were calculated for a wide range of parameter values which seem to be biologically meaningful. The results are presented in Table 2.

Table 2 Compilations of the means and standard deviations of D of (7), $H_A H_B$ of (8) and R of (9). $E(\ )$ indicates the mean and σ the standard deviation.

$2Nu$	$2Nc$	$E(D)$ $\times 10^2$	σ_D $\times 10^2$	$E(H_A H_B)$ $\times 10^2$	$\sigma_{H_A H_B}$ $\times 10^2$	$E(D)/E(H_A H_B)$	$E(R)$	σ_R
0.01	0	0.068	0.590	0.069	0.982	0.313	0.008	0.074
0.01	0.1	0.028	0.638	0.057	0.930	0.487	0.011	0.087
0.01	1	0.007	0.175	0.052	0.772	0.131	0.004	0.035
0.01	5	0.006	0.156	0.063	0.873	0.098	0.004	0.033
0.01	10	0.001	0.051	0.028	0.474	0.041	0.001	0.013
0.01	20	0.001	0.031	0.061	0.821	0.022	0.001	0.013
0.05	0	0.567	2.570	2.146	6.723	0.264	0.057	0.178
0.05	0.1	0.561	2.483	2.243	7.041	0.250	0.058	0.177
0.05	1	0.485	1.530	4.725	8.202	0.103	0.031	0.096
0.05	5	0.079	0.533	0.970	3.596	0.082	0.014	0.055
0.05	10	0.047	0.294	1.096	3.838	0.043	0.009	0.036
0.05	20	0.034	0.176	1.439	4.498	0.024	0.007	0.026
0.1	0	1.267	4.021	3.289	7.008	0.385	0.107	0.237
0.1	0.1	1.144	3.625	3.140	6.881	0.364	0.092	0.212
0.1	1	0.547	2.039	2.966	6.577	0.185	0.050	0.129
0.1	5	0.304	1.005	3.909	7.577	0.078	0.029	0.068
0.1	10	0.195	0.590	4.461	8.259	0.044	0.019	0.044
0.1	20	0.123	0.331	5.031	8.843	0.024	0.012	0.027
0.2	0	2.836	5.305	8.811	11.142	0.322	0.162	0.239
0.2	0.1	2.633	5.038	8.184	10.994	0.322	0.152	0.230
0.2	1	1.566	3.188	9.006	11.071	0.174	0.087	0.145
0.2	5	0.891	1.543	12.086	13.038	0.074	0.048	0.070
0.2	10	0.597	0.950	13.943	14.020	0.043	0.031	0.043
0.2	20	0.419	0.554	18.100	16.022	0.023	0.019	0.088

Table 2 (continued)

$2Nu$	$2Nc$	$E(D)$ $\times 10^2$	σ_D $\times 10^2$	$E(H_A H_B)$ $\times 10^2$	$\sigma_{H_A H_B}$ $\times 10^2$	$E(D)/E(H_A H_B)$	$E(R)$	σ_R
0.3	0	4.407	5.803	15.386	13.995	0.286	0.193	0.217
0.3	0.1	4.047	5.441	14.843	13.810	0.273	0.181	0.205
0.3	1	2.560	3.663	15.400	13.454	0.166	0.114	0.138
0.3	5	1.404	1.726	20.380	15.304	0.069	0.056	0.062
0.3	10	1.033	1.075	25.300	16.254	0.041	0.036	0.034
0.3	20	0.740	0.613	33.000	17.939	0.022	0.021	0.017

The present study reveals several important facts about the behavior of linked loci. Firstly the level of D is quite small in an equilibrium population even if $c = 0$ (complete linkage). Secondly, the ratio of the two means, namely $E(D)/E(H_A H_B)$ is often much larger than the mean of the ratio, $E(D/H_A H_B)$. Importantly, the experimentally observable quantity is usually the ratio R, rather than the ratio of the means. Finally, I want to mention that statistical distributions of linkage disequilibrium values are not normal, and have a large variances. In order to cope with future problems in linkage analysis, much further study will be needed, and the method based on Itô's stochastic integrals as used in this paper will be a valuable tool.

References

1. M. Kimura: Nature 217, 624 (1968)
2. J.L. King, T.H. Jukes: Science 164, 788 (1969)
3. M. Kimura, T. Ohta: Nature 229, 467 (1971)
4. T. Maruyama: *Stochastic Problems in Population Genetics*, Lecture Notes in Biomathematics, Vol. 17 (Springer, Berlin, Heidelberg, New York 1977)
5. T. Maruyama: Adv. in Appl. Prob. 12, 263 (1980)
6. T. Maruyama, N. Takahata: Heredity (in press, 1980)
7. T. Maruyama, M.Nei: Genetics (submitted)
8. L. Arnold: *Stochastic Differential Equations: Theory and Applications* (Wiley, New York 1974)
9. W. Rümelin: Report Nr. 12, Forschungsschwerpunkt Dynamische Systeme (Universität Bremen)
10. S. Wright: Encyclopedia Britanica, 14th ed., 10, 111 (1949)
11. M. Kimura, J.F. Crow: Genetics 49, 725 (1964)
12. F.M. Stewart: Theor. Popula. Biol. 9, 188 (1976)
13. F.M. Johnson, H.E. Schaffer: Biochem. Genetics 10, 149 (1973)
14. T. Mukai, R.A. Voelker: Genetics 86, 175 (1977)
15. W.F. Bodmer, J.G. Bodmer: Med. Bull. 34, 3o9 (1978)
16. W.G. Hill, A. Robertson: Theor. Appl. Genetics 38, 226 (1968)
17. T. Ohta, M. Kimura: Genetical Res. 13, 47 (1969)

Poisson Processes in Biology

Henry C. Tuckwell

Department of Mathematics, Monash University,
Clayton, Victoria, Australia 3168

1. Introduction

A simple Poisson process, $[N(t), t \geq 0]$, with parameter λ, is a random process with stationary independent increments such that if $0 \leq t_1 < t_2$, then

$$(1)\quad Pr[N(t_2) - N(t_1) = k] = [\lambda(t_2-t_1)]^k \exp[-\lambda(t_2-t_1)]/k!$$

$k = 0,1,2..$

As Khintchine [15] pointed out, it is more of a surprise than not that some naturally occurring processes look like simple Poisson processes. This extends into biology. An example arises at the junction between nerve and muscle called neuromuscular junction. Here little blips called miniature end plate potentials, representing changes in the electric potential difference across the muscle membrane, occur under certain conditions in accordance with the defining characteristics of a simple Poisson process [6 , 8]. The 'Poisson hypothesis' for these events has been looked at very closely with a battery of statistical tests [3 , 4]. Not surprisingly this has ultimately led to the refutation of the hypothesis. The subject is still of current interest in pharmacology because drugs may have a drastic effect on the sequence of times of occurrence of the miniature end plate potentials [20]. As yet the mechanisms involved in the events which culminate in the miniature end plate potentials have not been delineated, so it is a fertile area for modeling.

We will be concerned with the application of Poisson processes in models in neurobiology and population biology. Some of the work has appeared and some will appear in more detail in the near future. I am indebted to several people with whom the work has been joint or has benefited greatly from discussion - Davis K. Cope, Floyd B. Hanson, John B. Walsh and Frederic Y.M. Wan.

2. Modeling the activity of nerve cells

2.1. Biological Background

The fundamental unit of the nervous system is the nerve cell or neuron. Such cells have many different forms but the traditional paradigm is that a cell has the following important features. A branching structure called the dendritic tree, a cell body or soma which is often about spherical or pyramidal and an axon which starts somewhere around the cell body and extends possibly over large distances to often bifurcate and form junctions called synapses with 'target cells'.

Most neurons emit action potentials which usually tend to form at a region on the neuronal surface where there is a high density of sodium ion channels. This region is called a trigger zone and is often located just by the cell body. The occurrence of an action potential can be monitored by an extracellular or intracellular microelectrode and the times of occurrence recorded and stored in a computer.

Evidence for the variability of the time interval between action potentials even under 'steady' input conditions is now overwhelming. The first examples came from receptor cells called muscle spindles of frog [1 ,10]. Neurons in mammalian brain also exhibit various degrees of variability in their interspike intervals, the name often employed for the time between action potentials. Interest in this variability from an experimental point of view comes from classifying the pattern of activity in accordance with anatomical location [19] or with the behavioral state of the animal [2]. There is also the question of information transmission in the nervous system which seeks to explain how a sequence of spikes is read by target cells when the sequence is apparently random [18]. Here it is opportune to point out that variability of interspike times does not necessarily imply randomness in the mechanisms which produce the spikes. This can be seen by applying Scharstein's [23] novel graphical method of predicting a spike train from a given input current to a simple neuron model. The input may be completely regular (e.g. sinusoidal) yet the output train can be highly irregular. The distinction between this and the 'random spike train' can be found by computing serial correlation coefficients.

If the spike sequence from a neuron is random we address the question of how this randomness arises and attempt to elucidate the pattern of activity of cells with the aid of models for nerve cell activity. There are thought to be two main sources of randomness which contribute to the variability of the interspike interval. These are fluctuations in the properties of the spike generating mechanism and the random nature of the synaptic input. An example of a study of the former is that of Levine & Shefner [17]. Concerning the latter it should be noted that many neurons are covered with possibly thousands of synapses which are not all synchronized in their activity so that particularly for a 'spontaneously active' cell, the randomness of the synaptic input is a good candidate for producing variability in the spike train.

We will consider a few models of nerve cell activity with random input. The physiologist will realize how much reality is left out of these models, but unfortunately even the 'simple' models generate very difficult problems when we begin to ask for quantitative results. A review of the subject is contained in Holden's monograph [14].

2.2 Stein's model

In this model which was introduced by Stein [24], the spatial extent of the cell is ignored. Some justification for this comes from the existence of trigger zones which are usually quite small. Based on neurophysiological investigations with intracellular electrodes the electrical activity of some cells can be approximated as that of a resistor and capacitor in parallel. This leads to exponential decay of the depolarization in the absence of inputs. The model actually dates back to 1907 when it was used by Lapicque [16]. The input currents to this model cell are approximated by delta functions and their arrival times made to coincide with the times of occurrence of events in simple Poisson processes.

Let $X(t)$ be the depolarization at time t with initial value $X(0) = x$. Excitation increases $X(t)$ by a_1 units and has mean rate λ_1; inhibition arrives independently, decreasing $X(t)$ by a_2 and has mean rate λ_2. Taking account of the exponential decay we can write a stochastic differential equation for $X(t)$:

$$(2)\quad dX(t) = -X(t)dt + a_1 dN_1(t) - a_2 dN_2(t),\ X(0) = x,\quad X(t) < \theta,$$

where $N_1(t)$ and $N_2(t)$ are Poisson processes with parameters λ_1 and λ_2. Here time is measured in units of the membrane time constant. The equation is valid only for $X(t)$ less than the threshold for firing of action potentials which is assumed in the first instance to be constant. Let $T_\theta(x)$ be the random variable

$$(3)\quad T_\theta(x) = \inf\{t \mid X(t) \geqslant \theta \mid X(0) = x < \theta\}.$$

We identify this as the random time between action potentials, save for the addition of an assumed fixed absolute refractory period. $T_\theta(x)$ is a <u>first passage time</u> for $X(t)$ to level θ. The determination of the distribution function of $T_\theta(x)$ requires solution of a partial differential difference equation, whereas finding the moments of $T_\theta(x)$ requires solution of the recursion system of ordinary differential difference equations. With $M_n(x)$ the n-th moment, these are

$$(4)\quad -x\frac{dM_n}{dx} + \lambda_1 M_n(x + a_1) + \lambda_2 M_n(x - a_2) - (\lambda_1 + \lambda_2)M_n(x) = -nM_{n-1}(x),$$

$n = 0,1,2,\ldots,\ x < \theta.$, as can be found by applying theorems on first exit times for Markov processes [28]. Equations such as this have been investigated hardly at all, especially on account of the occurrence of both forward and backward differences. The boundary condition at θ is that $M_n(x) = 0,\ x \geqslant \theta,\ n = 1,2,\ldots$ but with $\lambda_2 \neq 0$ the condition at $x = -\infty$ is difficult to prescribe.

2.2.1 An exact calculation

When $\lambda_2 = 0$ (no inhibition) we can sometimes solve (4) for small θ and small n exactly. A simple illustrative example will suffice. For convenience set $\lambda_1 = a_1 = 1$ and consider the equation for the first moment:

$$(5)\quad -x\frac{dM_1}{dx} + M_1(x + 1) - M_1(x) = -1,\quad x < 2.$$

We can restrict the domain of M_1 to $[x_1,2)$ where $x_1 \leqslant 0$. $X(t)$ can only exit from $[x_1,2)$ at the right hand end point. It is sufficient for our problem to take $x_1 = 0$ because we are primarily interested in an initial condition at resting level, $x = 0$. On $[1,2)$ put $M_1(x) = F_1(x)$ and on $[0,1)$ put $M_1(x) = F_2(x)$. Then F_1 satisfies the simple ordinary differential equation

$$(6)\quad \frac{dF_1}{dx} + \frac{F_1}{x} = \frac{1}{x},$$

with solution $F_1(x) = x^{-1}[x + c_1]$, with c_1 a constant of integration. This is used on $[0,1)$:

$$(7) \quad \frac{dF_2}{dx} + \frac{F_2}{x} = \frac{2}{x} + \frac{c_1}{x(x+1)} .$$

This is solved to give

$$(8) \quad F_2(x) = x^{-1}[2x + c_1 \ell n(x+1) + c_2],$$

with c_2 a second constant of integration. The values of c_1 and c_2 are found by imposing the conditions that $M_1(x)$ is continuous and bounded on $[0,2)$. This leads to the solution

$$(9) \quad M_1(x) = \begin{cases} 1 + [1/(1 - \ell n2)]x, & x \in [1,2), \\ 2 + [1/(1 - \ell n2)][\ell n(x+1)]/x, & x \in [0,1). \end{cases}$$

This solution has a discontinuity at $x = 2$ where it jumps from a positive value to zero (see Figure 2 of [27]). Exact calculation is possible for other parameter values and for the second moment as long as the threshold θ is small (less than three). These calculations do reveal some interesting facts concerning the coefficient of variation of the interspike interval. Physiologists use coefficient of variation as a measure of the noisiness of the transmitted signal, as it is in fact the inverse of the 'signal to noise ratio'. It had been thought that the coefficient of variation was a monotonically increasing function of the mean interval but calculations reveal that in some parameter ranges the dependence is not monotonic but includes maxima and minima for a given threshold. However, it seems that the predicted structure may not be observed experimentally because it is based on a model which is not realistic enough and other sources of randomness would drown out the structure. Full details are in [32].

2.2.2 Numerical methods

For large θ and when there is inhibition we have solved the differential difference equation (4) for small n by two methods [5, 32]. In the first approach, devised by Wolfgang Richter, the differential difference equation is written as a coupled linear system of ordinary differential equations which was solved using Runge-Kutta techniques. It was possible to ascertain the dependence of the moments of the interspike time on θ and also on the input rate of excitation. An attempt was made by the method of moments to predict the three parameters of the model (the threshold to excitatory postsynaptic potential amplitude ratio, the time constant of the cell membrane circuit and the rate of arrival of excitatory inputs) for two cells of the cat cochlear nucleus. Though the predicted values are all reasonable, they cannot be taken seriously because of the great oversimplifications of the model and the assumptions made in its implementation.

When there is inhibition we have employed a different technique [5]. This relied on an asymptotic form for the solution for large negative x which contained an additive unknown constant. The solution is thus extended to positive values of x and the boundary condition at θ was employed to find the previously unknown constant. This method seemed to work although no boundary condition was employed at large negative x. The integration of such equations could benefit from theoretical investigations, with particular emphasis on the development of numerical methods.

2.2.3 Diffusion approximation

A diffusion approximation, $X^*(t)$, may be constructed for $X(t)$ given by equation (2), such that the first and second infinitesimal moments of the two processes are the same. This leads to

$$\text{(10)} \quad dX^*(t) = (-X^*(t) + \alpha)dt + \sigma dW(t),$$

where $\alpha = \lambda_1 a_1 - \lambda_2 a_2$, $\sigma = (\lambda_1 a_1{}^2 + \lambda_2 a_2{}^2)^{1/2}$ and $W(t)$ is a standard Wiener process. The sequence of approximating processes whose infinitesimal moments converge to those of $X^*(t)$ is the Ornstein-Uhlenbeck process which has mainly appeared in connection with modeling the phenomenon of Brownian motion. By computing solutions of the differential difference equation (4) and comparing them with solutions of the corresponding equation for $X^*(t)$:

$$\text{(11)} \quad \frac{\sigma^2}{2} \frac{d^2 M_n}{dx^2} + (\alpha - x)\frac{dM_n}{dx} = -nM_{n-1}(x), \quad n = 1,2,...,$$

[22] it is possible to see how the diffusion approximation fairs in estimating the threshold crossing time. This had been done for $n = 1$ and a general scheme for the error deduced. [31]. It is pointed out that overshoot of threshold can occur for the discontinuous process $X(t)$ but not the diffusion, $X^*(t)$.

2.2.4 Computer simulation

With package programs available for random number generation, it is straightforward to carry out computer simulations for Stein's nerve cell model. An example of first passage time densities for $X(t)$ obtained through approximating histograms is given in [29]. One feature of such densities is the presence of an extremely large tail in the case of inhibitory inputs and experimentalists' hypotheses about the nature of the density [2] were possibly explained. An interesting observation was that only when there was significant inhibition could the coefficient of variation of the interspike interval take values greater than units [30]. This leads to the conjecture that if the coefficient of variation is greater than one, then the cell under observation must be receiving significant amounts of inhibitory input. More recently Frederic Wan [36] has shown using singular perturbation methods for solving (11) with σ small, that this conjecture is true for the diffusion model $X^*(t)$.

2.2.5 Generalization of Stein's model

Stein's model may be generalized to include an arbitrary distribution of jump amplitudes [25]. Here $X(t)$ satisfies the stochastic differential equation

$$\text{(12)} \quad dX(t) = -X(t)dt + \int_R uN(du,dt),$$

where $N(.,.)$ is a Poisson random measure [9] such that

$$\text{(13)} \quad Pr[N(A,t) = k] = (t\lambda(A))^k \exp[-t\lambda(A)]/k!, \; A \in \mathcal{B}(R), \; k=0,1,2,....$$

$\lambda(\cdot)$ being the rate measure. The moment equations (4) become

$$\text{(14)} \quad -x\frac{dM_n}{dx} + \int_R M_n(x+u)\,\lambda(du) - \Lambda M_n(x) = -nM_{n-1}(x),$$

where Λ is the total jump rate. Apart from the case already considered, the case where there is an exponential ditribution of excitatory inputs has also been studied [26], in which case (14) becomes

$$(15)\qquad \frac{-x\,dM_n}{dx} + \alpha \int_R M_n(x+u)e^{-\beta u}du - \frac{\alpha}{\beta} M_n(x) = -nM_{n-1}(x).$$

A further generalization of the model equation (2) is the inclusion of smooth noise along with the discontinuities so that the process is a diffusion with jumps occuring at time intervals which are exponentially distributed:

$$(16)\qquad dX(t) = -X(t)dt + a_1 dN_1(t) - a_2 dN_2(t) + \sigma dW(t),$$

which leads to the moment equations

$$(17)\qquad \frac{\sigma^2}{2}\frac{d^2M_n}{dx^2} - x\frac{dM_n}{dx} + \lambda_1 M_n(x+a_1) + \lambda_2 M_n(x-a_2) - (\lambda_1+\lambda_2)M_n(x) = -nM_{n-1}(x), \quad n = 1,2,\ldots$$

This equation should yield some interesting problems in singular perturbation methods and is of interest to see how small jitter may influence the firing time in the presence of random inputs of larger magnitudes.

2.3. Modification of Stein's model

In Stein's model the amplitude of the postsynaptic potentials is independent of the value of the membrane potential at their time of occurrence. There is much physiological evidence that this is not true and that for excitation and inhibition there are reversal potentials [7] at which these amplitudes become zero. To include these reversal potentials is not difficult formally as we now have:

$$(18)\qquad dX(t) = -X(t)dt + (X_1 - X(t))a_1 dN_1(t) + (X_2 - X(t))a_2 dN_2(t), X(0)=x,$$

where X_1 and X_2 are the (constant) reversal potentials and a_1 and a_2 are further constants. The earlier form for the expectation of $X(t)$ [30] should read

$$(19)\qquad E[X(t)] = (k_2/k_1) + [x - (k_2/k_1)]\exp(-k_1 t),$$

where $k_1 = 1 + \lambda_1 a_1 + \lambda_2 a_2$, $k_2 = \lambda_1 a_1 X_1 + \lambda_2 a_2 X_2$, λ_1 and λ_2 being the rate parameters of the Poisson processes. I thank Charles Smith for pointing this out.

The first passage time moments now satisfy a slightly different set of differential difference equations which cannot be solved as a system of ordinary equations on intervals of constant size. Some analytic results were obtained for excitation only and it was found that for reasonable values of the reversal potentials, substantially different firing times occur [30].

2.4 Varying Threshold

In the above models the threshold for action potential generation was assumed to be constant. This is an approximation for most cells as there is intrinsic refractoriness as well as threshold elevation due to aftercurrents. These phenomena can be included in the above framework by making θ a function of time.

A method for handling time varying thresholds is available if the threshold satisfies a differential equation of first order. The following result for a general Markov process is obtained not by viewing first passage of a scalar process $X(t)$ to a time varying barrier $Y(t)$ but by considering the equivalent first exit problem in the phase plane for the vector valued process $(X(t), Y(t))$.

Theorem. Let $X(t)$ be a Markov process satisfying the stochastic differential equation

$$dX(t) = \alpha(X(t))dt + \beta(X(t))dW(t) + \int_R \gamma(X(t),u)N(du,dt), X(0) = x, \tag{20}$$

and let $Y(t)$ be the solution of the deterministic equation

$$dY(t)/dt = \delta(Y(t)), \quad Y(0) = y. \tag{21}$$

Let $T(x,y)$ be the time of first passage of $X(t)$ to $Y(t)$:

$$T(x,y) = \inf\{t | X(t) = Y(t) | X(0) = x, Y(0) = y\}, \tag{22}$$

and set

$$M_n(x,y) = E[T^n(x,y)], \quad n = 1,2,... \tag{23}$$

Then assuming $T(x,y)$ is finite with probability one, the moments satisfy the recursion system of partial-differential-integro equations,

$$\frac{1}{2}\beta^2(x)\frac{\partial^2 M_n(x,y)}{\partial x^2} + \alpha(x)\frac{\partial M_n}{\partial x} + \delta(y)\frac{\partial M_n}{\partial y} + \int_R M_n(x+\gamma(x,u))\lambda(du) - \Lambda M_n = -nM_{n-1}, \tag{24}$$

with boundary conditions that $M_n(x,y) = 0$ for $x > y$, assuming $X(t)$ starts below $Y(t)$.

Proof. The proof is simple upon considering the vector valued process and applying standard results for first exit times. Further details can be found in [34] where some applications are presented.

2.5 Spatial Models

In deterministic modeling of nerve cells spatial effects have long known to be important. The same stimulus delivered close to the cell body will have a different effect when delivered on distal dendrites. The pioneering work in this area was Rall's [21]. We can consider instead of a point model such as Stein's, a spatial model of the kind used in deterministic modeling. Such models represent portions of the cell as cylinders with cable structures containing resistors and capacitors. These are still passive elements as no intrinsic threshold properties arise. With suitable units for time and distance, if we consider a nerve cylinder of length L with a Poisson input at $x = x_0 \in (0,L)$, we then find that the depolarization satisfies the stochastic partial differential equation

(25) $$\frac{\partial V}{\partial t} = -V + \frac{\partial^2 V}{\partial x^2} + \delta(x - x_0)a_1 \frac{dN_1}{dt}, \quad 0 < x < L, \quad t > 0,$$

where $V(x,t)$ is the depolarization, a_1 is a constant, $\delta(.)$ is Dirac's delta function and $N_1(.)$ is a Poisson process with rate λ_1. We will assume for simplicity that the boundary conditions are

(26) $$V_x(0,t) = V_x(L,t) = 0,$$

and the initial data is

(27) $$V(x,0) = 0,$$

though the results extend to other cases. An eigenfunction expansion for $V(x,t)$,

(28) $$V(x,t) = \sum_{n=0}^{\infty} \phi_n(x)V_n(t),$$

is possible, where

(29) $$\phi_n(x) = \begin{cases} (1/L)^{1/2}, & n = 0, \\ (2/L)^{1/2} \cos(n\pi x/L), & n = 1,2,..., \end{cases}$$

are the spatial eigenfunctions. The random processes $V_n(t)$ satisfy ordinary stochastic differential equations of the kind in Stein's point model:

(30) $$dV_n(t) = -\mu_n^2 V_n(t)\, dt + a_1\phi_n(x_0)dN_1(t),$$

where the eigenvalues are,

(31) $$\mu_n^2 = 1 + n^2\pi^2/L^2, \quad n = 0,1,2,...$$

The moments of $V(x,t)$ are computable as infinite sums and their values in the steady state obtainable. Results are similar for the case of a white noise input for the first two moments and the covariance [33, 35]. First passage theory can be applied to truncated versions of the sum in (28) and lead to problems with vector-valued jump processes which have not been previously considered.

3. Modeling the growth of populations

There are many examples of population growth where sudden decreases occur apparently due to random disasters [12]. We have attacked the problem of estimating the persistence time of a population which is beset by occasional disasters of fixed magnitude by considering the population size as satisfying the stochastic differential equation

(32) $$dX(t) = rX(t)[1 - X(t)/K]dt - \alpha dN(t), \quad X(0) = x \in (0,K),$$

where K is a constant called carrying capacity, r is the intrinsic growth rate and α is the magnitude of the portion of the population removed by a disaster. In the absence of disasters the growth obeys the simple logistic law.

For the model described by (32) the extinction of the population is considered as the passage of $X(t)$ to 0 for the first time. Hence the problems in calculating extinction time are very similar to those in determining the firing time of the nerve cell model of

Stein. The moments $M_n(x)$ of the extinction time now satisfy the recursion system,

$$(33) \quad rx\left(1 - \frac{x}{K}\right) \frac{dM_n}{dx} + \lambda[M_n(x - \alpha) - M_n(x)] = -nM_{n-1}(x),$$

$n = 1,2,\ldots,$ with boundary condition $M_n(x) = 0$ for $x \leqslant 0$. The equations for $n = 1,2$ have been solved by numerical methods employing a singular decomposition [11]. Probably the most interesting feature of these results, apart from their quantitative values, is the appearance of plateaus in the extinction time moments as functions of x for small disaster rates relative to the growth rates. The implication is a safety zone for populations in low disaster rate environments where the expected survival time is not very sensitive to how close the population size is to carrying capacity. That is, no ecological advantage is obtained by maintaining a high level population size when disasters are relatively infrequent.

We have also made the disasters a function of population size. The simplest problem to consider is that when the amount of the population removed is proportional to the present magnitude of the population. The stochastic differential equation for this case is

$$(34) \quad dX(t) = rX(t)[1 - X(t)/K]dt - \alpha X(t)dN(t), \; X(0) = x, \; 0<\Delta<x<K.$$

Here the population cannot, if $\alpha < 1$, ever be driven to level zero, so a small level population size must be chosen as the extinction level. The equations corresponding to (33) become

$$(35) \quad rx\left(1 - \frac{x}{K}\right) \frac{dMn}{dx} + \lambda[M_n((1 - \alpha)x) - M_n(x)] = -nM_{n-1}(x),$$

and constant steps cannot be employed in the integration procedure. Numerical results as well as some computer simulations which enable a comparison to be made of the models (32) and (34) are contained in a forthcoming article [12]. We are currently extending these results to incorporate a distribution of disaster amplitudes in both density-dependent and density independent situations [13].

There are two kinds of problems in this area that would profit from attention. The first is the inclusion in a model such as (34) the presence of smaller amplitude noise by means of a diffusion term:

$$(36) \quad dX(t) = rX(t)[1 - X(t)/K]dt - \alpha X(t)dN(t) + \sigma dW(t),$$

where σ is small. The effects of this extra term will be to make zero accessible to $X(t)$ and the appropriate moment equations will be

$$(37) \quad rx\left(1 - \frac{x}{K}\right) \frac{dMn}{dx} + \frac{\sigma^2}{2} \frac{d^2Mn}{dx^2} + \lambda[M_n((1 - \alpha)x) - M_n(x)] = - nM_{n-1}(x).$$

Again, such equations should lead to some interesting applications of singular perturbation methods with boundary layers at $x = 0$ and $x = K$. A second problem worth looking at in this context is a vector valued version of the above discontinuous model equations so that extinction in coupled systems of, say, predator and prey may be considered via their first exit times. However, both of these as yet untackled problems will present some very difficult computational tasks.

4. References

1. Brink, F., Bronk, D.W. & Larrabee, M.G., Ann.N.Y.Acad.Sci. 47, 457 (1946).
2. Burns, B.D. & Webb, A.C., Proc.Roy.Soc. Lond.B. 194, 211 (1976).
3. Cohen, I., Kita, H. & Van der Kloot, W., Brain Res. 54, 318 (1973).
4. Cohen, I., Kita, H. & Van der Kloot, W., J. Physiol. 326, 327 (1974).
5. Cope, D.K. & Tuckwell, H.C., J.Theor.Biol. 80, 1 (1979).
6. Cox, D.R. & Lewis, P.A.W. The statistical analysis of series of events. Methuen, London (1966).
7. Eccles, J.C. The physiology of synapses. Academic, New York (1964).
8. Fatt, P. & Katz, B. J.Physiol. 117, 109 (1952).
9. Gihman, I.I. & Skorohod, A.V. Stochastic differential equations. Ergebnisse der Mathematik und ihrer Grenzgebiete, Vol. 72. Springer, Berlin, Heidelberg, New York (1972).
10. Hagiwara, S., Jap.J.Physiol. 4, 234 (1954).
11. Hanson, F.B. & Tuckwell, H.C., Theor.Pop.Biol. 14, 46 (1978).
12. Hanson, F.B. & Tuckwell, H.C., Theor.Pop.Biol. in press (1980).
13. Hanson, F.B. & Tuckwell, H.C., in preparation.
14. Holden, A.V. Models of the stochastic activity of neurones. Lecture Notes in Biomathematics, Vol. 12. Springer, Berlin, Heidelberg, New York (1976).
15. Khintchine, A.Y. Mathematical methods in the theory of queueing. Griffin, London (1960).
16. Lapicque, L. J.Physiol. Pathol.Gen. 9, 620 (1907).
17. Levine, J.W. & Shefner, J.M., Biophys.J. 19, 241 (1977).
18. MacGregor, R.J. & Lewis, E.R. Neural modeling. Plenum, New York (1977).
19. O'Brien, J.H., Packham, S.C. & Brunnhoelzl, W.W., J.Neurophysiol. 36, 1051.
20. Quastel, D.M.J. Synaptic transmission and neuronal interaction, p23. Raven, New York (1974).
21. Rall, W., Exp.Neurol. 1, 491 (1959).
22. Roy, B.K. & Smith, D.R., Bull.Math.Biophys. 31, 341 (1969).
23. Scharstein, H., J.Math.Biol. 8, 403 (1979).
24. Stein, R.B., Biophys.J. 5, 173 (1965).
25. Stein, R.B., Biophys,J. 7, 37 (1967).
26. Tsurui, A. & Osaki, S., Stoch.Proc.Appl. 4, 79 (1976).
27. Tuckwell, H.C., Biol.Cybernetics 18, 225 (1975).
28. Tuckwell, H.C., J.Appl.Prob. 13, 39 (1976).
29. Tuckwell, H.C., Biophys.J. 21, 289 (1978).
30. Tuckwell, H.C. J. Theor. Biol. 77, 65 (1979).
31. Tuckwell, H.C. & Cope, D.K., J.Theor.Biol. 83, 377 (1980).
32. Tuckwell, H.C. & Richter, W., J.Theor.Biol. 71, 167 (1978).
33. Tuckwell, H.C. & Wan, F.Y.M., J.Theor.Biol. in press (1980).
34. Tuckwell, H.C. & Wan, F.Y.M., in preparation.
35. Wan, F.Y.M. & Tuckwell, H.C., Biol.Cybernetics 33, 39 (1979).
36. Wan, F.Y.M. & Tuckwell, H.C., in preparation.

Part VII

Space-Time Processes and Stochastic Partial Differential Equations

Linear Stochastic Evolution Equation Models for Chemical Reactions*

Ludwig Arnold and P. Kotelenez

Fachbereich Mathematik, Universität Bremen, D-2800 Bremen, Fed. Rep. of Germany

Ruth F. Curtain

Mathematics Institute, University of Groningen,
9700 AV Groningen, The Netherlands

0. Introduction

Recently there has been interest in modelling random physical phenomena using stochastic partial differential equations [5], [2]. Various mathematical models together with existence and uniqueness results have been known for some years, and we refer the reader to the survey [8] for an account of a semigroup description and references to other mathematical formulations. Although a stochastic calculus and some stability results for stochastic p.d.e.'s have been developed [7],[12], [13], until recently, little attention had been paid to Markovian properties of the solutions. In fact, although the Markov property of solutions of stochastic p.d.e.'s has been postulated in [5], the first proof appeared in [3].

This paper is an account of Markovian and other statistical properties of mild solutions of a class of linear abstract evolution equations and is based on the work in [10] and [3], where detailed proofs can be found. To illustrate the theory we have chosen a model for a chemical reaction arising from [2], which has provided the motivation for the study of abstract evolution equations in [3] and [10].

1. Deterministic Abstract Evolution Equations

We wish to formulate the following linearized local deterministic model for a chemical reaction as an abstract evolution equation on a Hilbert space

---

* Presented by Ruth F. Curtain

$$(1.1)\quad \begin{cases} \frac{\partial y}{\partial t}(r,t) = D\Delta y(r,t) + f'(\varphi(r,t))y(r,t) \\ y(r,0) = y_0(r) \ ; \ 0 \le r \le 1; \quad t \ge 0 \end{cases}$$

$$(1.2)\quad y(0,t) = 0 = y(1.t)$$

$$(1.3)\quad \frac{\partial y}{\partial r}(0,t) - \alpha y(0,t) = 0 = \frac{\partial y}{\partial r}(1,t) + \beta y(1,t) = 0$$

where D, α, β are given constant, Δ is the Laplacian operator, $\varphi(r,t)$ is a given smooth function, $y_0 \in L_2(0,1)$ and $f'(x) = \lambda'(x) - \mu'(x)$; μ and λ being known polynomials, and ' denotes differention. We shall need the concepts of semigroups and mild evolution operators.

Definition 1.1 Strongly continuous semigroup

A strongly continuous semigroup T_t on a Banach space Z is a map $T_t : R^+ \to \mathcal{L}(Z)$, which satisfies

$$(1.4)\quad T_{t+s} = T_t T_s \quad \text{for all } t, s \in R^+$$

$$(1.5)\quad T_0 = I$$

$$(1.6)\quad ||T_t h-h|| \to 0 \quad \text{as} \quad t \to 0+ \quad \forall\, h \in Z .$$

Useful consequences of the defintion are the following

$$(1.7)\quad ||T_t|| \le Me^{wt}$$

for some constant m and w, $M \ge 0$.

T_t uniquely defines a closed, linear, densely defined operator A on Z by

$$(1.8)\quad Ah = \lim_{\delta \to 0+} \frac{T_\delta h-h}{\delta}$$

for all h such that $T_t h$ is differentiable. A is called the infinitesimal generator of T_t and (1.8) implies that

$$(1.9)\quad \frac{d}{dt}(T_t h) = AT_t h = T_t Ah \quad \forall h \in D(A).$$

Conversely, certain classes of closed linear operators generate semigroups.

To illustrate these ideas let us consider $A = \Delta$, the Laplacian operator acting on the Hilbert space, $L_2(0,1)$, and suppose its domain is

$$D_1(\Delta) = \{h \in L_2(0,1): \Delta h \in L_2(0,1) \text{ and } h \text{ satisfies } (1.2)\} .$$

Then Δ generates the strongly continuous semigroup T_t on $L_2(0,1)$ given by

$$(1.10)\quad (T_t h)(x) = \sum_{n=1}^{\infty} e^{-n^2\pi^2 t} \sin n\pi x \int_0^1 h(x) \sin n\pi x dx$$

for $h \in L_2(0,1)$.

Property (1.9) implies that $z(t) = T_t z_0$ is the unique solution of the abstract differential equation

$$(1.11)\quad \dot{z} = Az \ ; \ z(0) = z_0 \in D(A) \subset Z.$$

In particular, for $A = \Delta$, (1.10) is the unique solution of the equation

$$(1.12)\quad \begin{cases} \frac{\partial y}{\partial t} = \Delta y \ ; \ y(0,t) = 0 = y(1,t) \\ y(r,0) = h(r) \end{cases}$$

provided $h \in D(\Delta)$.

In fact, the way one arrives at the explicit formula (1.10) for T_t is to find an explicit solution for (1.12) by separation of variables. Similar remarks apply for Δ defined on $L_2(0,1)$ but with the domain

$$D_2(\Delta) = \{h \in L_2(0,1);\ \ \Delta h \in L_2(0,1) \text{ and } h \text{ satisfies } (1.3)\}.$$

In general many linear partial differential equations can be expressed as an abstract evolution equation of type (1.11) for a suitable choice of A and Z and their unique solution is then given in terms of the strongly continuous semigroup T_t generated by A. This can be a very useful way of studying problems associated with partial differential equations (see [14]) and will be the approach followed here. Here we need to consider a more general class of abstract evolution equation, namely

$$(1.13) \qquad \dot{z} = (A + E(t))z\ ;\quad z(t_0) = z_0$$

where $E(t) \in B_\infty((0,T);\ \mathcal{L}(Z))$, the class of strongly measurable $\mathcal{L}(Z)$ valued functions with $\operatorname{ess\,sup}_{0\le t\le T} ||E(t)|| < \infty$. The solution of (1.13) can be expressed in terms of a mild evolution operator (see [6]).

Definition 1.2

$U(t,s)$ for $(t,s) \in \Delta(T) = \{(t,s);\ 0 \le s \le t \le T\}$ is a mild evolution operator on Z if $U(t,s) \in \mathcal{L}(Z)$ satisfies

$$(1.14) \qquad U(t,r)\,U(r,s) = U(t,s)\ \ .\ \ 0 \le s \le r \le t \le T$$

(1.15) $\quad U(t,\cdot)$ is strongly continuous on $[0,t)$ and $U(\cdot, s)$ is strongly continuous on $(s,T]$.

A special class of mild evoltuion operators is $U(t,s) = T_{t-s}$ and those generated by operators like $A + E(t)$, where A is the infinitesimal generator of a semigroup. In general, $U(t,s)$ is not differentiable in t, but if $E(t)$ is C^1 in t in the uniform topology, this is true and we call $U(t,s)$ is a strong evolution operator, with the extra property

$$(1.16) \qquad \begin{cases} U(t,s) : Z \to D(A) \text{ and} \\ \dfrac{\partial}{\partial t}(U(t,s)h) = (A+E(t))U(t,s)h \quad \text{for } h \in D(A). \end{cases}$$

In this case, $z(t) = U(t,t_0)\,z_0$ is differentiable in t for $z_0 \in D(A)$ and is a (strong) solution of (1.13). Otherwise $z(t) = U(t,t_0)z_0$ is a mild solution of (1.13) by which we mean that

$$(1.17) \qquad z(t) = z_0 + A\int_{t_0}^{t} z(s)ds + \int_{t_0}^{t} E(s)z(s)ds.$$

To illustrate this concept we consider (1.1) and let T_t be the strongly continuous semigroup generated by $D\Delta$ on $L_2(0,1)$ under either (1.2) or (1.3).
Define $E(t) : L_2(0,1) \to L_2(0,1)$ by

$$(1.18) \qquad (E(t)h)(r) = f'(\varphi(r,t))h(r).$$

Now from [1], we know that $0 \le \varphi(r,t) \le \rho$ on $[0,1] \times [0,\infty)$ and $\varphi \in C^1([0,\infty), L_2(0,1)) \cap C([0,\infty), H^2(0,1))$.
Thus $E(t) \in \mathcal{L}(L_2(0,1))$ for each t and $E(t)$ is C^1 in t in the uniform topology in $\mathcal{L}(L_2(0,1))$. So $D\Delta + E(t)$ generates a strong evolution operator $U(t,s)$ on $L_2(0,1)$ (1.1) can be reformulated as the following abstract evolution equation on $L_2(0,1)$

$$(1.19) \qquad \begin{cases} \dot{z} = (D\Delta + E(t))z \\ z(0) = y_0 \end{cases}$$

and if $y_0 \in D(\Delta)$, (1.19) has the solution $z(t) = U(t,0)y_0$. If $y_0 \notin D(\Delta)$, then this is still the mild solution of (1.19). It is also of interest to consider inhomogeneous abstract evolution equations

$$(1.20) \quad \begin{cases} \dot{z} = (A + E(t))z + f(t) \\ z(t_0) = z_0 . \end{cases}$$

If $A + E(t)$ generates a strong evolution operator, $z_0 \in D(A)$ and $f(t) \in C^1(0,T;Z)$, then (1.20) has the strong solution

$$(1.21) \quad z(t) = U(t,t_0)z_0 + \int_{t_0}^{t} U(t,s)f(s)ds.$$

However if any of these conditions fail, $z(t)$ may not be differentiable and (1.21) is interpreted as the mild solution of (1.20). In fact for abstract evolution equations representing partial differential equations the mild solution is just the weak solution defined in terms of distributions (see [5] chapter 2).

The long term behaviour on stability of (1.21) is determined by the behaviour of $||U(t,s)||$ as $t \to \infty$. In the semigroup case, we have the bound (1.7) which tells us that of $w < 0$ we have exponential asymptotic stability of the zero solution of (1.11) and instability if $w > 0$. ($w = 0$ is uncertain as always). For some classes of semigroup, the spectrum of A gives a testable stability criterion via

$$(1.22) \quad w(A) = \sup \{\text{Re } \lambda : \lambda \in \sigma(A)\}.$$

If $w(A) = w$, then $w(A) < 0$ ensures asymptotic stability, $w(A) = w$ for parabolic systems, which includes the case $A = D\Delta$. For the time dependent case, we quote a perturbation result from [6].

Lemma 1.1

If A generates T_t and $||T_t|| \leq Me^{-\lambda t}$, then if $E(\cdot) \in B_\infty(0,\infty; \mathcal{L}(Z))$ and $A + E(t)$ generates the mild evoltuion operator $U(t,s)$, we have

$$(1.23) \quad ||U(t,s)|| \leq Me^{(MC-\lambda)t} \quad ; \quad C = \underset{0 \leq t \leq \infty}{\text{ess sup}} ||E(t)||.$$

For our example, $D\Delta$ generates a contraction semigroup, so that $M = 1$. Under (1.2), $\lambda = \pi^2/D$ and under (1.3), $\alpha = \beta = 0$; $\lambda = 0$. (1.23) Is a fairly crude estimate and for better results you need te make a more careful analysis, especially in the time dependent case. Of special interest are the steady state solutions of the local deterministic model, that is, solutions of

$$(1.24) \quad \frac{d^2\varphi}{dx^2} = (\lambda - \mu)\varphi$$

which are independent of t and x. Then if $f'(\varphi)$ is negative, all eigenvalues of $D\Delta + f'(\varphi)$ are < 0 and we have stability, but if $f'(\varphi) < 0$, we get instability under (1.3) and stability under (1.2) only if $|f'(\varphi)| < \pi^2/D$.

2. Abstract Stochastic Integrals

We survey here briefly the fundamentals of stochastic integration in Hilbert spaces as developed in ([6] chapter 5). Let $(\Omega,\mathcal{P},p)$ be a complete probability space and let H,K be real, separable Hilbert spaces. Then a Wiener process on K has the representation

$$(2.1) \quad w(t) = \sum_{i=0}^{\infty} \beta_i(t)e_i$$

where $\{e_i\}$ is a complete orthonormal basis for K and $\beta_i(t)$ are mutually independent scalar Wiener processes such that

$$(2.2) \quad E\{(\beta_i(t) - \beta_i(s))^2\} = \lambda_i|t - s| \quad ; \quad \sum_{i=0}^{\infty} \lambda_i < \infty .$$

It is interesting to note that if we let $K = R^n$ and identify $e_i = (0, , 1,...0)'$, then (2.1) agrees with the definition. The interesting point to notice in the infinte dimensional definition is that the variance of $\beta_i(t) \to 0$ as $i \to \infty$; in other words, we cannot have an incremental variance equal to the identity. The covariance operator turns out to be trace class. We list here the most important properties of $w(t)$

(2.3) $\quad w(t) \in L_2(\Omega, p; K)$ and

$$E\{||w(t) - w(s)||^2\} = (t-s)\sum_{i=0}^{\infty}\lambda_i \ (=(t-s) \text{ trace } W)$$

(2.4) $\quad \mathrm{Cov}\{w(t) - w(s)\} = W(t-s)$

where $W \in \mathcal{L}(H)$ is positive, self adjoint, and trace class and defined by

(2.5) $\quad We_i = \lambda_i e_i \quad i = 0, .., \infty$.

We define F_t to be sigma field generated by $w(s)$; $0 \leq s \leq t$. For $\Phi \in B_2((0,T); \mathcal{L}(K,H))$, the space of strongly measurable $\mathcal{L}(K, H)$ - valued functions, with $\int_0^T ||\Phi(t)||^2 dt < \infty$, we can define the following stochastic integral for $[t_0,t_1] \subset [0,T]$.

(2.6) $$\int_{t_0}^{t_1} \Phi(s)dw(s) = \sum_{i=0}^{\infty}\int_{t_0}^{t_1} \Phi(s)e_i \, d\beta_i(s).$$

It has the following properties

(2.7) $$E\{\int_{t_0}^{t_1} \Phi(s)\, dw(s)\} = 0$$

(2.8) $$E\{||\int_{t_0}^{t_1} \Phi(s)dw(s)||^2\} \leq \text{trace } W \, E\int_{t_0}^{t_1}\{||\Phi(s)||^2\}ds.$$

Finally we quote the infinite dimensional version of Itô's lemma [7].

Lemma 2.1 Itô's lemma

Suppose that $\Phi \in B_2((0,T), \mathcal{L}(K,H))$, f is adapted to F_t and $\in L_1((0,T)\times\Omega)$, $z_0 \in L_2(\Omega, p; H)$, then

$$z(t) = z_0 + \int_0^t f(s)ds + \int_0^t \Phi(s)dw(s)$$

is an H-valued stochastic process and we write its stochastic differential

(2.9) $\quad dz(t) = f(t)dt + \Phi(t)dw(t)$.

Then if $g: [0,T] \times H \to R$ is twice Fréchet differentiable on H for each $t \in T$ and once differentiable in t and these derivatives are continuous on $[0,T] \times H$, then $y(t) = g(t, z(t))$ has the following stochastic differential

(2.10) $$dy(t) = (g_t(t,z(t)) + \langle \delta g(t,z(t)), f(t)\rangle + \tfrac{1}{2}\text{ trace }\{\Phi^*(t)\delta^2 g(t,z(t))\Phi(t)W\})dt + \langle\delta g(t,z(t)), \Phi(t)dw(t)\rangle.$$

3. Stochastic Evoluation Equations. The Markov Property

The class of stochastic evolution equations we consider is the stochastic analogue of (1.13), namely

(3.1) $$\begin{cases} dz(t) = (A + E(t)z(t))dt + \Phi(t)dw(t) \\ z(t_0) = z_0 \end{cases}$$

where A and $E(t)$ satisfy the assumptions in §1, $\Phi(t) \in B_2((0,T); \mathcal{L}(K,H))$,

$z_0 \in L_2(\Omega, p; H)$ and is measurable relative to $\mathcal{F}_0$ and $w(t)$ is a Wiener process on K. (K and H are assumed to be real, separable Hilbert spaces). It is useful here to remark that (3.1) is reminiscent of a finite dimensional Itô stochastic differential equation which has the unique solution

$$(3.2) \qquad z(t) = U(t,t_0)z_0 + \int_{t_0}^{t} U(t,s)\,\Phi(s)dw(s)$$

where $U(t,s)$ is the fundamental matrix solution of $\dot{x} = Ax + E(t)x$. This analogy is very useful and is one of the reasons for the usefulness of the semigroup approach. However, in the infinite dimensional case, the results are not as nice and all we can say is that (3.2) is always a well-defined stochastic process on H, is in $C((0,T); L_2(\Omega, p; H))$ and satisfies (3.1) in the following weak sense.

$$(3.3) \qquad z(t) = z_0 + A\int_{t_0}^{t} z(s)ds + \int_{t_0}^{t} E(s)z(s)ds + \int_{t_0}^{t}\Phi(s)dw(s).$$

The problem lies in the fact that $z(t)$ defined by (3.2) is not in $D(A)$ in general. Sufficient conditions for $z(t)$ to be a strong solution are given in [3], but these are even more restrictive than in the deterministic case and are rarely satisfies in applications. Here we find it sufficient to consider mild solutions of (3.1), which do have the Markov property and if z_0 is Gaussian so is $z(t)$ given by (3.2). In [10] it is established that if $E(t)$ and $\Phi(t)$ are strongly continuous in t, the mild solution defined by (3.2) has the generator $\mathcal{A}(t)$ defined by

$$(3.4) \qquad \mathcal{A}(t)g(t,x) = \frac{\partial g}{\partial t}(t,x) + \langle (A^* + E^*(t)\delta g(t,x), x\rangle + \tfrac{1}{2}\operatorname{trace}_k\{\Phi^*(t)\delta^2 g(t,x)\Phi(t)W\}$$

where $g \in B((0,T) \times H)$ is such that g_t, δg. $\delta^2 g$ and $A^*\delta g$ exist, are continuous in x and t and uniformly bounded in norm in $[0,T] \times H$.

In an attempt to make this expression more transparent, we choose the class of functionals $g(t,x)$ given by

$$(3.5) \qquad g(t, x) = \gamma(t, \langle x,h\rangle)$$

where $\gamma(t,u)$ is C^1 in t, C^2 in u and has uniformly bounded derivatives; $h \in D(A^*)$ is fixed. Then we have

$$(3.6) \qquad \begin{aligned} \delta g(t,x) &= \gamma'(t, \langle x,h\rangle)\ \langle h,\cdot\rangle \\ A^*\delta g(t,x) &= \gamma''(t, \langle x,h\rangle)\ \langle A^*h,\cdot\rangle \\ \delta^2 g(t,x) &= \gamma''(t, \langle x,h\rangle)\ h \circ h \end{aligned}$$

and (3.4) becomes

$$(3.7) \qquad \begin{aligned} \mathcal{A}(t)g(t,x) &= \frac{\partial\gamma}{\partial t}(t,\langle x,h\rangle) + \gamma'(t,\langle x,h\rangle)\langle (A^* + E^*(t))h, x\rangle \\ &\quad + \tfrac{1}{2}\gamma''(t, \langle x,h\rangle)\sum_{i=0}^{\infty}\langle \Phi(t)W^{\frac{1}{2}}e_i, h\rangle_H^2 . \end{aligned}$$

These results are quite general and provide a means for modelling linear stochastic phenomena with a distributed nature, for example, stochastic turbulence as postulated in [5] and chemical reactions as we shall discuss in §4. The essential restrictions are that the deterministic driving term be linear and of hyperbolic or parabolic type and that the stochastic disturbance enter linearly and additively.

4. A Stochastic Evolution Equation Model for Chemical Reactions

In [2] the following central limit theorem was obtained

$$(4.1) \qquad \lim L(x(t) - \varphi(t)) = y(t) \quad \text{weakly}$$

where $\varphi(t)$ is the solution of the local deterministic model and $x(t)$ is the Markov process obtained for the local stochastic model, and $y(t)$ is an $L_2(0,1)$ valued Gaussian Markov process with generator given formally by

(4.2) $$A(t)\, g(t,x) = \frac{\partial g}{\partial t}(t,x) + \langle D\Delta + f'(\varphi(r,t))\, \delta g(t,x),\, x \rangle_{L_2(0,1)}$$

$$+ \tfrac{1}{2}\, \mathrm{trace}\, \{P(t)\, \delta^2 g(t,x)\}$$

where $P(t)$ is a differential operator given by

(4.3) $$P(t) = -D\frac{\partial}{\partial r}(\varphi(t,r)\frac{\partial}{\partial r}) + (\lambda + \mu)(\varphi(t,r))\ ;\ D(P) = H_0^2(0,1)$$

and $f = \lambda - \mu$ as in the deterministic model (1.1).

The qualification formal has been made because so far (4.2) was only been established for the class of functionals satisfying (3.5)and the trace term depends on the spaces chosen. However in light of the theory described in § 3, this suggests that $y(t)$ could be represented as the mild solution of a stochastic evolution equation which has a weak generator of the form (4.2).
For the formulation of a stochastic evolution equation a careful choice of the state space H and the space K on which the Wiener process is defined is necessary. An initial try with $H = L_2(0,1)$ proved to be unsuccessful as it was incompatible with the physical requirement that the noise be "white" on $L_2(0,1)$ (this translates into a variance operator equal to the identity which is not allowed in this formulation - see 2). Intuitively our choice of the state space should be the Sobolev-space $H^{-1}(0,1)$, but for technical simplicity we have used the scaled space H_{-1}.
We define the linear operator B by

(1.4) $$Be_i = ie_i \ ;\ i = 1, 2, \ldots, \infty.$$

where $e_i = \sin \pi i x$

and we define the scaled spaces, H_n, to be the inner product spaces

(4.5) $$H_n = \{h \in L_2(0,1) : ||h||_n = ||B^n h||_{L_2(0,1)} < \infty\}$$

$$n = 0, \pm 1, \pm 2, \ldots .$$

In §1 we stated that $D\Delta$ under (1.2) or (1.3) generates a strongly continuous semigroup on the space $L_2(0,1)$. Now because $L_2(0,1) \subset H_{-1}$ with continuous and dense inclusion, and T_t is linear, we can easily show that T_t extends to a linear and bounded operator on H-1 with the semigroup property and moreover T_t is strongly continuous on H_{-1}. So T_t also defines a strongly continuous semigroup in H_{-1}, but what is changed is that its infinitesimal generator on H_{-1} will have a larger domain than Δ. This will not concern us, as we shall only consider mild solutions which are fully specified in terms of T_t.

We now define our Wiener process on H_1 by

(4.6) $$w(t) = \sum_{i=1}^{\infty} \beta_i(t)\, e_i^1$$

where $e_i^1 = \frac{1}{i} e_i$ forms an orthonomal basis on H_1. and $\beta_i(t)$ are mutually independent Wiener processes with incremental variance λ_i and $\Sigma\lambda_i < \infty$.

Considered as a closed linear operator on $L_2(0,1)$, P is self adjoint and positive (the latter follows since $\varphi(r,t) \geq 0$). Hence P has a unique square root, $P^{\frac{1}{2}}$, which is also self adjoint and negative and P can be factorized

(4.7) $$P(t) = G^2(t).$$

$G(t)$ is self adjoint and nonnegative as a closed linear operator on $L_2(0,1)$, but can also be regarded as being in $\mathcal{L}(H_1, L_2(0,1))$. We now define $\Phi(t) \in (H_1, H_{-1})$ by

(4.8) $$\Phi(t) = G(t)B.$$

Then the following is a well-defined Markov process with state space H_{-1}:

(4.9) $$z(t) = U(t,t_0)z_0 + \int_{t_0}^{t} U(t,s)\, \Phi(s)dw(s).$$

$U(t,s)$ is the evolution operator on H_{-1} with generator $D\bar{\Delta} + E(t)$, where $\bar{\Delta}$ is the infinitesimal generator of T_t on H_{-1} and $E(t)$ is (1.13) interpreted as a bounded operator on H_{-1}. According to the theory reviewed in §3, the weak generator of (4.9) is given by

$$(4.10)\qquad A(t)g(t,x) = \frac{\partial g}{\partial t}(t,x) + \langle (D\bar{\Delta}^* + E^*(t))\delta g(t,x),\ x\rangle_{H_{-1}} + \tfrac{1}{2}\mathrm{trace}_{H_1}\{\Phi^*(t)\delta^2 g(t,x)\phi(t)W\}$$

where $g \in (B(0,T)\times H_{-1})$.

In order to make a comparison with the generator in (4.2), we suppose that g has the form (3.5). The first terms in (4.10.) and (4.2) clearly agree, so let us evaluate the trace term in (4.10)

$$\tfrac{1}{2}\,\mathrm{trace}_{H_1}\{\Phi(t)\delta^2 g(t,x)\ \Phi(t)W\}$$

$$= \tfrac{1}{2}\,\gamma''(t, \langle x, h\rangle)\ \mathrm{trace}_{H_1}\{\Phi^*(t)h \circ h\ \Phi(t)W\} \qquad \text{from (3.6)}$$

$$= \tfrac{1}{2}\,\gamma''(t, \langle x, h\rangle)\ \mathrm{trace}_{H_1}\{W^{\frac{1}{2}}\Phi^*(t)(h \circ h)\Phi(t)W^{\frac{1}{2}}\}$$

$$= \tfrac{1}{2}\,\gamma''(t, \langle x, h\rangle)\ \mathrm{trace}_{H_1}\{\Phi(t)W\ \Phi^*(t)(h \circ h)\}$$

using properties of trace class operators from [4]

$$= \tfrac{1}{2}\,\gamma''(t, \langle x, h\rangle)\ \mathrm{trace}_{H_1}\{G(t)G(t)\ h \circ h\} \quad \text{using (4.8)}$$

$$= \tfrac{1}{2}\,\gamma''(t, \langle x, h\rangle)\ \mathrm{trace}_{H_{-1}}\{P(t)\ h \circ h\} \qquad \text{from (4.7)}$$

$$= \tfrac{1}{2}\,\mathrm{trace}_{H_{-1}}\{P(t)\delta^2 g(t, x)\}.$$

So we obtain agreement, if we evaluate the trace on H_{-1}, or equivalently if we choose the state space for our process to be H_{-1} and not $L_2(0,1)$. On $L_2(0,1)$ the trace term becomes infinite and mathematically (4.1) is only a generalized stochastic process on $L_2(0,1)$ (see [11],[4]). An intuitive interpretation is that the noise term is modelled as $G(t)d(Bw(t))$, with $G(t) \in (H_1, L_2(0,1))$ and $Bw(t) = \sum_{i=1}^{\infty} \beta(t)\ e_i$ is "white noise" on H_1, with covariance operator the identity which is what one would expect from physical reasoning. So our model (4.9) is a generalized stochastic process on $L_2(0,1)$ and a well-defined Markov process on H_{-1} whose generator agrees with that obtained in the limiting procedure in [2] at least for functions g satisfying (3.5)

5. Regularity Properties of Mild Solutions

We have already stated that $z(t)$ given by (3.2) $\in C((0,T); L_2(\Omega,p;H_{-1})$ and is a Markov process. Of interest is when $z(t)$ has continuous sample paths, for then it is a strong Markov process. In general $z(t)$ does not have continuous sample paths, but several types of sufficient conditions for sample path continuity are given in [3], [9]. In particular, $z(t)$ has continuous sample paths if $\Phi(t)$ is C^1 in t or if A generates a group, or if A generates an analytic semigroup, $\Phi(t)$ is weakly sample continuous and $W^{\frac{1}{2}}$ is trace class. The first set of conditions apply to our chemical reaction model (4.8), since the coefficients in T (4.3) are polynomials in $\varphi(r,t)$ or $\frac{\partial}{\partial r}\varphi(r,t)$ and it is known [1] that $\varphi \in C^1([0,\infty), L_2(0,1))$. So $z(t)$ is in fact a strong Markov process on H_{-1}.

If we regard $z(t)$ as a function of its initial state z_0, we can write $z(t) = g(z_0)$ where $g: Z \to Z$ and $Z = C((t_0,T); L_2(\Omega,p;H_{-1}))$. In [3] it is shown that g is a continuously Fréchet differentiable map and so $z(t)$ depends continuously differentiably upon its initial state. Regarding $z(t)$ as a function of the initial time t_0, we have sample continuity in t_0.

6. Moment Equations for Mild Solutions [3]).

Using the properties of the stochastic integral in §2, we can calculate the first and second moments of $z(t)$ of (3.2)

$$(6.1)\qquad E\{z(t)\} = U(t,t_0)\ E\{z_0\}$$

$$(6.2) \quad P(t) = Var(z(t)) = U(t,t_0)P_0 \, U^*(t,t_0) + \int_{t_0}^{t} U(t,s)\Phi(s)W\Phi^*(s)U^*(t,s)ds$$

where $P_0 = Var\{z_0\}$.

$$(6.3) \quad Cov(z(t), z(s)) = \begin{cases} U(t,s) \, Var\{z(s)\} & \text{for } t > s \\ Var\{z(t)\}U(s,t) & \text{for } t < s. \end{cases}$$

If $U(t,s)$ is a strong evolution operator, we can obtain a differentiated version of (6.2)

$$(6.4) \begin{cases} \frac{d}{dt} \langle P(t)h,k\rangle = \langle P(t)h, (A^* + E^*(t))k\rangle + \langle P(t)k, (A^* + E^*(t))h\rangle + \langle \Phi(t)W\phi^*(t)h,k\rangle \\ P(0) = P_0 \qquad \text{for } h, k \in D(A^*) \end{cases}$$

where the inner products are in H_{-1}.

In our chemical reaction model (4.9), $U(t,s)$ is a strong evolution operator since $E(t)$ is C^1 on H_{-1} and so for the second moment equation we obtain

$$(6.5) \quad \frac{d}{dt} \langle P(t)e_i^{-1}, e_j^{-1}\rangle_{H^{-1}} = \langle P(t)e_i^{-1}, (D\Delta + f'(\varphi))e_j^{-1}\rangle_{H^{-1}} + \langle P(t)e_j^{-1}, (D\Delta + f'(\varphi)e_i^{-1}\rangle_{H^{-1}})$$
$$+ \langle \Phi(t)W\Phi^*(t)R_i^{-1}, e_j^{-1}\rangle_{H^{-1}}$$

$$P(0) = P_0 .$$

Writing

$$(6.6) \quad P(t) = \sum_{i,j=1}^{\infty} p_{ij}(t)e_i^{-1} \langle \cdot, e_j^{-1}\rangle_{H^{-1}} \; ; \quad P_{ij} = P_{ji}$$

$$= \sum_{i,j=j}^{\infty} p_{ij}(t)e_i \langle \cdot, e_i\rangle_{L_2(0,1)}$$

we see that (6.5) yields equation for the Fourier coefficients $p_{ij}(t)$ of the variance function.

7. Itô's Lemma for Mild Solutions. The Hopf Equation ([10]).

The Itô's lemma stated in §2 can be directly applied to strong solutions of (3.1), that is, provided $z(t) \in D(A)$ w.p.1. In general this is not true, and for mild solutions a more restricted form of Itô's lemma was developed in [10], which requires that in addition to the assumptions in lemma 2.1, g must have uniformly bounded derivatives and $\delta g(t,u) \in D(A^*)$ and $A^*\delta g(t,u)$ is continuous and uniformly bounded. Functionals of the class (3.5) satisfy these conditions, for example. This modified Itô's lemma can then be applied to obtain the Hopf equation for the characteristic functional of the mild solution.

$$(7.1) \quad Q_t(h) = E\{\exp i \langle h,z(t)\rangle\}$$

$$(7.2) \quad dQ_t(h) = \langle (A^* + E^*(t))h, \partial_h Q_t(h)\rangle - \tfrac{1}{2}Q_t(h) \sum_{k=0}^{\infty} \langle h,\Phi(t)W^{\frac{1}{2}}e_k\rangle^2$$

$$Q_0(h) = E\{\exp\{i \langle h,\eta\rangle\}\} \qquad \forall h \in D(A^*).$$

For our chemical reaction model we can define

$$(7.3) \quad q_j(t) = Q_t(e_j^{-1}) \quad \text{and obtain}$$

$$(7.4) \quad dq_j(t) = \langle (D\Delta + f'(\varphi))e_j^{-1}, \partial_h Q_t(e_j^{-1})\rangle_{H_{-1}}$$
$$- \tfrac{1}{2} q_j(t) \sum_{k=1}^{\infty} \frac{1}{k^4} \langle G(t)e_k^{-1}, e_j^{-1}\rangle^2_{H_{-1}} .$$

8. Kolmogorov's Equation

We consider now the mild soltuion at time T as a function of its initial conditions.

$$(8.1) \qquad z(T) = U(T,t)x + \int_t^T U(T,s)\ \Phi(s)dw(s)$$

As stated in §5, z(T) depends continuously on (t,x) and for functionals $\Psi : H \to R$, which are twice continuously Fréchet differentiable and uniformly bounded in norm

$$(8.2) \qquad F(t,x) = E\ \{\psi(z\ (T)) \mid z(t) = x\}$$

enjoys the same properties. If in addition, $\Phi(t)$ and E(t) are strongly continuous in t and either

$$(8.3) \qquad A\,U(t,s) \text{ in bounded for all } t > s$$

or

(8.4) $\quad A * \delta\Psi(x)$ is well-defined and continuous and bounded in x, then F(t,x) is a solution of Kolmogorov's equation.

$$(8.5) \qquad \frac{\partial F}{\partial t} + \mathcal{A} F = 0 \quad ; \quad F(T,x) = \Psi(x).$$

In our chemical reaction model (8.3) holds and so Kolmogorov's equation has F(t,x) as a solution.

9. Stability and Long Term Behaviour of Mild Solutions

Following [13], we say that the system (3.2) is mean square stable if

$$(9.1) \qquad \lim_{T\to\infty} \frac{1}{T} \int_0^T E\{\|\, z(t)\|^2\}dt < \infty .$$

From the moment equations in §6, one obtains for (3.2)

$$(9.2) \qquad E\{\|z(t)\|^2\} \leq \|U(t,t_0)\|^2 \text{ trace } P_0 + \text{trace } W \int_{t_0}^t \|U(t,\alpha)\|^2 \|\Phi(\alpha)\|^2\, d\alpha$$

and so if U(t,s) is exponentially stable (cf (1.20)), (3.2) is mean square stable. Thus stochastic mean square stability depends on the stability of the deterministic equation.
If $\Phi(t) = \Phi$ is time invariant, then the limit in (9.1) equals trace $\{\Phi w \Phi^*\}$, which defines an invariant measure associated with the Markov process (3.2).

10. References

1. L. ARNOLD, On the consistency of mathematical models of chemical reactions. Forschungsschwerpunkt Dynamische Systeme, Universität Bremen.

2. L. ARNOLD and P. KOTELENEZ, Central limit theorem for the stochastic model of chemical reactions with diffusion. (in preparation)

3. L. ARNOLD, R.F. CURTAIN and P. KOTELENEZ, Nonlinear stochastic evolution equations in Hilbert space. Forschungsschwerpunkt Dynamische Systeme, Universität Bremen, Report Nr. 17 (1980).

4. A. BENSOUSSAN, Filtrage optimal des systemes lineaires. Dunod 1971.

5. P. CHOW, Stochastic partial differential equations in turbulence related problems. Probabilistic analysis and related topics. Vol. 1. Ed. Barucha-Reid pp.1 - 44.

6. R.F. CURTAIN and A.J. PRITCHARD, Infinite Dimensional Linear Systems Theory, Lecture Notes in Control and Information Sciences, Vol. 8 (Springer, Berlin, Heidelberg, New York 1978).

7. R.F. CURTAIN, Itô's lemma in infinite dimensions. J. Math. Anal. and Appl. (1970) 31 , pp. 434 - 448.

8. R.F. CURTAIN, "Linear Stochastic Itô Equations in Hilbert Space", in: Stochastic Control Theory and Stochastic Differential Systems, Lecture Notes in Control and Information Sciences, Vol. 16 (Springer, Berlin, Heidelberg, New York 1980) pp. 61-85

9. P. KOTELENEZ and R.F. CURTAIN, Local behaviour of Hilbert space valued stochastic integrals and the continuity of mild solutions of stochastic evolution equations. Report Forschungsschwerpunkt Dynamische Systeme, 21 Universität Bremen. 1980.

10. R.F. CURTAIN, Markov processes generated by linear stochastic evolution equations, 1980. T.W. Report Nr. 219, University of Groningen.

11. J.M. GELFAND and N. Ya. VILENKIN, Generalized functions, Vol. 4, (Translation) Academic Press 1964.

12. U.G. HAUSSMANN, Asymptotic Stability of the linear Itô equation in infinite dimensions. J. Math. Anal. of Appl. 65 (1978) pp. 219-235.

13. A. ICHIKAWA, Dynamic programming approach to stochastic evolution equations. SIAM J. Control and Opt. 17 (1979) pp. 152 - 174

14. S.G. KREIN, Linear differential equations in Banach space, Amer. Math. Soc.1971.

Stochastic Measure Processes

Donald A. Dawson

Department of Mathematics and Statistics, Carleton University,
Ottawa, Canada, K1S5B6

1. The Role of Stochastic Measure Processes

Stochastic measure processes arise as mathematical models of the evolution of spatially distributed populations under conditions in which fluctuations are of importance. Models of this type have arisen in several fields including nonequilibrium statistical physics [30], chemical kinetics [30], population genetics [16], ecology [31], epidemiology and human geography [1].

A detailed description of a spatially distributed system can be described as follows:

(i) there is a total poluation of N(t) individuals at time t,

(ii) each individual in the population belongs to one of a finite number, s, of distinct species,

(iii) at time t, the ith individual is located at a point, $\xi_i(t)$, in the environmental space, R^d, and

(iv) the individuals undergo spatial motion, transformation, reproduction and interaction; these processes are frequently random in nature so that only a statistical description is possible.

Mathematical models at this "microscopic" level of description are usually too unwieldy to be of use and consequently the most frequently used mathematical models involve various mathematical idealizations and simplifications which ignore certain aspects of the actual population. Some standard classes of mathematical population model are the following:

(a) finite systems of ordinary differential equations (continuous, deterministic, aggregated model),

(b) stochastic birth and death models (discrete, stochastic, aggregated model),

(c) finite systems of stochastic differential equations (continuous, stochastic aggregated model),

(d) finite systems of partial differential equations (continuous, deterministic, distributed model),

(e) interacting particle systems (discrete, stochastic, distributed),

(f) finite systems of stochastic partial differential equations (continuous, stochastic, distributed).

The last two classes of models, namely, distributed stochastic models are subsumed under the study of stochastic measure processes. The stochastic measure processes to be described in detail in this paper incorporate the following three principal phenomena:

(i) the spatial motion and dispersion of the population,

(ii) the inherent fluctuations in the population due to environmental and demographic stochasticity, and

(iii) nonlinear interaction effects such as competition, predation and contagion.

In many situations deterministic models are perfectly adequate and stochastic effects can be safely neglected. However in other situations the inclusion of stochastic effects are essential to the understanding of the qualitative behavior of a system. The same can be said for the question of spatial effects. One of the principal objectives of the study of stochastic measure processes is the characterization of those situations in which both stochastic and spatial effects play an essential role. Heuristically we can recognize the following general paradigms:

(a) stochastic effects are limited to microscopic linearized fluctuations about a deterministic behavior,

(b) in population models at low population levels stochastic effects must be introduced to model the phenomena of extinction and small scale spatial heterogeneity,

(c) in nonlinear systems the parameters which measure noise levels, that is, "temperature-like" parameters, are essential in the description of the macroscopic (thermodynamic) behavior of a system,

(d) in certain special situations fluctuations can grow to macroscopic scale; these situations occur at conditions "near criticality" and have been extensively studied in the context of phase transitions in statistical physics [37].

One of the major problems in the study of stochastic population models is to make these notions precise. The basic methods of attack on this problem include asymptotic analysis [32], limit theorems and self-similar random fields [33], [35], and the diffusion approximation [15], [24]. In the context of stochastic measure processes the diffusion approximation leads to stochastic partial differential equations and stochastic measure diffusions. Diffusion processes are used in the study of stochastic population models for many of the same reasons that ordinary differential equations are used to model deterministic (discrete) population systems. Moreover, in the light of the fundamental work of W. FELLER [15], the nature of the singularities in the diffusion limit provides insight into the basic qualitative behavior of the underlying stochastic population model.

A stochastic measure diffusion process is obtained as a measure-valued solution, $\{X(t,.):t \geq 0\}$, of a stochastic evolution equation of the form:

$$\partial X(t,.)/\partial t = A^*(X)X(t,.) + (Q(X(t)))^{1/2}W'; \tag{1.1}$$

$X(.,.)$ is a random function from $[0,\infty)\times R^d$ into R^s. For each t, $X(t,.)$

denotes the spatial distribution of the s species present over the environmental space, R^d, on which the population is assumed to live; $A^*(.)$ is a non-anticipating operator-valued function which describes the spatial motion and deterministic interaction effects of the population. The expression $(Q(X(t)))^{1/2}W'$ describes the population dependent stochastic fluctuations. In this expression, W' can be viewed as a space-time white noise. As it stands, (1.1) is no more than a suggestive formal equation. In order to lead up to a mathematically precise reformulation of (1.1) the next section is devoted to an outline of the theory of measure-valued Markov processes.

2. Basic Notions of Stochastic Measure Processes

2.1 Environmental Space D

D is assumed to be a compact subset of R^d and is the space on which the population is assumed to live. (The extension to the case $D = R^d$ requires some technical modifications and is omitted for expository simplicity.)

2.2 State Space M(D)

The set of possible states of a measure process is the set M(D) of all (nonnegative) measures on D. In order to discuss the continuity of such a process we introduce the weak topology on M(D). The sequence of measures μ_n is said to <u>converge weakly</u> to the measure μ, $\mu_n \overset{W}{\to} \mu$, if for every $f \in C(D)$, the space of continuous real-valued functions on D,

$$\int f(x)\mu_n(dx) \to \int f(x)\mu(dx), \text{ as } n \to \infty. \tag{2.1}$$

2.3 Set of Histories of a Measure Process: Canonical Construction

Let Ω denote the set of continuous histories, that is, Ω is the set of continuous functions from $[0,\infty)$ into M(D). [Similarly, Ω^D denotes the set of right continuous functions from $[0,\infty)$ into M(D).] The <u>canonical process</u> is defined by: for $\omega \in \Omega$,

$$X(t,\omega) \equiv \omega(t) \in M(D). \tag{2.2}$$

2.4 Alternate Description of a Measure Process

A stochastic measure process, $\{X(t):t \geq 0\}$ is characterized as follows:

(i) for every $t \geq 0$, X(t) is a random measure, that is, for every Borel subset, A, of D, X(t,A) is a random variable which denotes the (random) population "mass" in the set A at time t,

(ii) if $\{A_n:n \geq 1\}$ is a sequence of disjoint Borel subsets of D, then

$$X(t, \bigcup_n A_n) = \sum_n X(t,A_n), \text{ with probability one,} \tag{2.3}$$

(iii) for a fixed Borel subset, A, of D, X(.,A) is a real-valued stochastic process.

2.5 Atomic and Continuous Processes

There are two special classes of stochastic measure processes which have simple representations. The first is the class of atomic measure-valued processes which can be represented as

$$X(t) = \sum_{i=1}^{N(t)} a_i(t)\delta_{\xi_i(t)}, \tag{2.4}$$

where $\xi_i(t)$ denotes the location and $a_i(t)$ the mass of the ith atom at time t and N(t) denotes the total number of atoms. The second is the class of continuous processes having a density. In this case there is a (non-negative) function-valued process, $Y(t,.)$, such that

$$X(t,A) = \int_A Y(t,x)dx, \text{ for every Borel set A.} \tag{2.5}$$

It might be expected that these two classes would cover all situations of interest in applications. One of the surprises in the study of stochastic measure processes is that this is not the case. Examples of "singular" processes and their intuitive significance are discussed below.

2.6 Distribution of a Stochastic Measure Process

The distribution of a continuous [respectively right continuous] stochastic measure process is given by a family of probability measures $\{P_\mu : \mu \in M(D)\}$ on Ω [respectively Ω^D]. The probability measure, P_μ, is the probability law of the process subject to the initial condition $X(0) = \mu$.

2.7 The Markov Property

The past information is contained in the σ-algebra, $\underline{\underline{F}}_s$, which is the smallest σ-algebra which contains all events of the form $\{\omega : X(t,\omega;A) \in B\}$ where A is a Borel subset of D, B is a Borel subset of R^1_+ and $t \leq s$. A stochastic measure process is (stationary) Markov if for $s \leq t$, and B, a Borel subset of D,

$$P_\mu(X(t) \in B | \underline{\underline{F}}_s) = P_{X(s)}(X(t-s) \in B), \tag{2.6}$$

P_μ-almost surely. [Warning: in general the Markov property (2.6) does not imply that the real-valued stochastic processes $X(.,A)$, A fixed, are Markov.]

2.8 Transition Mechanism and Generator

Recall that simple Markov processes are often specified in terms of the Kolmogorov equations satisfied by their transition functions. In the general theory of Markov processes the specification is usually in terms of the infinitesimal generator of the process. To define it consider a function, f, in $C(M(D))$, the space of bounded continuous real-valued functions defined on $M(D)$ and let

$$T_t f(\mu) \equiv E_\mu(f(X(t))) \tag{2.7}$$

where E_μ denotes expectation with respect to the probability law P_μ. The process is said to be a Feller process if T_t maps $C(M(D))$ into itself. In this case, $\{T_t : t \geq 0\}$ forms a semigroup of contraction operators on $C(M(D))$. A function $f \in C(M(D))$ is said to belong to the domain, $D(\underline{\underline{L}})$, of the (weak) infinitesimal generator if for every

$\mu \in M(D)$,

$$\lim_{t \downarrow 0} t^{-1}[T_t f(\mu) - f(\mu)] = \underline{\underline{L}} f(\mu), \tag{2.8}$$

where $\underline{\underline{L}}f(.) \in C(M(D))$, and for all sufficiently small t,

$$\| t^{-1}[T_t f(\mu) - f(\mu)] \| \leq M < \infty, \tag{2.9}$$

where $\|.\|$ denotes the supremum norm. The semigroup $\{T_t : t \geq 0\}$ is uniquely determined by its infinitesimal generator, $\underline{\underline{L}}$.

2.9 Example: The Birth, Death and Migration Process

Let D be a sphere in R^d, A be an elliptic operator of the form

$$Ag(x) = \sum_{i,j}^{d} b_{ij}(x) \partial^2 g / \partial x_i \partial x_j \ , \ x \in D, \tag{2.10}$$

and absorbing boundary conditions be imposed on the boundary, ∂D. The operator, A, can be interpreted as the infinitesimal generator of a Markov process in D which describes the motion of individual members of the population. We now prescribe an operator which can be uniquely extended to be the infinitesimal generator of an atomic measure-valued process which corresponds to a particle system in D. Let $f \in C(M(D))$ be a function of the form $f(\mu) = F(\int \phi(x)\mu(dx))$ where F is a smooth function on R^1 and $\phi \in D(A)$, the domain of A. For

$$\mu = \sum_k \delta_{x_k} \ , \quad x_k = (x_{k,1}, \ldots, x_{k,d}) \in R^d,$$

$$\underline{\underline{L}}f(\mu) \equiv F'\Big(\sum_k \phi(x_k)\Big)\Big[\sum_k A\phi(x_k)\Big] + \tfrac{1}{2}F''\Big(\sum_k \phi(x_k)\Big) \sum_k \sum_{i,j}^{d} b_{ij}(x_k)\phi_i(x_k)\phi_j(x_k)$$
$$+ \sum_{\tilde{\mu}} \int c(\mu;\tilde{\mu},d\nu)\,[F(\int\phi(x)\mu(dx) - \int\phi(x)\tilde{\mu}(dx) + \int\phi(x)\nu(dx)) - F(\int\phi(x)\mu(dx))] \tag{2.11}$$

where $c(\mu:\tilde{\mu},\nu)$ denotes the infinitesimal rate at which a subset of atoms, $\tilde{\mu}$, of μ is removed and replaced by a new set of atoms, ν, given that the current mass distribution is given by μ. This corresponds to a particle system in which particles are moving according to the motion process in D given by the infinitesimal generator A and subject to death and birth of particles or groups of particles. Under appropriate regularity conditions on $c(.;.,.)$ it can be verified that there is a unique extension of $\underline{\underline{L}}$ which is the infinitesimal generator of a semigroup of contraction operators on $C(M(D))$.

2.10 The Measure-Valued Martingale Problem

For many stochastic measure processes it is difficult to directly establish the existence of a Feller semigroup. It turns out that it is more convenient to specify a stochastic measure process in terms of a "martingale problem" in the sense of STROOCK and VARADHAN [34].

A <u>martingale problem</u> on Ω (or Ω^D) is assigned by a pair $(\underline{\underline{L}}, D(\underline{\underline{L}}))$ where $\underline{\underline{L}}$ is a linear operator defined on a linear subspace, $D(\underline{\underline{L}})$, of $C(M(D))$. A <u>solution</u> to the martingale problem associated with the pair $(\underline{\underline{L}}, D(\underline{\underline{L}}))$ is a mapping $\mu \to P_\mu$ from $M(D)$ to $\Pi(\Omega)$, the set of probability laws on Ω, which satisfies the two conditions: for each $\mu \in M(D)$,

$$P_\mu(X(0) = \mu) = 1, \tag{2.12}$$

$Y_{f,t} \equiv f(X(t)) - \int_0^t \underline{\underline{L}}f(X(s))ds, \quad f \in D(\underline{\underline{L}})$, is a P_μ-martingale (2.13).

The reason for the usefulness of the martingale problem is the following basic result which shows that a martingale problem which has a unique solution specifies a measure-valued Markov process.

Theorem 2.1

Assume that the martingale problem on Ω associated with $(\underline{\underline{L}},D(\underline{\underline{L}}))$ has a unique solution, $\{P_\mu:\mu \in M(D)\}$, which satisfies the following regularity conditions:

(i) for each compact subset, K, of M(D) the set $\{P_\mu:\mu \in K\}$ is compact, and

(ii) for each $\phi \in C(M(D))$, $\lim_{M\to\infty} \sup_{\mu \in K} E_\mu(k_M(\int\phi(x)X(t,dx))) = 0$

where $k_M(x) = |x|$ if $|x| > M$,

$= 0$, otherwise.

Then $\{P_\mu:\mu \in M(D)\}$ is the probability law of a unique Feller Markov stochastic measure process whose generator, $\underline{\underline{L}}_0$, is an extension of $\underline{\underline{L}}$.

The proof of Theorem 2.1 is found in [7]; for a complete development of the theory of martingale problems, refer to [34].

2.11 Measure Diffusion Processes

In this section we reformulate the stochastic evolution equation (1.1) (under the assumption that A(.) = A) as a measure-valued martingale problem. Assuming that (1.1) has a solution and following the standard rules of stochastic calculus, we obtain the following natural candidate for the infinitesimal generator of the solution:

$$\underline{\underline{L}}f(\mu) \equiv A\ \delta f(\mu)/\delta_{\mu,x} + \tfrac{1}{2}\iint(\delta^2 f(\mu)/\delta_{\mu,x}\delta_{\mu,y})(Q(\mu:dx\times dy)) \qquad (2.14)$$

where $\delta f(\mu)/\delta_{\mu,x}$ represents a function of x which denotes the variational derivative

$$\delta f(\mu)/\delta_{\mu,x} \equiv \lim_{\varepsilon\downarrow 0}[f(\mu+\varepsilon\delta_x)-f(\mu)]/\varepsilon\ ,\ \mu \in M(D),\ x \in R^d, \qquad (2.15)$$

and δ_x denotes a unit mass at the point x. By analogy with finite dimensional diffusions, $Q(\mu:dx\times dy)$ represents the infinitesimal rate of change of population mass covariance at the points x and y when the current mass distribution is given by μ. To avoid technical problems concerning variational derivatives we restrict our attention to a family of functions whose derivatives can be explicitly computed. In particular, let $D(\underline{\underline{L}})$ denote the subspace of C(M(D)) generated by functions of the form $f(\mu) = \psi(\int\phi_1(x)\mu(dx),\ldots,\int\phi_n(x)\mu(dx))$, where $\psi \in C^2(R^n)$, the space of bounded twice differentiable functions and $\phi_1,\ldots,\phi_n \in C(D) \cap D(A)$. For $f \in D(\underline{\underline{L}})$, $\underline{\underline{L}}f$ is defined as follows:

$$\underline{\underline{L}}f(\mu) \equiv \sum_{i=1}^{n}\psi_{x_i}(\mu)(\int A\phi_i(x)\mu(dx)) + \tfrac{1}{2}\sum_{i,j=1}^{n}\psi_{x_i x_j}(\mu)[\iint\phi_i(x)\phi_j(y)Q(\mu:dx\times dy)] \qquad (2.16)$$

For each μ, the fluctuation quadratic form, $Q(\mu:dx\times dy)$ is assumed to be a symmetric signed measure on D×D which satisfies the following conditions:

(i) $\iint\phi(x)\phi(y)Q(\mu:dx\times dy) \geq 0$, for each $\phi \in C(D)$,

(ii) $\mu(A) = 0$ implies that $Q(\mu:A\times A) = 0$, and

(iii) $Q(\mu:D\times D) < \infty$.

Condition (ii) is a consequence of the fact that the diffusion coefficient of a diffusion on $[0,\infty)$ must vanish at 0 and is imposed to insure that the solution to the martingale problem is non-negative measure-valued.

The measure-valued martingale problem is specified by either the pair $(\underline{\underline{L}},D(\underline{\underline{L}}))$ or the pair (A,Q). A solution to this martingale problem is called a <u>stochastic measure diffusion</u> (process) with spatial dispersion generator, A, and fluctuation quadratic form, Q. For a detailed discussion of the foundations of stochastic measure diffusions, refer to [3],[5],[7].

3. Some Basic Measure Diffusions and Diffusion Approximations

3.1 Branching Systems

Branching systems are models of populations in which individuals reproduce by splitting into a random number of offspring with the basic assumption that individuals reproduce independently of each other. A <u>branching random field</u> is informally described as follows. At each instant, t, the process, $X(t)$, is described by a random field of particles located in the subset, D, of R^d. Each particle has a random lifetime which is exponentially distributed with mean $1/V$. At the end of its lifetime each particle branches into n identical particles with probability, p_n, $n=0,1,2,\ldots$. The mean offspring size, $m_1 \equiv \sum np_n$, is assumed to be finite. In addition, during its lifetime each particle moves in D according to a Markov process with infinitesimal generator, A. Finally, immigration into D is allowed to occur according to a space-time Poisson process with rate $r(dx)$. This process is characterized by an integer-valued measure-valued martingale problem on Ω^D. The operator, $\underline{\underline{L}}$, is a special case of (2.11) with coefficient

$$\begin{aligned} &c(\mu:0,\delta_x) = r(dx), \\ &c(\mu:\delta_x,n\delta_x) = Vp_n\mu(dx), \\ &c(\mu:\tilde{\mu},d\nu) = 0, \text{ otherwise.} \end{aligned} \tag{3.1}$$

3.1.1 Theorem [10]

The martingale problem associated with $(\underline{\underline{L}},D(\underline{\underline{L}}))$ with $\underline{\underline{L}}$ defined as in (2.11), (3.1) has a unique solution on Ω^D. The resulting stochastic measure process can be characterized in terms of its Laplace functional,

$$L_{X(t)}(\phi) \equiv E[\exp(-\int\phi(x)X(t,dx))],\ \phi \geq 0,\ \phi \in C(D). \tag{3.2}$$

Assuming that $X(0)$ is a Poisson random field on D with intensity measure $\lambda(dx)$, it is obtained as follows:

$$L_{X(t)}(\phi) = \exp[-\int H_t(1-\exp(-\phi);x)\lambda(dx) - \int_0^t \int H_{t-s}(1-\exp(-\phi);x)r(dx)ds],$$

where $H_t(1-\xi;x) \equiv 1-G_t(\xi;x)$, and $G_t(.;,)$ satisfies the Skorokhod equation

$$G_t(\xi;x) = \exp(-Vt)\int\xi(y)p(t;x,dy) + \sum_{n=0}^{\infty} p_n \int_0^t V\exp(-Vs)T_s[G_{t-s}(\xi;.)]^n ds, \tag{3.3}$$

where $\{T_s : s \geq 0\}$ and $p(s;x,dy)$ denote the semigroup of operators and the probability transition function of the spatial motion process with infinitesimal generator, A, respectively.

To study the high density (coarse scale) behavior of this process a mass, w, is assigned to each particle and the initial condition is assumed to correspond to a high density of particles. In particular, given $\varepsilon > 0$, let $w_\varepsilon = \varepsilon w$, $\lambda_\varepsilon = \lambda/\varepsilon$. $r_\varepsilon = r/\varepsilon$, $m_1^\varepsilon = 1+\nu(\varepsilon)$ and $V_\varepsilon = V(\varepsilon)$. Let $\{X_\varepsilon(t) : t \geq 0\}$ denote the resulting $M(D)$-valued Markov process.

3.1.1 Deterministic Limit and Microscopic Fluctuations

Consider the situation in which the branching occurs at a rate which is slow in comparison with the Malthusian growth of the population, that is, far from criticality. This corresponds to the choice $V_\varepsilon = V$ and $\nu(\varepsilon) = \nu$. Then $\{X_\varepsilon(.)\}$ converges (in the sense of weak convergence of stochastic processes) to a deterministic, continuous measure-valued process which is characterized as the solution of the evolution equation

$$\partial X(t,.)/\partial t = A^* X(t,.) + V\nu X(t,.) \tag{3.4}$$

where A* denotes the adjoint of the generator, A. In addition, for small ε, the fluctuation process $X_\varepsilon(t,.)-X(t,.)$ is asymptotically described by a space-time Gaussian process. Thus far from criticality the random effects are confined to Gaussian fluctuations around a deterministic behavior.

3.1.2 Theorem: Existence of the Diffusion Limit [5],[36]

Consider the situation of nearly critical branching, that is, $\nu(\varepsilon) = \nu\varepsilon$, and consider a speeded up version of the process, that is, $V_\varepsilon = V/\varepsilon$. Let m_2, m_3 denote the second and third factorial moments of the offspring distribution. Assume that

$$m_2^\varepsilon \to m_2,\ m_3^\varepsilon \to M < \infty, \text{ as } \varepsilon \to 0. \tag{3.5}$$

Then the stochastic measure processes, $\{X_\varepsilon(t,.)\}$ converge (in the sense of weak convergence of stochastic processes) to a measure diffusion process, $\{X(t,.) : t > 0\}$. The latter process, called the <u>Branching Measure Diffusion</u>, is characterized as the unique solution of the martingale problem (A,Q) with fluctuation quadratic form

$$Q(\mu : dx \times dy) = \gamma\delta_x(dy)\mu(dx) \tag{3.6}$$

where $\gamma = m_2 V$ denotes the <u>demographic temperature</u>.

3.1.3 Theorem: Transition Functional for the B.M.D. [10]

The B.M.D. process, $\{X(t) : t \geq 0\}$ is characterized by its Laplace transiotion functional

$$\begin{aligned} L_{t,\mu}(\phi) &\equiv E_\mu[\exp(-\int\phi(x)X(t,dx))],\ \mu \in M(D),\ \phi \in C(D),\ \phi \geq 0, \\ &= \exp[-\int H_t(\phi;x)\mu(dx) - \int_0^t \int H_{t-s}(\phi;x)r(dx)ds] \end{aligned} \tag{3.7}$$

where $H_t(.,.)$ is obtained as the solution of the equation

$$H_0(\phi;x) = \phi(x),$$

$$\partial H_t(\phi;x)/\partial t = AH_t(\phi;x) + V\nu H_t(\phi;x) - \tfrac{1}{2}\gamma(H_t(\phi;x))^2. \tag{3.8}$$

In this case the fluctuations remain of macroscopic size in the limit.

3.2 Sampling Systems

A sampling system is a model of a population of fixed size in which individuals belong to one of a finite number of classes. The population evolves according to a mechanism in which individuals are removed and replaced by new individuals of a randomly chosen type. The probability that the replacement individual is of a given type depends on the relative proportion of the total population represented by that type. We begin with a simple model which has been extensively studied in population genetics.

Consider a finite population of N individuals. To each individual is assigned d numerical characteristics measured in discrete units. Hence the "type" of an individual is represented by an element of the set $\{ku : k \in Z^d\}$. (Z denotes the set of integers.) Let $n_k(t)$ denote the number of individuals of type k at time t. Then

$$\sum_{k \in Z^d} n_k(t) = N, \text{ for all } t \geq 0. \tag{3.9}$$

The process $N(t) = \{n_k(t) : k \in Z^d\}$ is assumed to be a continuous time Markov chain with transition rule:

$$P(x_j \to x_k \text{ in } [t,t+\Delta t)) = g_{jk}(t)\Delta t + o(\Delta t),\ j,k \in Z^d, \tag{3.10}$$

where $x_j \to x_k$ denotes the event that an individual of type j is replaced by an individual of type k. Let $p_j \equiv n_j/N$; then $\underline{p} = \{p_k : k \in Z^d\}$ belongs to the infinite simplex $\Sigma \equiv \{\underline{p} : p_i \geq 0, \sum p_i = 1\}$. The generator of the Markov chain is given by:

$$\underline{\underline{G}}f(\underline{p}) = \sum_{i \neq j} g_{ij}(\underline{p})[f(\underline{p}^{ij}) - f(\underline{p})], \tag{3.11}$$

$$\begin{aligned} p_k^{ij} &= p_k - 1/N, \text{ if } k=i, \\ &= p_k + 1/N, \text{ if } k=j \\ &= p_k, \text{ if } k \neq i,j. \end{aligned}$$

The process, n(t) can also be represented as a measure-valued process as follows:

$$X_N(t,A) \equiv \sum_{ku \in A} n_k(t). \tag{3.12}$$

3.2.1 Random Mating and Mutation

A model of random mating originally introduced by MORAN and mutation leads to the choice:

$$g_{ij}(\underline{p}) = [\gamma p_i p_j + p_i \theta_{ij}],\ \gamma \geq 0,\ \theta_{ii} = -1,\ \theta_{ij} = 1/2d \text{ if } |i-j| = 1. \tag{3.13}$$

The term $\gamma p_i p_j$ represents random sampling and θ_{ij} represents a mutation rate per individual of a $x_i \to x_j$ mutation. We next look at the process speeded up and in coarse spatial scale. This leads to:

$$Y_N(t,A) \equiv \sum_{uj/N^{1/2} \in A} N^{-1} X_N(N^2 t, \{x_j\}). \tag{3.13}$$

The generator of the stochastic measure process $Y_N(t,.)$, $\underline{\underline{G}}_N$ has the following form. Let $\psi(\underline{p}) = f(\langle\phi,\underline{p}\rangle)$ where $f \in C^3(R^1)$, the space of functions with bounded third derivatives, $\phi \in C_K(R^d)$, the space of continuous functions with compact support in R^d and $\langle\phi,\underline{p}\rangle = \sum\phi(x_j)p_j$.

$$\underline{\underline{G}}_N\psi(\underline{p}) = \sum_{i,j} N^2[f(\langle\phi,\underline{p}\rangle-\phi(x_i)/N+\phi(x_j)/N)-f(\langle\phi,\underline{p}\rangle)][\gamma p_i p_j+p_i\theta_{ij}]$$
$$= N^2 f'(\langle\phi,\underline{p}\rangle)(\sum_i p_i N^{-1}[\sum_{|j-i|=1}\phi(x_j)-2d\phi(x_i)])$$
$$+\tfrac{1}{2}\gamma f''(\langle\phi,\underline{p}\rangle)(\sum_{i,j}p_i p_j[\phi^2(x_i)+\phi^2(x_j)-2\phi(x_i)\phi(x_j)]) + N^{-1}R(\phi) \quad (3.14)$$

where $R(\phi)$ is a bounded remainder term if $|f^{(3)}| \leq M$. Hence

$$\underline{\underline{G}}_N\psi(\underline{p}) = f'(\langle\phi,\underline{p}\rangle)(\langle\Delta\phi,\underline{p}\rangle) + \tfrac{1}{2}\gamma f''(\langle\phi,\underline{p}\rangle)[\langle\phi^2,\underline{p}\rangle-\langle\phi,\underline{p}\rangle^2] + N^{-1}R(\phi). \quad (3.15)$$

3.2.2 Theorem: Existence of the Diffusion Limit [16],[17]

(a) The stochastic measure processes $\{Y_N(.,.)\}$ converge in the sense of weak convergence of stochastic processes, as $N \to \infty$, to a measure diffusion process, $\{X(t,.):t \geq 0\}$. The process $\{X(.,.)\}$ takes its values in $M_1(R^d)$, the space of probability measures on R^d.

(b) The measure diffusion $X(t,.)$, called the Fleming-Viot Process, is characterized as the unique solution of the martingale problem on Ω, (Δ,Q), where Δ represents the d-dimensional Laplacian operator and

$$Q(\mu:dx\times dy) = \mu(dx)\delta_x(dy)-\mu(dx)\mu(dy). \quad (3.16)$$

3.2.3 Weighted Random Sampling

The basic random sampling mechanism of the Fleming-Viot model can be modified to allow for weighted sampling which leads to measure diffusions which need not be probability-measure-valued. The fluctuation quadratic form in this case is given by:

$$Q(\mu:dx\times dy) = [\int\theta(x,y)\mu(dy)]\delta_x(dy)\mu(dx) - \theta(x,y)\mu(dx)\mu(dy), \quad (3.17)$$

where $\theta(.,.) \in C_K(R^d\times R^d)$, and $\theta(x,y) = \theta(y,x) \geq 0$. $\theta(x,y)$ represents the rate at which an individual of type x is replaced by one of type y. In the case of a finite initial measure the total measure is preserved under the evolution. However this model can be extended to the case of a spatially homogeneous initial condition if $\theta(x,y)$ is replaced by a function of the form $\theta(|x-y|)$ and $\theta \in C_K(R^d)$. In the latter case this model is a relative of the stepping stone model of population genetics.

3.3 Population Growth in a Randomly Varying Environment

The final example of measure diffusion process to be discussed is a model for a the growth of a population in a spatially homogeneous but randomly varying environment in R^d. The basic assumptions of the model are as follows:

(a) individuals move independently in R^d according to a Brownian motion process,

(b) the population reproduces at replacement rate at average environmental conditions, and

(c) the environment varies randomly in space, R^d, and time (short range dependence in time, possibly long range dependence in space).

These assumptions motivate the spatially homogeneous stochastic evolution equation

$$\partial X(t,dx)/\partial t = \Delta X(t,dx) + \sigma X(t,dx)W'(t,dx), \quad (3.18)$$

where $W(t,dx)$ is a spatially homogeneous Brownian motion, that is, a Gaussian space-time process with covariance function

$$\mathrm{Cov}\left(\int\phi(x)W(t,dx),\int\phi(y)W(s,dy)\right) = \min\{s,t\}\iint\phi(x)\phi(y)\tilde{Q}(x-y)\,dx\,dy \tag{3.19}$$

where $\tilde{Q}(.)$ is a symmetric positive definite function on R^d which has a rapidly decreasing tail. The process $W(t,.)$ has a spectral form:

$$W(t,dx) = \int\exp(i\langle x,\lambda\rangle)M(t,d\lambda)\,dx, \tag{3.20}$$

where $M(t,.)$ is a Gaussian random generalized function with orthogonal increments.

3.3.1 Theorem [11]

(a) The stochastic evolution equation (3.18) has a unique solution $X(t,.)$ which can be represented as

$$X(t,dx) = \sum_{n=0}^{\infty} X_n(t,x)\,dx, \tag{3.21}$$

$$X_n(t,x) = \sigma\int_0^t\iint p(t-s;x,y)X_{n-1}(s,y)\exp(i\langle\lambda,y\rangle)\,dy\,M(ds\times d\lambda). \tag{3.22}$$

(b) The process $X(.,.)$ is characterized as the unique solution of the martingale problem on Ω associated with the pair (Δ,Q) where

$$Q(\mu:dx\times dy) = \tilde{Q}(x-y)\mu(dx)\mu(dy). \tag{3.23}$$

4. Clustering at Microscopic Scales

At a fixed time, t, a stochastic measure process, $X(t)$, is described by a random measure which may deviate significantly from uniformity [i.e. Poisson, Lebesgue in the atomic, continuous cases, respectively]. Spatial disorder, the degree of nonuniformity, depends on the "scale" at which the process is observed. In this section we investigate the phenomena of clustering at the smallest scales.

The idea of clustering can be intuitively described as follows. Consider a population of N individuals distributed in a cube, $V\subset R^d$, which is subdivided into Γ^d equal subcubes. We assume that $N \ll \Gamma^d$ and count the number of subcubes which are occupied. If the distribution is nearly Poisson, then the number of occupied subcubes is of the "order" of Γ^d. If the number of subcubes occupied is much less than Γ^d, then the population is said to be highly clustered. The phenomena of clustering is seen dramatically by considering the diffusion limit and identifying the sets on which the measure is concentrated.

Consider a unit cube, $V\subset R^d$, which for each $n > 1$ is subdivded into Γ_n^d equal subcubes of volume Γ_n^{-d} where $\{\Gamma_n : n \geq 1\}$ is an increasing sequence of positive integers. Consider a set, B, which is constructed as follows: $B_0 \equiv V$, $B_n \supset B_{n-1}$, $n \geq 1$, B_n is a union of N_n subcubes of volume Γ_n^{-d}, and $B = \bigcap_{n=0}^{\infty} B_n$. B constructed in this way is a "generalized Cantor set" and the Hausdorff dimension (c.f. [27]) can be shown to be:

$$\dim B = \liminf_{n\to\infty}\,[\log N_n/\log\Gamma_n]. \tag{4.1}$$

Now let X denote a random measure on V. Given $\varepsilon > 0$, let

$$N_n^{\varepsilon}(X) \equiv \min\{m: \sum_{i=1}^{m} X(v_i) \geq X(V)-\varepsilon\}, \tag{4.2}$$

$$K_n^{\varepsilon} \equiv \bigcup_{i=1}^{N_n^{\varepsilon}(X)} v_i, \tag{4.3}$$

where $\{v_i : i=1,2,\ldots,N_n^{\varepsilon}(X)\}$ is a collection of disjoint subcubes of volume Γ_n^{-d} achieving the minimum in (4.2).

4.1 Lemma [12]

Assume that

$$P([\log N_n^{\varepsilon_n}(X)/\log \Gamma_n] \leq D(1+\eta_n)) \geq 1-\varepsilon_n', \tag{4.4}$$

where $\varepsilon_n \downarrow 0$, $\eta_n \downarrow 0$, $\varepsilon_n' \downarrow 0$, as $n \to \infty$. Then there exists a random set, $B(\omega)$, such that

$$X(\omega,B(\omega)) = X(\omega,V), \text{ a.e. } \omega, \text{ and } \dim B(\omega) \leq D \text{ for all } \omega. \tag{4.5}$$

4.2 Theorem [12],[13]

Let $\{X(t,.):t \geq 0\}$ denote the measure diffusion process associated with either a branching system (i.e. Theorem 3.1.2) or a sampling system (i.e. Theorem 3.2.2). Then for $d > 2$, the random measure, $X(t)$, $t > 0$, is carried by a random set of Hausdorff dimension 2.

Theorem 4.2 implies that the empirical measure corresponding to a population of individuals in R^3 is highly clustered. For a discrete particle system, this has the following implication. If the distance an individual Brownian particle travels in a mean life time is small compared to the mean interparticle distance, then the type of clustering described above will appear at the scale of mean interparticle distance.

5. Fluctuations at Large Scales

At large scales a spatially homogeneous system is "nearly uniform"; this is a consequence of the ergodic theorem. Fluctuation limit theorems are concerned with the rate of convergence to uniformity as we look at increasingly large scales and the nature of the deviations. Two significantly different situations can arise: short range fluctuations or long range fluctuations. We first consider the former.

5.1 Spatial Mixing

To illustrate the notion of spatial mixing we consider the example of a spatially homogeneous subcritical branching measure diffusion, $X(t,.)$, with constant immigration. It can be demonstrated [10] that $X(t)$ approaches a steady state random measure, X, as $t \to \infty$. The Laplace functional of the random measure, X, can be represented as:

$$\log L_X(\phi) = \sum_{n=1}^{\infty} (1/n!)\int\ldots\int\phi(x_1)\ldots\phi(x_n)q_n(x_1,\ldots,x_n)dx_1\ldots dx_n. \tag{5.1}$$

The functions, $q_n(.,\ldots,.)$ are called the cumulant density functions.

5.1.1 Lemma [19],[20]

The random field, X, is Brillinger mixing, that is, for $n \geq 2$,

$$\int\ldots\int|q_n(x_1,\ldots,x_{n-1},x_n)|dx_1\ldots dx_{n-1} < \infty. \tag{5.2}$$

(This condition represents a form of asymptotic independence of fluctuations at large distances.)

As a consequence of spatial mixing we have the following result.

5.1.2 Theorem: Invariance Principle [2],[19],[20]

Consider the rescaled random field:

$$Y_K(\phi) \equiv [\int \phi(x/K)X(dx) - E(\int \phi(x/K)X(dx))]/K^{d/2}. \tag{5.3}$$

Then Y_K converges, in the sense of weak convergence of random fields, as $K \to \infty$, to a Gaussian random field on R^d characterized by the covariance kernel:

$$\Gamma(\phi,\psi) = \sigma \int \phi(x)\psi(x)dx, \tag{5.4}$$

where σ is a positive constant.

5.2 Scaling Limit Theorems for Long Range Fluctuations

The spatial mixing condition of 5.1.1 breaks down at criticality, that is, $\nu = 0$, leading to long range dependence and fluctuations which remain important at macroscopic scales. The basic tool in the study of this phenomena is the group of scaling transformations of space and time. Let $\{X(t,.):t \geq 0\}$ denote a space-time process. The group of scaling transformations $\{R^K_\eta : K > 0\}$ is defined by

$$\langle R^K_\eta X(t,.),\phi\rangle \equiv K^{-\eta}[\int X(K^2t,dx)\phi(x/K)]. \tag{5.5}$$

Since the space-time process $\{X(t,.)\}$ is associated with a probability measure, P, on $\Omega^* \equiv C([0,\infty),\mathcal{S}'(R^d))$, (5.5) induces a group, $\{\mathcal{R}^K_\eta : K > 0\}$, on $\Pi(\Omega)$. ($\mathcal{S}'(R^d)$ denotes the space of tempered distributions on R^d.) The probability measure, P*, is said to be the scaling limit of, P, if

$$P^* = \lim_{K\to\infty} \mathcal{R}^K_\eta P, \tag{5.6}$$

in the sense of weak convergence of probability measures. A measure, P*, obtained in this way is self-similar in the sense that

$$\mathcal{R}^K_\eta P^* = P^*, \quad \text{for all } K > 0. \tag{5.7}$$

5.2.1 Scaling Limit Theorem [8],[11],[18]

Let $X(t,.)$ denote either

(a) a critical branching measure diffusion, $\nu = 0$, on R^d, or
(b) the solution of the stochastic evolution equation (3.18) under the assumption that $\int |\tilde{Q}(x)| dx < \infty$.

Then, if $d > 2$,

$$R^K_\eta [X(.,.)-E(X(.,.))] \to\to Y(.,.), \text{ as } K \to \infty,\ \eta = (d+2)/2, \tag{5.8}$$

in the sense of weak convergence of probability measures. The limit process, $Y(.,.)$ is a "generalized Ornstein-Uhlenbeck process" which can be characterized as the solution of the linear stochastic evolution equation

$$\partial Y(t,.)/\partial t = \Delta Y(t,.) + \sigma W'(t,dx), \tag{5.9}$$

where $W'(.,.)$ is a space-time white noise. In addition, the generalized Ornstein-Uhlenbeck process possesses a steady state random field which can be characterized as a Gaussian random field on R^d with covariance kernel

$$\gamma(x) = 1/|x|^{d-2}, \quad x \in R^d. \tag{5.10}$$

(The long range dependence associated with the kernel (5.10) is typical of those obtained in scaling limit theorems.)

6. Spatial Disorder and Nonlinear Interaction

The types of spatial disorder discussed above have important implications for the study of nonlinear systems. For example, consider a distributed population in which individuals compete for a limited resource. If significant spatial clustering occurs at the scale of the range of competitive interaction, then a decrease in the equilibrium population would be expected. This is indeed the case and a detailed study of this phenomena has been carried out in [9]. More generally, the effects of spatial disorder (heterogeneity) are important in the study of coexistence and related topics in ecology (refer to LEVIN [25]).

7. References

1. M.S. Bartlett, The statistical analysis of spatial pattern, Chapman and Hall, London, 1975.

2. D.R. Brillinger, Statistical inference for stationary point processes. In Stochastic Processes and Related Topics, M.L. Puri, Ed., Academic Press, New York and London, 1974.

3. A. Bose and D.A. Dawson, "A class of measure-valued Markov processes, in: Measure Theory. Applications to Stochastic Analysis, ed. by G. Kallianpur, D. Kölzow, Lecture Notes in Mathematics, Vol. 695, Springer, Berlin, Heidelberg, New York, 1978, pp. 115-125.

4. P.L. Chow, Stochastic partial differential equations in turbulence related problems. In Probabilistic Analysis and Related Topics, Vol. 1, Academic Press, New York, San Francisco, London, 1978.

5. D.A. Dawson, Stochastic evolution equations and related measure processes, J. Mult. Anal. 5(1975), 1-52.

6. ______, The critical measure diffusion process, Z. Wahr. verw Geb. 40(1977), 125-145.

7. ______, Geostochastic calculus, Can. J. Stat. 6(1978), 143-168.

8. ______, Limit theorems for interaction free geostochastic systems, Seria Coll. Math. Janos Bolyai 24(1980).

9. ______, Qualitative behavior of geostochastic systems, J. Stoch. Proc. Appl. 10(1980), 1-32.

10. D.A. Dawson and G. Ivanoff, Branching diffusions and random measures, Adv. in Prob. 5(1978), 61-104.

11. D.A. Dawson and H. Salehi, Spatially homogeneous random evolutions, J. Mult. Anal. 10(1980), 1-40.

12. D.A. Dawson and K.J. Hochberg, The carrying dimension of a stochastic measure diffusion, Ann. Prob. 7(1979), 693-703.

13. ______, Wandering random measures in the Fleming Viot model, in preparation.

14. R.L. Dobrushin, Gaussian and their subordinated self-similar random generalized fields, Ann. Prob. 7(1978), 1-28.

15. W. Feller, Diffusion processes in genetics, Proc. Second Symp. Math. Stat. Prob., University of California Press, Berkeley, 227-246, 1951.

16. W. Fleming and M. Viot, Some measure-valued population processes. In Stochastic Analysis, A.Friedman and M. Pinsky, Eds, 97-108, Academic Press, New York, San Francisco, London, 1978.

17. ______, Some measure-valued Markov processes in population genetics theory, Indiana Univ. Math. J. 28(1979), 817-843.

18. R.A. Holley and D.W. Stroock, Generalized Ornstein-Uhlenbeck processes and infinite particle Branching Brownian motion, Publ. R.I.M.S., Kyoto 14(1978), 741-788.

19. B.G. Ivanoff, Branching diffusions with immigration, J. Appl. Prob. 17(1980), 1-15.

20. ______, Branching random fields, Adv. Appl. Prob. (1980).

21. P. Jagers, Aspects of random measures and point processes, Adv. in Prob. 3(1974), 179-238.

22. O. Kallenberg, Random Measures, Academic Press and Akademie-Verlag, 1976.

23. R. Kulperger, Parametric estimation for a simple branching diffusion process, J. Mult. Anal. 9(1979), 101-115.

24. T.G. Kurtz, Approximation of Population Processes, to appear.

25. S.A. Levin, Ed., Studies in Mathematical Biology, Part II, Populations and community, Math. Assoc. Amer., 1978.

26. P.A.W. Lewis, Ed., Stochastic point processes: statistical analysis, theory and applications, J. Wiley and Sons, New York, 1972.

27. B.B. Mandelbrot, Fractals: form chance and dimension, W.H. Freeman, San Francisco, 1977.

28. Y. Miyahara, White noise analysis and an application to stochastic differential equations in Hilbert space, to appear.

29. S. Mizuno, On some infinite dimensional martingale problems and related stochastic evolution equations, Ph.D. thesis, Carleton University, 1978.

30. G. Nicolis and I. Prigogine, Self-organization in non-equilibrium systems, Wiley-Interscience, New York, London, Sydney, Toronto, 1977.

31. A. Okubo, Diffusion and ecological problems: mathematical models, Springer-Verlag, Berlin, Heidelberg, New York, 1980.

32. G.C. Papanicolaou, Asymptotic analysis of stochastic equations. In Studies in Probability Theory, M. Rosenblatt, Ed., 111-179, Math. Assoc. Amer., 1978.

33. Ya. G. Sinai, Self-similar probability distributions, Theor. Prob. Appl. 21(1976), 64-80.

34. D.W. Stroock and S.R.S. Varadhan, Multidimensional diffusion processes, Grundlehren der mathematischen Wissenschaften, Vol. 233, Springer, Berlin, Heidelberg, New York, 1979.

35. M.S. Taqqu, Self-similar processes and related ultraviolet and infrared catastrophies, Seria Coll. Math. Janos Bolyai, to appear.

36. S. Watanabe, A limit theorem of branching processes and continuous state branching processes, J. Math. Kyoto Univ. 8(1968), 141-167.

37. J.M. Ziman, Models of disorder, Cambridge University Press, 1979.

Part VIII

Phase Transitions and Irreversible Thermodynamics

Stochastic Methods in Non-Equilibrium Thermodynamics

Robert Graham

Fachbereich Physik, Universität Essen – GHS, D-4300 Essen, Fed. Rep. of Germany

1. Macroscopic systems

The thermodynamics of non-equilibrium steady states [1] and of equilibrium states should be closely related. Just how close this relationship can be made on a formal level will be the subject of this talk. Thermodynamics is, of course, only applicable to macroscopic systems, i.e. systems which can be described by a well defined set of macroscopic variables q^{ν} $(\nu=1...n)$, widely separated in time scale from all other microscopic variables. For simplicity of notation, only discrete variables are considered in this talk.

Equilibrium thermodynamics applies to macroscopic systems subject to equilibrium boundary conditions. Such systems are either closed or interact with an environment which is itself in complete thermodynamic equilibrium.

Non-equilibrium thermodynamics, on the other hand, applies to macroscopic systems subject to boundary conditions which prevent the system to reach thermodynamic equilibrium. We restrict attention henceforth to non-equilibrium boundary conditions which are stationary at least in a wide sense (i.e. boundary conditions oscillating or fluctuating in a stationary way are still allowed).

The natural statistical description of such macroscopic systems is the Fokker Planck equation which is, of course, not exact but is a good approximation in as much as the assumptions defining macroscopic systems are satisfied.

$$\frac{\partial P}{\partial \tau} = -\frac{\partial}{\partial q^{\nu}} K^{\nu}(q,\varepsilon)P + \frac{1}{2}\varepsilon \frac{\partial^2}{\partial q^{\nu}\partial q^{\mu}} Q^{\nu\mu}(q)P \tag{1.1}$$

Let us suppose the variables chosen in such a way that eq. (1.1) is subject to zero boundary conditions at infinity. Eq. (1.1) is satisfied by the conditional probability density $P(q|q_o,\tau)$ to observe (q,dq) if a time τ earlier q_o was realized with probability 1. It involves the drift $K^{\nu}(q)$ and the diffusion matrix $Q^{\nu\mu}(q)$. The parameter ε appearing in (1.1) is a measure of the strength of the fluctuations. In equilibrium thermodynamics ε may formally be identified with Boltzmann's constant. More generally ε may be identified with the ratio of a microscopic and a macroscopic scale.

The limit $\varepsilon \to 0$ is the deterministic limit, in which the macroscopic variables satisfy deterministic differential equations.

$$\dot{q}^{\nu} = K^{\nu}(q,0) \tag{1.2}$$

In equilibrium thermodynamics, statistical mechanics puts strong restrictions [2-4] on the allowed form of the deterministic transport laws (1.2). Such restrictions are generally absent in the non-equilibrium case. Thus, for equilibrium thermodynamics we have the form first given by ONSAGER [5]

$$K^{\nu}(q,0) = -L^{\nu\mu}(q)\frac{\partial\Phi}{\partial q^{\mu}} \tag{1.3}$$

where Φ is the thermodynamic potential required by the particular equilibrium boundary conditions chosen ($\Phi=-S$ for closed systems). The transport matrix $L^{\nu\mu}(q)$ satisfies

$$\frac{1}{2}\left(L^{\nu\mu}(q) + L^{\mu\nu}(q)\right) \equiv D^{\nu\mu}(q) = \frac{1}{2}Q^{\nu\mu}(q). \tag{1.4}$$

If time reversal is defined by

$$q^{\nu} \to \tilde{q}^{\nu} = \varepsilon^{\nu}q^{\nu} \qquad \varepsilon^{\nu} = \begin{cases} 1 & \text{for even variables} \\ -1 & \text{for odd variables} \end{cases} \tag{1.5}$$

one may split

$$K^{\nu}(q,0) = d^{\nu}(q) + r^{\nu}(q) \tag{1.6}$$

with

$$\begin{aligned} r^{\nu}(\tilde{q}) &= -\varepsilon^{\nu}r^{\nu}(q) \\ d^{\nu}(\tilde{q}) &= \varepsilon^{\nu}d^{\nu}(q) \end{aligned} \tag{1.7}$$

where $r^{\nu}(q)$ and $d^{\nu}(q)$ are, respectively, the reversible and irreversible deterministic drift, given by

$$\begin{aligned} r^{\nu}(q) &= -A^{\nu\mu}(q)\frac{\partial\Phi}{\partial q^{\mu}} \\ d^{\nu}(q) &= -D^{\nu\mu}(q)\frac{\partial\Phi}{\partial q^{\mu}} \end{aligned} \tag{1.8}$$

where

$$\begin{aligned} L^{\nu\mu}(q) &= A^{\nu\mu}(q) + D^{\nu\mu}(q) \\ A^{\nu\mu}(q) &= -A^{\mu\nu}(q) \\ D^{\nu\mu}(q) &= D^{\mu\nu}(q) \\ L^{\nu\mu}(q) &= \varepsilon^{\nu}\varepsilon^{\mu}L^{\mu\nu}(\tilde{q}). \end{aligned} \tag{1.9}$$

Let us emphasize again that within the assumptions stated eqs. (1.1),

(1.2) are general for macroscopic systems, but eqs. (1.3) - (1.9) follow from statistical mechanics only for macroscopic systems subject to equilibrium boundary conditions.

2. Non-equilibrium thermodynamic potential and irreversible drift in the steady state

We may be interested in the time independent macroscopic ensemble describing a non-equilibrium steady state. In equilibrium thermodynamics we know that the equilibrium ensemble is given by[1)]

$$W(q) \sim \exp\left(-\frac{\Phi(q)}{\varepsilon}\right) \tag{2.1}$$

where $\Phi(q)$ is a coarse grained thermodynamic potential, defined by an average over the microscopic variables in which the fluctuating macroscopic variables q are kept fixed. This coarse grained thermodynamic potential is given by the transport law (1.3) if $L^{\nu\mu}(q)$ there is taken as the matrix of non-renormalized transport coefficients. The latter differ from the measured transport coefficients by the renormalization they receive due to the fluctuations of the macroscopic variables. The requirement that (2.1) is indeed a time independent solution of (1.1) leads to the relation [4]

$$K^{\nu}(q,\varepsilon) = K^{\nu}(q,0) + \varepsilon \, \frac{\partial L^{\nu\mu}(q)}{\partial q^{\mu}} \tag{2.2}$$

for systems interacting with an environment in thermodynamic equilibrium.

We now take a generalizing step and look for a coarse grained non-equilibrium thermodynamic potential $\Phi_o(q)$ for steady states which is associated to the deterministic transport law (1.2) and generalizes (2.1) for steady states. To this end we insert

$$W(q,\varepsilon) \sim \exp\left(-\frac{\Phi_o(q)}{\varepsilon} + \ldots\right) \tag{2.3}$$

as an ansatz into (1.1) and take the deterministic limit $\varepsilon \to 0$.

We obtain

$$K^{\nu}(q,0)\,\frac{\partial \Phi_o}{\partial q^{\nu}} + \frac{1}{2}\,Q^{\nu\mu}(q,0)\,\frac{\partial \Phi_o}{\partial q^{\nu}}\,\frac{\partial \Phi_o}{\partial q^{\mu}} = 0 \tag{2.4}$$

which, indeed, relates Φ_o to $K^{\nu}(q,0)$ and $Q^{\nu\mu}(q,0)$, however, in a much more complicated way as encountered before in equilibrium thermodynamics. In order to make a closer comparison let us define [9,10] in analogy to (1.4), (1.6), (1.8)

$$d^{\nu}(q) = -\frac{1}{2}\,Q^{\nu\mu}(q,0)\,\frac{\partial \Phi_o}{\partial q^{\mu}} \tag{2.5}$$

$$r^{\nu}(q) = K^{\nu}(q,0) - d^{\nu}(q) \tag{2.6}$$

---

1) We assume that we have chosen natural variables in the sense of [4] in order to avoid complications associated with the metric in state space [4].

with

$$\Phi_o(q) = - \lim_{\varepsilon \to 0} \varepsilon \ \ln W(q,\varepsilon). \tag{2.7}$$

The quantity $r^\nu(q)$ is important to characterize both equilibrium and non-equilibrium steady states and is sometimes called 'circulation' [8]. Eq. (2.4) may now be rewritten as the simple requirement

$$r^\nu(q) \frac{\partial \Phi_o}{\partial q^\nu} = 0 \tag{2.8}$$

i.e. the circulation only takes place on equipotential surfaces of Φ_o. Generally, the circulation $r^\nu(q)$ may have sources and sinks, $\frac{\partial r^\nu(q)}{\partial q^\nu} \neq 0$. It can be shown, however, that near a stable fixed point attractor of the deterministic transport law (1.2) the circulation is free of sources and sinks (cf. the next section).
It is obvious that in equilibrium thermodynamics eq. (2.5) just defines the irreversible deterministic drift, eq. (2.6) the reversible one, and that (2.8) (and therefore (2.4)) is satisfied automatically because of (1.8) and the asymmetry of $A^{\nu\mu}(q)$. In non-equilibrium thermodynamics, on the other hand, $d^\nu(q)$ and $r^\nu(q)$ defined by (2.5), (2.6) still exist but do not posess any special time reversal properties. Therefore, they cannot be easily determined from $K^\nu(q,0)$; indeed, the only method known is to determine (2.7) as a solution of (2.4), and determine d^ν and r^ν from (2.5), (2.6) respectively. The circulation $r^\nu(q)$ has a simple statistical meaning: Eq. (1.1) defines the probability current

$$g^\nu(q,\varepsilon) = K^\nu(q,\varepsilon)\ P + \frac{1}{2}\ \varepsilon \frac{\partial}{\partial q^\mu} Q^{\nu\mu}(q,\varepsilon)P. \tag{2.9}$$

In the steady state for small ε, $P \to W \sim \exp\ (- \Phi_o | \varepsilon)$ and we obtain

$$r^\nu(q) = \lim_{\varepsilon \to 0} \frac{g^\nu(q,\varepsilon)}{W(q,\varepsilon)} \tag{2.10}$$

i.e. $r^\nu(q)$ is the steady state drift rate associated with the probability current in the steady state. Our result may therefore be summarized succinctly: A steady state generally (not always) differs, from an equilibrium state in that its steady state drift rate (circulation) has an irreversible component. Thus a steady state is (generally) dissipative, an equilibrium state never.

Due to the definition (2.7) eq. (2.4) has to be satisfied by Φ_o with the boundary condition that Φ_o approaches infinity for q^ν going to infinity (for a stable system). Further requirements on the function Φ_o are single valuedness, continuity and at least piecewise first order differentiability. Unfortunately, the general restrictions on $K^\nu(q,\varepsilon)$ and $Q^{\nu\mu}(q,\varepsilon)$ under which the limit (2.7) exists and has the required analytical properties are not known. Mathematical papers pertaining to problems of this kind are [11,12].

Many specific examples are by now known, where Φ_o could be determined explicitely [9,10], the physical applications ranging from quantum optics to hydrodynamic instabilities and chemical reactions. These examples are distinguished by the fact that they satisfy a detailed balance condition [13], which is a special feature not satisfied in most steady states. Two examples <u>without</u> detailed balance are discussed in sections 3, 4.

Assuming the existence of $\Phi_o(q)$ one may gain some insight by looking at its properties:

i) Φ_o gives an asymptotic expression of the steady state distribution function in the limit of small noise via eq. (2.3). This relation generalizes the usual Boltzmann formula of equilibrium states to steady states.

ii) Φ_o is a function which decreases under the deterministic time evolution (1.2) since

$$\frac{d\Phi_o}{dt} = \frac{\partial \Phi_o}{\partial q^\nu} K^\nu(q,0) = -\frac{1}{2} Q^{\nu\mu}(q,0) \frac{\partial \Phi_o}{\partial q^\nu} \frac{\partial \Phi_o}{\partial q^\mu} \tag{2.11}$$

and $Q^{\nu\mu}(q,0)$ is a non-negative matrix. This generalizes the usual H-theorem satisfied by thermostatic potentials. Eq. (2.10) shows that Φ_o is a Lyapunoff function of eq. (1.2).

iii) As a consequence of the property (2.11), Φ_o has a local minimum for any stable attractor of eq. (1.2) in q-space.
On all connected subsets of the attractor Φ_o is constant. Unstable steady state solutions of eq. (1.2) correspond to maxima or saddle points of $\Phi_o(q)$.

We will now proceed to study examples where these features can be exhibited explicitely. Before doing so it may be worth recalling that (2.3) is just the leading contribution to $W(q,\varepsilon)$ for small ε. Provided $\Phi_o(q)$ is twice differentiable, corrections may be obtained by calculating the next term in the exponent of (2.3), which is independent of ε. However, in the following we want to concentrate on Φ_o alone.

3. Local potential of a monostable steady state [6]

Let us assume that the system has a unique, stable steady state and let us chose coordinates such that $q^\nu=0$ is the deterministic description of that state. Let then $\dot{q}^\nu=A^\nu{}_\mu q^\mu$ and $Q^{\nu\mu}=L^{\nu\mu}(0) + L^{\mu\nu}(0)$ be the linear transport law and the constant transport coefficients in the vicinity of that state. Eq. (2.4) then acquires the form

$$A^\nu{}_\mu\, q^\mu \frac{\partial \Phi_o}{\partial q^\mu} + \frac{1}{2} Q^{\nu\mu} \frac{\partial \Phi_o}{\partial q^\nu} \frac{\partial \Phi_o}{\partial q^\mu} = 0 . \tag{3.1}$$

Eq. (3.1) is solved by the ansatz

$$\Phi_o = \frac{1}{2} \sigma^{-1}_{\nu\mu} q^\nu q^\mu ; \qquad \sigma^{-1}_{\nu\mu} = \sigma^{-1}_{\mu\nu} \tag{3.2}$$

which satisfies the required boundary condition. From (3.1) we obtain (with $\sigma^{\alpha\nu}\sigma^{-1}_{\nu\beta}=\delta^\alpha_\beta$)

$$A^\alpha{}_\mu\, \sigma^{\mu\beta} + A^\beta{}_\mu\, \sigma^{\mu\alpha} = - Q^{\alpha\beta} \tag{3.3}$$

from which $\sigma^{\mu\nu}$ can be determined, provided the determinant of this system of linear equations does not vanish. From (3.3) it may be inferred that $\sigma^{\alpha\beta}$ is a positive definite matrix, if the real parts of the eigenvalues of $A^{\nu}_{\ \mu}$ are all negative, i.e. if the steady state is indeed stable as was assumed. Regardless of this condition we find

$$\frac{d\Phi_o}{dt} = -\frac{1}{2}\,\sigma^{-1}_{\nu\varkappa}\,Q^{\nu\mu}\,\sigma^{-1}_{\mu\lambda}\,q^{\varkappa}q^{\lambda} < 0 \tag{3.4}$$

i.e. the system in any case moves downhill in the potential Φ_o. Thus it approaches the steady state if it is indeed stable, while it moves to infinity if it is unstable. Because of the linearization of the transport laws around the steady state, Φ_o may be called a local potential; it only applies in the local vicinity of the steady state $q^{\nu}=0$. For $r^{\nu}(q)$, defined by (2.6), we obtain

$$r^{\nu}(q) = (A^{\nu}_{\ \mu} + \frac{1}{2}\,Q^{\nu\lambda}\,\sigma^{-1}_{\lambda\mu})\,q^{\mu}. \tag{3.5}$$

The orthogonality relation (2.7) is indeed satisfied due to eq. (3.3). Furthermore $r^{\nu}(q)$ in the present example also satisfies

$$\frac{\partial r^{\nu}(q)}{\partial q^{\nu}} = 0 \tag{3.6}$$

as may again be veryfied with the help of (3.3). Thus $r^{\nu}(q)$ has no sources or sinks and leaves Φ_o unchanged. This part of the drift $K^{\nu}(q,0)$ is therefore indifferent i.e. neither stabilizing nor destabilizing the steady state. The stabilization is entirely due to the drift $d^{\nu}(q)$ defined by (2.5).

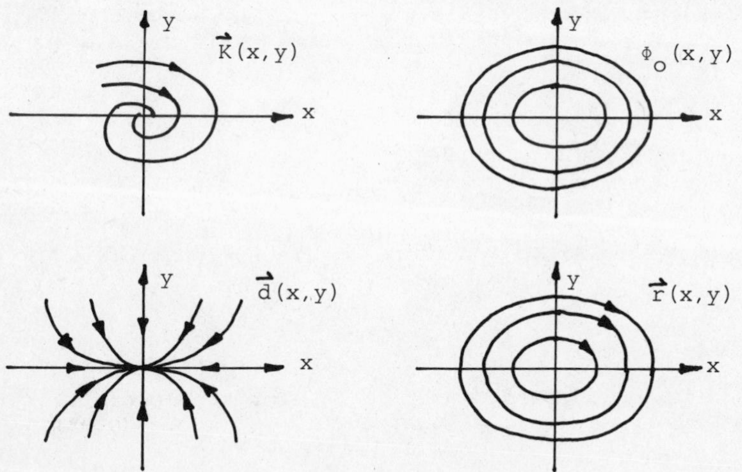

Fig. 1

In fig. 1 we give schematic plots of the drift $K^{\nu}(q)$, the potential $\Phi_o(q)$ and the two parts $d^{\nu}(q)$ and $r^{\nu}(q)$ for a monostable steady state of a two variable system in coordinates x, y in which $Q^{\nu\mu}=\begin{pmatrix} Q & 0 \\ 0 & Q \end{pmatrix}$, $\sigma^{\nu\mu}=\begin{pmatrix} \sigma^{xx} & 0 \\ 0 & \sigma^{yy} \end{pmatrix}$. It is clear that $r^{\nu}(q)$ carries the oscillatory part (if there is any) of the relaxation into the steady state and may thereby be observed. The drift $d^{\nu}(q)$ describes the monotonic part of the approach of the steady state via the drift down the potential Φ_o.

Besides this deterministic dynamic aspect, Φ_o also has its probabilistic aspect. The steady state distribution in the limit of weak noise is given by (2.3) which, in the present case, is the exact steady state solution of the Fokker Planck equation (1.1) with drift and diffusion given by $A^\nu_\mu q^\mu$ and $Q^{\nu\mu}$, respectively. This confirms that the solution (3.2) of eq. (3.1) is indeed the potential Φ_o defined by (2.7).

4. Non-equilibrium thermodynamic potential of a bistable steady state [14]

In collaboration with A. Schenzle I have applied the preceding general theory to the physically very interesting example of optical bistability [14]. This is a nonlinear problem without detailed balance, except in special cases. There is no place here to go into the physics of that problem. What is important in the present context is that the complex amplitude E of the transmitted electromagnetic field under certain conditions can be shown to satisfy a deterministic transport law of the form (1.2) with

$$\dot{E} = (-1 - i\delta)E + E_o - \Gamma^2 \frac{(1+i\Delta)}{1+|E|^2} \cdot E \tag{4.1}$$

and the constant transport coefficient $Q_{E^*E}=2\varepsilon Q$. E_o is the amplitude of an external driving field, δ and Δ are linear and nonlinear frequency mismatches, respectively, and Γ^2 measures the strength of nonlinearity in the response of the employed medium to the driving field. For sufficiently strong Γ^2, the steady state solution of (4.1) as a function of E_o has a bistable domain.

For this system we have determined the non-equilibrium thermodynamic potential Φ_o from (2.4) exactly to first order in the driving field E_o. In this case a necessary condition to be satisfied by any acceptable solution of (2.4) is that it reduces to the correct result for $E_o \to 0$, given by the equilibrium thermodynamic potential. The solution obtained under this condition and the requirement of single valuedness is

$$\begin{aligned} Q\ \Phi_o = {} & |E|^2 - EE_o/(1-i\delta) - E^*E_o/(1+i\delta) + \Gamma^2 \ln (1+|E|^2) \\ & - 2E_o\ \Gamma^2\ (\delta-\Delta)\ \mathrm{Im} \left[\frac{E\ F\ (-\ |E|^2/(1+\Gamma^2))}{(1-i\delta)(1+\Gamma^2-i(\Delta)\Gamma^2+\delta)} \right] \end{aligned} \tag{4.2}$$

where F(z) is the hypergeometric function (in the notation of [15])

$$F(z) = {}_2F_1\left(1, \frac{1}{2} - \frac{i\delta}{2}; \frac{3}{2} - \frac{i}{2}\ \frac{\delta+\Delta\Gamma^2}{1+\Gamma^2}; z\right). \tag{4.3}$$

The result (4.2) is exact to first order in E_o. For the special case $\delta=\Delta$, eq. (4.2) even solves eq. (2.4) exactly, i.e. it then determines Φ_o to all orders in E_o.

$$Q\Phi_o\ (E,E^*) = |E|^2 - \frac{EE_o}{1-i\delta} - \frac{E^*E_o}{1+i\delta} + \Gamma^2 \ln (1+|E|^2) \tag{4.4}$$

It is possible to check expliciteIy, that Φ_o indeed satisfies all the general properties derived in section 2. In particular it has minima on the two stable branches in the bistable domain and a maximum on the unstable state in between. Φ_o can also be used like a thermodynamic potential to find the generalization of the thermodynamic Maxwell rule for the steady state, which determines the value of E_o where absolute stability is exchanged between the two locally stable branches. The reader is referred to our original paper [14] for a detailed discussion of these results and a plot of the probability distribution $W(q) \sim \exp(-\Phi_o/\varepsilon)$ in the bistable domain, exhibiting two local maxima, as expected. Another non-equilibrium thermodynamic potential for a class of hydrodynamic instabilities without detailed balance can be found in ref. [10] section 4.4.

5. Extremum principles and non-equilibrium thermodynamic potentials

We saw in section 2 that Φ_o determines the attractors in the deterministic description. In this section we want to remark that Φ_o is itself the result of a minimum principle. This principle was developed in the work of ONSAGER [5] and ONSAGER and MACHLUP [16] for the case of linear systems interacting with equilibrium environments; since then it has been generalized considerably, both, for nonlinear systems and non-equilibrium environments (cf. [9,17,18]).

One way to arrive at this minimum principle is to observe that eq. (2.4) has the form of a Hamilton Jacobi equation if Φ_o is put into analogy with the action along the classical path of a system in classical mechanics. The "Hamiltonian" of that system may be read off eq. (2.4). It is of the form

$$H(p,q) = \frac{1}{2} Q^{\nu\mu}(q,0)p_\nu p_\mu + K^\nu(q,0)p_\nu \tag{5.1}$$

The characteristics of the Hamilton Jacobi equation then satisfy the canonical equations

$$\dot{p}_\nu = -\frac{\partial H}{\partial q^\nu}, \qquad \dot{q}^\nu = \frac{\partial H}{\partial p^\nu}. \tag{5.2}$$

Eqs. (5.2) may be obtained from the 'principle of least action'

$$\delta \int_{t_o}^{t} d\tau \, (p_\nu \dot{q}^\nu - H(p,q)) = 0 \tag{5.3}$$

where q^ν is prescribed at the end points $q^\nu(t_o) = q_o^{\ \nu}$, $q^\nu(t) = q^\nu$. During the motion along the characteristics described by eq. (5.2) the Hamiltonian $H(p,q)$ is conserved

$$H(p,q) = E. \tag{5.4}$$

Since we are here only interested in the solution of the time independent Hamilton Jacobi equation, only characteristics with E=0 (or near E=0) have to be considered. The non-equilibrium thermodynamic potential may now be represented by the minimum principle

$$\Phi_o(q) = \min_{(q_o)} \int_{-\infty}^{t} d\tau \, (p_\nu \dot{q}^\nu - H(p,q)) + \text{const}. \tag{5.5}$$

In (5.5) the time integral is carried out along the minimizing trajectory $p(\tau)$, $q(\tau)$ satisfying (5.2) and (5.4) with E=0 and connecting $q^\nu(-\infty) = q_o^\nu$ and $q^\nu(t) = q^\nu$. The minimum is also taken with respect to $q_o^{\ \nu}$, keeping the value of $\Phi_o(q)$ at some reference point, e.g. $q^\nu = 0$, fixed.

Because of the requirement of an infinite time interval the trajectory has to go from q_o^ν through a stationary point of (5.2), where it spends an infinitely long time until it goes to the final point q^ν within a finite time. In this way the initial point q_o^ν is forgotten. However, the minimizing with respect to q_o is still necessary, since the trajectory may pass through different stationary points of (5.2) depending on the region of q-space in which q_o is chosen. Of all the values of $\Phi_o(q)$ resulting for different choices of q_o always the lowest lying one has to be chosen, keeping $\Phi_o(q)$ fixed at some reference point and requiring continuity.

Let us see how this works in two simple one-dimensional examples defined by

$$K(x) = -x, \quad Q = 2 \tag{5.6}$$

and

$$K(x) = x - x^3, \quad Q = 2. \tag{5.7}$$

The initial value problem of the time dependent Hamilton Jacobi equation for these examples has been discussed by KITAHARA [19]. Fig. 2 shows the trajectories at E=0 in the x,p-plane and one choice of x_o

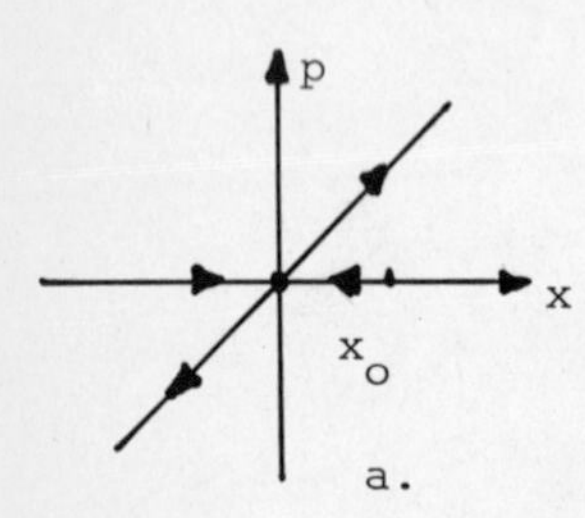

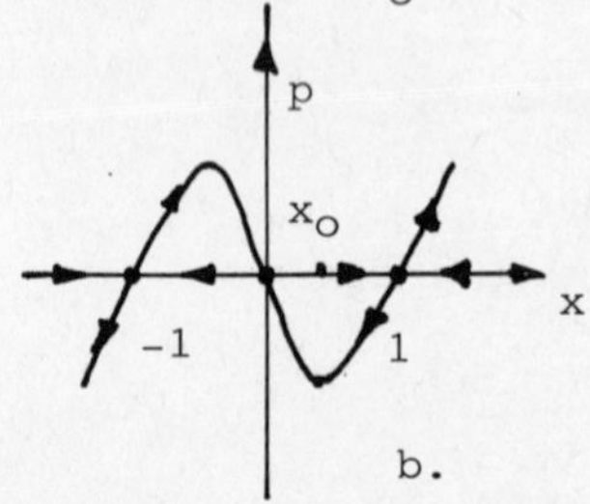

Fig. 2

In fig. 2a all points x>0 and x<0 are reached from x_o for infinite times by passing through the stationary point at x=0, p=0. Therefore, $\Phi_o(x)$ as given by (5.5) is independent of x_o even without minimizing with respect to x_o.

In fig. 2b all points x>0 are reached by passing through the stationary point at (1,0). However, all points $-1 < x < 0$ are reached by passing first through (1,0) then through (0,0), while all points $x < -1$ are reached via (1,0), (0,0), (-1,0). The resulting potential for x_o positive and $\Phi_o(0)=0$ is

$$\Phi_o(x, x_o^+) = \begin{cases} \frac{1}{4}x^4 - \frac{1}{2}x^2 & x > 0 \\ 0 & -1 < x < 0 \\ \frac{1}{4}x^4 - \frac{1}{2}x^2 + \frac{1}{4} & x < -1 \, . \end{cases} \tag{5.8}$$

For negative x_o we obtain

$$\Phi_o(x,x_o^-) = \begin{cases} \frac{1}{4}x^4 - \frac{1}{2}x^2 + \frac{1}{4} & x > 1 \\ 0 & 0 < x < 1 \\ \frac{1}{4}x^4 - \frac{1}{2}x^2 & x < 0 . \end{cases} \tag{5.9}$$

Taking the minimum with respect to x_o with $\Phi_o(0)=0$ we obtain

$$\Phi_o(x) = \frac{1}{4} x^4 - \frac{1}{2} x^2 \tag{5.10}$$

which, indeed, satisfies the definition (2.7) of the coarse grained potential.

Due to the condition $H(p,q)=0$ on the minimizing path eq. (5.5) may be simplified to assume the form

$$\Phi_o(q) = \min_{(q_o)} \int_{q_o}^{q} p_\nu dq^\nu \tag{5.11}$$

where the integral is taken along the minimizing trajectory and the minimum is taken with respect to q_o.

Another way to obtain these results is to start from the stochastic description (1.1), derive its path integral solution [7,9,17,18] and to go to the deterministic limit. Asymptotically, for small ε we obtain

$$W(q) \sim \exp\,(-\frac{\Phi_o}{\varepsilon}) \sim \int DpDq \exp\,(-\frac{1}{\varepsilon}\int_{-\infty}^{t} d\tau\ (\dot{q}^\nu p_\nu - H(p,q))) \tag{5.12}$$

where the path integral has to be done for fixed $q(t)=q$. Since only the asymptotic behaviour for $\varepsilon \to 0$ is of interest here, the saddle point approximation may be used on the right hand side and we immediately obtain eq. (5.5), where, indeed, the minimum with respect to the initial condition has also to be taken. One may then go on to infer eq. (2.4) from (5.5).

The existence of these extremum principles even for thermodynamic potentials of non-equilibrium steady states shows us how surprisingly close the formal connections between equilibrium thermodynamics and the non-equilibrium theory proposed here really are. However, we are now also able to appreciate the basic difference between both, which explains why equilibrium thermodynamics is so much more usefull than non-equilibrium thermodynamics:

In equilibrium thermodynamics reversible and irreversible processes are separated in a clean way, while in non-equilibrium thermodynamics, as we have seen they are not. A non-equilibrium thermodynamic potential therefore incorporates reversible and irreversible phenomena in a complicated way described by eq. (2.4), which usually makes it difficult to determine, and tedious to use. Testimony to these facts is given by the explicit example (4.2).

However, in view of the wealth of information carried by Φ_o about the steady state probability distribution function and the attractors of the deterministic equations it is an invaluable function, if it can be determined.

References

1 De Groot, S.R., Mazur, P., *'Nonequilibrium thermodynamics'*, Amsterdam, North Holland 1962

2 Green, M.S., J. Chem. Phys. 20, 1281 (1952)

3 Grabert, H., Green, M.S., Phys. Rev. A19, 1747 (1979)

4 Grabert, H., Graham, R., Green, M.S., Phys. Rev. A21, 2136 (1980)

5 Onsager, L., Phys. Rev. 37, 405 (1931); 38, 2265 (1931)

6 Graham, R., in *'Coherence and Quantum Optics'*, ed. L. Mandel, E. Wolf, Plenum, New York 1973, p. 851

7 Kubo, R., Matsuo, K., Kitahara, K., J. Stat. Phys. 9, 51 (1973)

8 Tomita, K., Tomita, H., Progr. Theor. Phys. 51, 1731 (1974)

9 Graham, R., "Statistical Theory of Instabilities in Stationary Nonequilibrium Systems with Applications to Lasers and Nonlinear Optics", in Quantum Statistics, Springer Tracts in Modern Physics, Vol. 66, Springer, Berlin, Heidelberg, New York 1973

10 Graham, R., in *'Fluctuations, Instabilities and Phase Transitions'*, ed. T. Riste, Plenum, New York 1975, p. 215

11 Ventsel, A.D., Freidlin, M.I., Usp. Mat. Nauk. 25, 3 (1970) = Russ. Math. Surveys 25, 1 (1970)

12 Ludwig, D., SIAM Rev. 17, 605 (1975)

13 Graham, R., Haken, H., Z. Physik 243, 289 (1971); 245, 141 (1971)

14 Graham, R., Schenzle, A., in Proceedings of the International Optical Bistability Conference, Asheville, USA, 1980, to be published Graham, R., Schenzle, A., Phys. Rev. A, to be published

15 Gradshteyn, I.S., Ryzhik, I.M., *'Table of Integrals, Series, and Products'*, Academic Press, New York 1965

16 Onsager, L., Machlup, S., Phys. Rev. 91, 1505 (1953); 91, 1512 (1953

17 Graham, R., "Path-Integral Methods in Nonequilibrium Thermodynamics and Statistics" in *'Stochastic Processes in Non-equilibrium Systems'* ed. by L. Garrido, P. Seglar, P.J. Shepherd, Lecture Notes in Physics, Vol. 84, Springer, Berlin, Heidelberg, New York 1978, p. 82

18 Graham, R., in Proceedings of the Workshop on Functional Integratior Louvain, Belgium 1979, ed. J.P. Antoine, to be published

19 Kitahara, K., Adv. Chem. Phys. 29, 85 (1975)

Fluctuation Dynamics Near Chemical Instabilities

S. Grossmann*

Fachbereich Physik der Philipps-Universität,
Renthof 6, D-3550 Marburg/Lahn, Fed. Rep. of Germany

1. Introduction

Chemical reaction systems are usually described by a master equation for the probability density $P(N_i,t)$ to find N_i particles of a certain species in a certain spatial cell, both characterized by i. This description takes care of all fluctuations due to reactions as well as to diffusion.

Generally the fluctuations are small since many particles are involved. Taking an appropriate limit $V\to\infty$ one obtains macroscopic rate equations. These allow to find the different macroscopic phases of the system, its stationary states, multiple state phases, limit cycles, and permanent chaotic time-dependent states. Phase boundaries and bifurcations as functions of the external parameters can be deduced from the macroscopic equations by use of linear stability analysis.

The latter also gives information about the decay time of the fluctuations if these are small. Then they are of Gaussian type and can be deduced in quasi-linear approximation from the basic master equation [3],[4] in order $V^{1/2}$. Van Kampen has developed this linearization also for time-dependent states, in particular near limit cycles [4].

Linearization breaks down near instabilities. The fluctuations are enhanced. The time correlation functions $\langle\delta N_i \delta N_j(t)\rangle$ which represent a lot of relevant information about the internal dynamics of the system have to be calculated beyond the diverging linear analysis. Perturbative approaches or path-integral techniques must be renormalized with respect to propagators as well as to vertices [5],[6].

A rather simple method to study the dynamics of fluctuations in the non-linear regimes near instabilities as well as within phases with non-Gaussian fluctuations is provided by a continued fraction analysis of the time correlation functions. Its basic idea is to connect the dynamics with the properties of the stationary state, i.e. the equal time moments. Together with a mode-decoupling approximation of the memory kernel this method (introduced by ZWANZIG [7] and MORI [8] via projector techniques) has proved as useful for many physical systems in equilibrium (some examples are cited in [9]).

In open systems far from equilibrium the fluctuation dynamics of course reflects the different physical situation: non-Hermitian time development, loss of detailed balance, fluxes through the system, etc. Nevertheless, the continued fraction analysis, properly adjusted, has proved to be useful, too (some examples are [10-12]). Applied to chemical reaction systems [1],[2] one gets information, for instance, about the dynamics of autocatalytic reactions or fluctuations in limit cycle states.

In this lecture this general idea to study fluctuations in non-linear chemical reaction systems is reviewed (sect.2) and applied to a discontinuous (first order) instability (sect.3) and to the continuous onset of a limit cycle (sect.4).

---

*The lecture is based on work done together with Rudi Schranner and Peter H. Richter [1], [2].

Although spatially inhomogeneous systems can be treated as well, I shall restrict myself to homogeneous well-stirred systems. Large, medium, and small particle numbers are considered. Macroscopic, sharp instabilities show up in the large N limit. The fluctuation physics with small and medium particle numbers may shed some light on the chemical behaviour under disturbances with medium or large wave number k in inhomogeneous systems.

2. Memory Kernel Representation

The basic equation of motion is the master equation

$$\partial_t P(t) = \mathcal{D} P(t) \quad .$$

$\mathcal{D}$ is a matrix operator $W_{x|x'}$. It is $x_i = N_i/N$. N measures the total particle number and serves as an extensivity parameter.

The (slightly modified) SCHLÖGL [13] reaction

$$\begin{aligned} &B + 2X \underset{k_2}{\overset{k_1}{\rightleftharpoons}} 3X \quad , \\ &B \xrightarrow{k_3} X \quad , \qquad X \xrightarrow{k_4} C \quad , \end{aligned} \tag{1}$$

is represented by

$$W_{x'|x} = N \begin{cases} bx(x-1/N) + b\kappa & x' = x + 1/N \\ \kappa x(x-1/N)(x-2/N) + x & x' = x - 1/N \end{cases} \quad . \tag{1'}$$

$N=(k_3k_4/k_1k_2)^{1/4}$ measures the size of the system. The parameter $\kappa=(k_2k_3/k_1k_4)^{1/2} \propto N^0$ describes the ratio of the spontaneous X production net rate k_3/k_4 to the autocatalytic one k_1/k_2. In equilibrium $\kappa=1$. Strong influence of the autocatalytic reaction is characterized by small κ. $b=[B](k_1^3k_3/k_2k_4^3)^{1/4} \propto N^0$ scales the externally controlled bath particle density. Varying b changes the external stress or the particle flux.

Another example is the Brusselator [14] introduced by PRIGOGINE and LEFEVER as a simple model for a chemical system capable of oscillations as observed e.g. in the Belousov-Zhabotinski reaction.

$$\begin{aligned} A &\xrightarrow{k_1} X \quad , \\ B + X &\xrightarrow{k_2} Y + C \quad , \\ 2X + Y &\xrightarrow{k_3} 3X \quad , \\ X &\xrightarrow{k_4} D \quad . \end{aligned} \tag{2}$$

The corresponding transition matrix is again $\propto N$.

$$W_{x'y'|xy} = N \begin{cases} a & x' = x + 1/N \ , \quad y' = y \\ yx(x-1/N) & x' = x + 1/N \ , \quad y' = y - 1/N \\ bx & x' = x - 1/N \ , \quad y' = y + 1/N \\ x & x' = x - 1/N \ , \quad y' = y \end{cases} \tag{2'}$$

$a=(k_1/k_4)([A]/N)$ scales the fixed A-density, $b=[B]k_2/k_4$ is the external driving parameter, $N=(k_4/k_3)^{1/2} \propto V$ serves as extensivity parameter. The dimensionless time is measured again in units of k_4^{-1}.

The master equation operator $\mathcal{D}$ determines the stationary state by $\mathcal{D}P_S=0$. Time-dependent dynamical variables $A(N_i)$ are introduced by

$$\langle A\rangle_t = \sum A(e^{\mathcal{D}t}P(0)) = \sum (e^{\mathcal{L}t}A)P(0) \quad .$$

$\mathcal{L}$ is the adjoint operator to $\mathcal{D}$ and $\partial_t A(t)=\mathcal{L}A(t)$. We are interested in the steady state time dependent correlation matrix of $\delta A=A-\langle A\rangle_S$,

$$C(t) = (\delta A|e^{\mathcal{L}t}\delta A) = \langle\delta A^{*}\delta A(t)\rangle_S \quad .$$

Its Laplace transform

$$C(w) = \int_0^\infty dt\ e^{-wt}\ C(t) = (\delta A|(w-\mathcal{L})^{-1}\delta A)$$

is the resolvent matrix taken in the subspace of variables A_k of interest. $(|)$ has the properties of an inner product, defined on the Hilbert space of dynamical variables with the stationary state P_S as a weight function, $\langle\ldots\rangle_S=\Sigma\ldots P_S$.

In a closed Hamiltonian system $\mathcal{L}$ is purely anti-Hermitian. In open systems far from equilibrium this is not the case. In the SCHLÖGL model $\mathcal{L}$ turns out to be Hermitian, describing a purely damped decay of the fluctuations without oscillations. The proof uses detailed balance which happens to hold in every single variable one step master equation. For the Brusselator $\mathcal{L}$ is complex. Its symmetric part contributes to the damping, its anti-Hermitian part gives rise to the limit cycle oscillations.

Since only the projection of the resolvent into the subspace $P=|\delta A)\Gamma^{-1}(\delta A|$ is of interest, it is natural to separate the macroscopic part $P\mathcal{L}P$ of the Liouvillian. Using an exact resolvent identity one gets the representation

$$C(w) = \Gamma\ [w\Gamma + \Lambda - K(w)]^{-1}\ \Gamma \quad . \tag{3}$$

$\Gamma=(\delta A|\delta A)$ is the matrix of stationary equal time correlations. $\Lambda=-(A|\mathcal{L}A)$ is the macroscopic part of the Liouvillian, also a stationary correlation matrix, though of higher order. $K(w)=(A|\mathcal{L}Q(w-Q\mathcal{L}Q)^{-1}Q\mathcal{L}A)$ is the resolvent in the space $Q=1-P$ perpendicular to that of the relevant variables A_k. Its w-dependence describes the non-Markovian memory, i.e. the deviation of the fluctuation decay from a Lorentzian line shape,

$$d_t\ C(t) = -\ (\Lambda\Gamma^{-1})\ C(t) + \int_0^t dt'\ [K(t')\Gamma^{-1}]\ C(t-t') \quad .$$

Applying the resolvent identity to the memory kernel K(w), etc., one generates the continued fraction representation of C(w). In [1],[2] we have checked the memory effects. Either a higher order fraction or a mode decoupling has been used. In order to simplify the description of the fluctuation dynamics let me omit K(w) in this lecture. Then

$$C(t) = e^{-\Lambda^0 t} \quad . \tag{4}$$

The physics of the correlation decay is contained in the normalized decay matrix

$$\Lambda^0 = \Lambda\Gamma^{-1} = -\ (A|\mathcal{L}A)\cdot(\delta A|\delta A)^{-1} \quad . \tag{5}$$

The macroscopic static moment of the Liouvillian $\mathcal{L}$ (adjoint to the master equation operator $\mathcal{D}$) and the stationary fluctuations Γ jointly govern the fluctuations.

In the linear regime $\mathcal{L}A\cong\lambda\delta A$. Then Λ^0 coincides with the results of the linear stability analysis. In the non-linear regime near instabilities or in limit cycle states the linearization breaks down. However, Λ^0 remains finite and useful throughout the non-linear regime, since the static moments Γ and Λ take proper care of the fluctuation enhancement. Eq. (5) is an approximate but rather deep connection between the dynamical and the static properties of the fluctuations. In sect.4 it will give rise, for instance, to a relation between the limit cycle's frequency and its amplitude and shape.

It should be remarked that (5) can also be used in the Fokker-Planck limit. $\mathcal{D}$ then is a differential operator and $\mathcal{L}$ its adjoint differential operator. As an example of an application the calculation of the laser line width near the threshold is mentioned [12].

3. Fluctuations at a 1st-Order Transition

Consider a chemical reaction system showing an instability similar to a first order phase transition. An experimental realization is the $N_2O_4 \rightleftharpoons 2NO_2$ reaction under laser stimulation [15]. SCHLÖGL's reaction (1) is a simple model.

Its macroscopic rate equation ($N \to \infty$) reads

$$\dot{x} = -\kappa x^3 + bx^2 - x + \kappa b \quad . \tag{6}$$

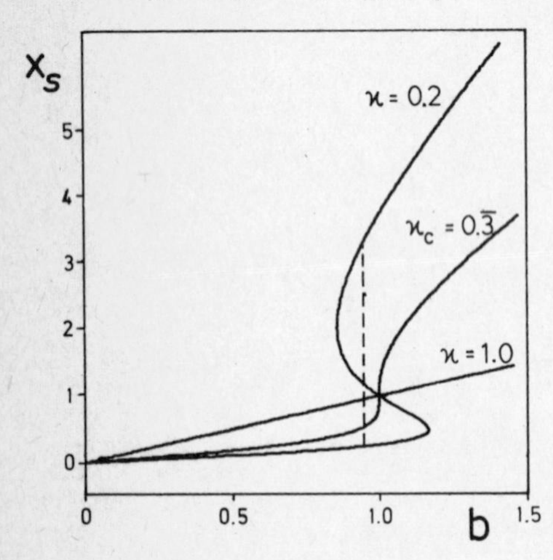

Fig.1 Steady state density x_s vs b; the Maxwell construction yields $b_{---} \cong 0.94$

There is no instability near $\kappa=1$ (mass equilibrium), a critical situation if $\kappa=\kappa_c$ with $\kappa_c=1/3$, and a first order transition if $\kappa<\kappa_c$.

The fluctuation decay as described by (5) is given by

$$\Lambda^0 = b\left[\frac{1}{N} + \frac{\langle X\rangle}{\Gamma}\left(\langle x\rangle - \frac{1}{N} + \frac{\kappa}{\langle x\rangle}\right)\right] \quad . \tag{7}$$

The intermediate's mean value and its variance are sufficient to discuss the line width. (All higher order moments have been eliminated using identities like $(1|\mathcal{L}\delta x^k)=0$.) At mass equilibrium, $\kappa=1$, we have $\Gamma=\langle X\rangle=Nb$, therefore $\Lambda^0=1+b^2$, i.e. a monotonous rapid increase of the decay constant with no peculiarity near $b_c \approx 1$. If κ is small, there are enhanced fluctuations (Fig.2) giving rise to a slowing down of Λ^0 (Fig.3). Away from the transition it is $\Gamma/\langle X\rangle \cong 1$ and $x_s \cong \kappa b$ or b/κ for small or large b resp. This yields

$$\Lambda^0 \cong \begin{cases} 1 + \kappa b^2 & b \ll 1 \quad , \\ \dfrac{b^2}{\kappa} + \kappa^2 & b \gg 1 \quad . \end{cases} \tag{7'}$$

In leading order this coincides with the eigenvalue of the linear stability analysis of (6)

$$\lambda = 3\kappa x_s^2 - 2bx_s + 1 \quad .$$

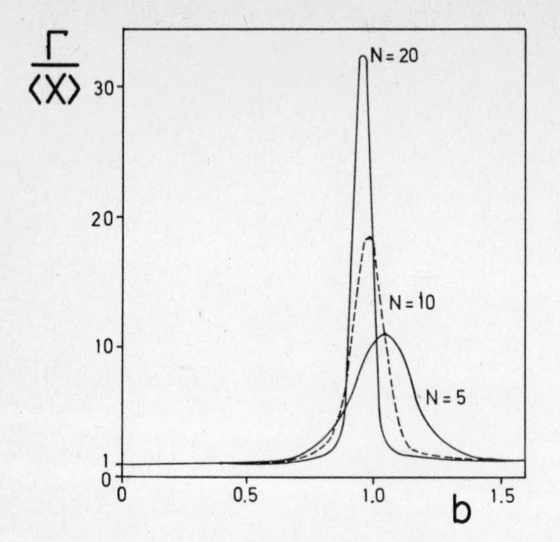

Fig.2 Reduced variance $\Gamma/\langle X\rangle$ vs b for κ=0.2

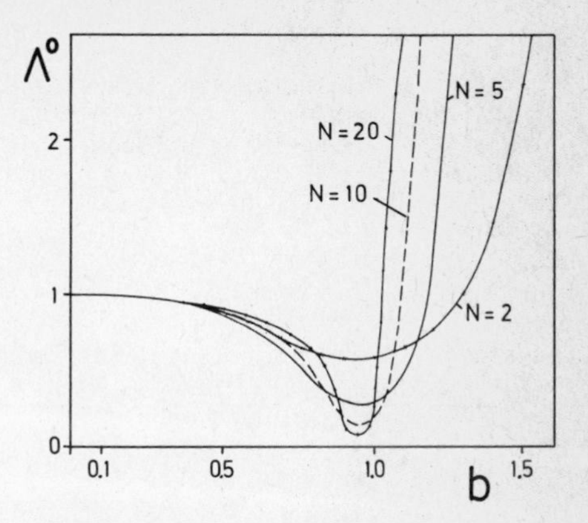

Fig.3 Relaxation frequency Λ^0 vs b for κ=0.2

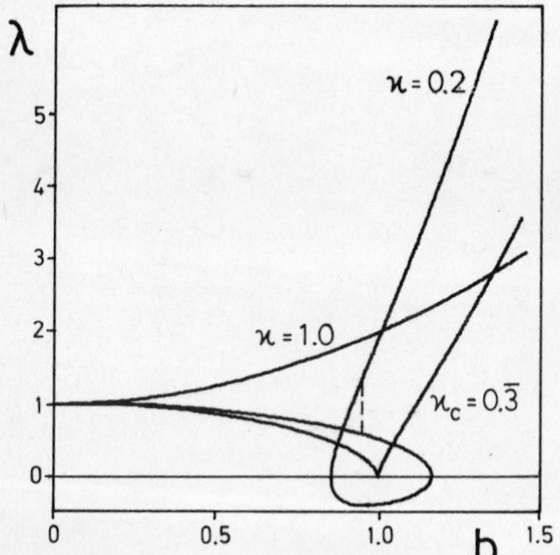

Fig.4 Breakdown of linear stability analysis in the transition region

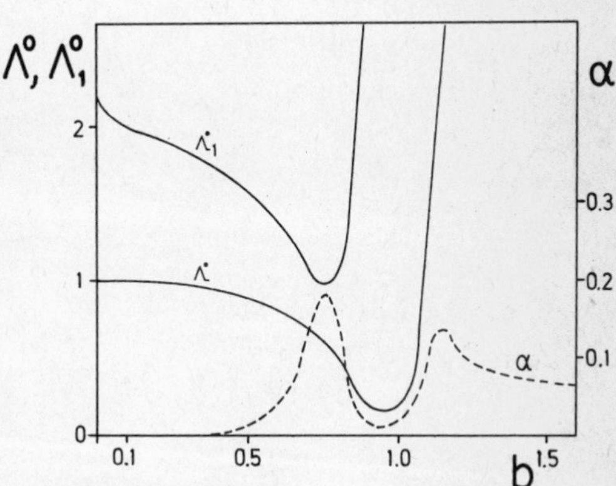

Fig. 5 Λ^0, Λ_1^0, and memory strength α for κ=0.2 and N=10

But evidently this approximation breaks down near the transition, see Fig.4.

It is of interest to check the memory decay time Λ_1^0 and the deviation from a Lorentzian line shape, see Fig.5. The calculation of C(t) with stochastic dynamics [16] yields qualitatively similar results.

4. Fluctuations at the Onset of a Limit Cycle

Sustained oscillations in open chemical systems have widely been observed [17],[18]. As a model system the Brusselator (2) shall be considered.

The macroscopic rate equations for the two intermediates are

$$\dot{x} = a - bx + x^2y - x \quad ,$$
$$\dot{y} = \quad bx - x^2y \quad . \tag{8}$$

For fixed a and increasing b the stationary point $x_s=a$, $y_s=b/a$ is monotonous stable if $b \leq (1-a)^2$, oscillatory stable if $(1-a)^2<b<b_c=1+a^2$, and a limit cycle develops continuously above b_c (Hopf bifurcation). These conclusions are drawn from the discussion of the matrix

$$\lambda = \begin{pmatrix} 1-b & -a^2 \\ b & a^2 \end{pmatrix}$$

obtained by linear stability analysis. They are summarized in Fig.6.

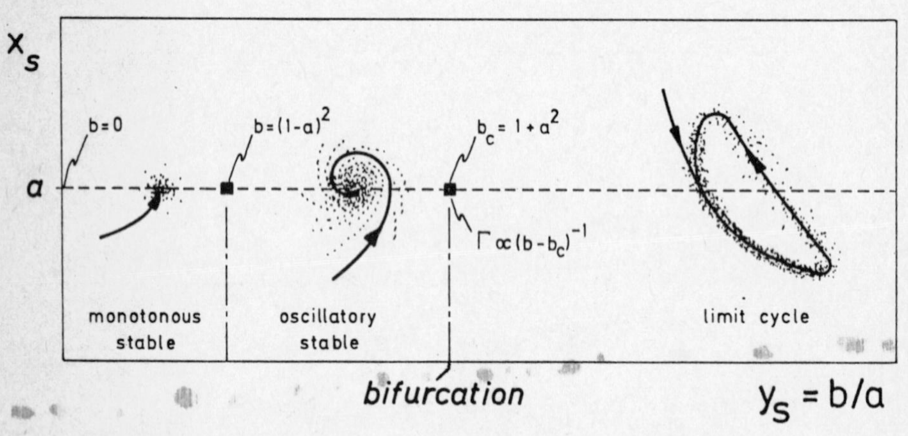

Fig.6 Macroscopic states of the Brusselator with increasing b

In quasi-linear approximation [19],[3] of the master equation one can correct for the effect of fluctuations for finite N, using $\lambda\Gamma+\Gamma\lambda^T=2D$. The critical point of bifurcation as well as the cycle's frequency are reduced, see [2]:

$$b_c^{(N)} = 1 + a^2 - 2\sqrt{\frac{a^4-1}{aN}} \quad , \qquad \omega_c^{(N)} = a\sqrt{1-\frac{1}{aN}} \quad .$$

In the immediate neighbourhood of the phase transition the fluctuations are enhanced, the quasi-linear approximation breaks down.

Due to the fluctuations the Hopf-bifurcation is smoothened, in fact it does not take place. A transition similar to that of the laser near threshold occurs. As an order parameter the finite amplitude and shape of the limit cycle can be used, more precisely $\Gamma_{xx}=\langle\delta X\delta X\rangle$ together with Γ_{yy}, while $\Gamma_{xy}=aN-\Gamma_{xx}$. Similar to the laser, fluctuations *along* the summit contour of P_s decay very slowly,

$$\eta \propto 1 / \langle\delta x\delta x\rangle N \quad , \tag{9}$$

due to phase decorrelation (phase diffusion). In addition, and different from the laser, the fluctuations rotate around the closed contour line of maximal probability with a period

$$\omega^{-1} = \frac{T}{2\pi} = \sqrt{\frac{\langle\delta y\delta y\rangle - \langle\delta x\delta x\rangle}{\langle\delta x\delta x\rangle}} \quad . \tag{10}$$

In contrast to the damping the period is of order N^0. In general ω is *smaller* than the Hopf-bifurcation frequency a, which is also found from (10) if the variances are calculated by reductive perturbation theory [20],[2]: $\langle\delta y\delta y\rangle=(b_c/a^2)\langle\delta x\delta x\rangle$. Using this approach the phase decorrelation of limit cycle fluctuations has already been found by TOMITA et al. [21].

The fluctuation behaviour just described can easily be derived and understood from the basic expression (5) throughout the entire transition region, not only for small $\varepsilon=(b-b_c)/b_c$. It also holds for small and medium size systems, not only in the $N\to\infty$ limit.

Using exact relations like $(1|\mathscr{L}\delta x)=0$ most moments needed for Λ can be calculated exactly. E.g. $\langle x\rangle=a$, $\langle yx(x-1/N)\rangle=ab$, $\langle\delta x\delta x\rangle+\langle\delta x\delta y\rangle=a/N$, etc. This yields

$$\Lambda = N\begin{pmatrix} a(b+1) & -ab \\ -ab & ab \end{pmatrix} + (\Gamma_{xx}-aN)\begin{pmatrix} 0 & 1 \\ -1 & 0 \end{pmatrix} . \tag{11}$$

Below threshold $\Gamma_{xx}\propto N$, the anti-symmetric term is of the order of the symmetric one and does not destroy its damping property. Above threshold, on the contrary, the main term is $\Gamma_{xx}=N^2\langle\delta x\delta x\rangle$, leading to eigenvalues $\propto\pm iN^2$ of Λ. The normalized decay matrix $\Lambda\Gamma^{-1}$ thus yields a frequency $\propto N^0$, see (10), and corrections $\propto N^{-1}$ as well as a damping of the same order, see (9). The decrease of the damping, roughly proportional to $b_c/(b-b_c)=\varepsilon^{-1}$, is a consequence of the increase of Γ, a measure of the limit cycle's amplitude.

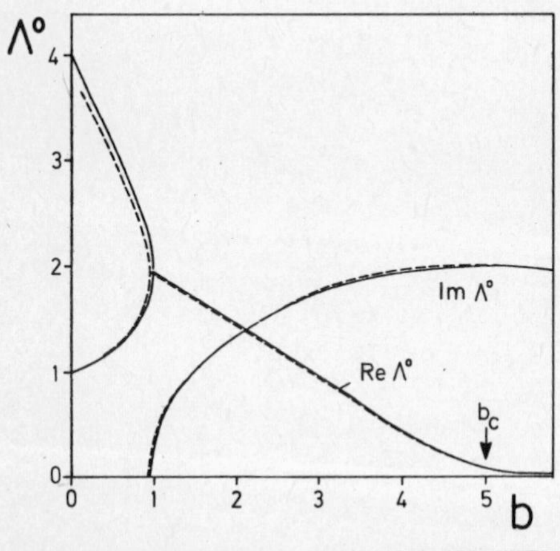

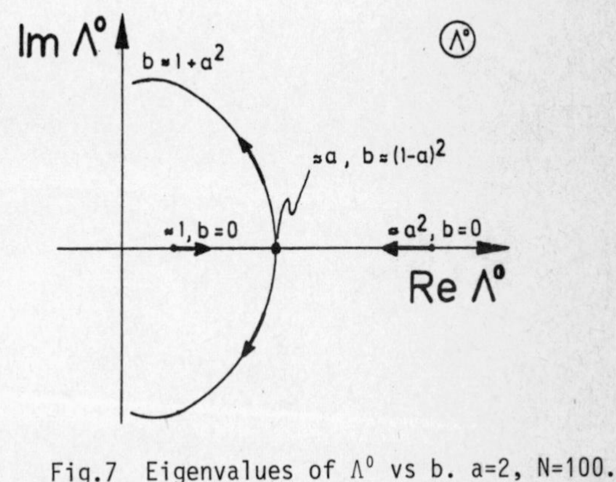

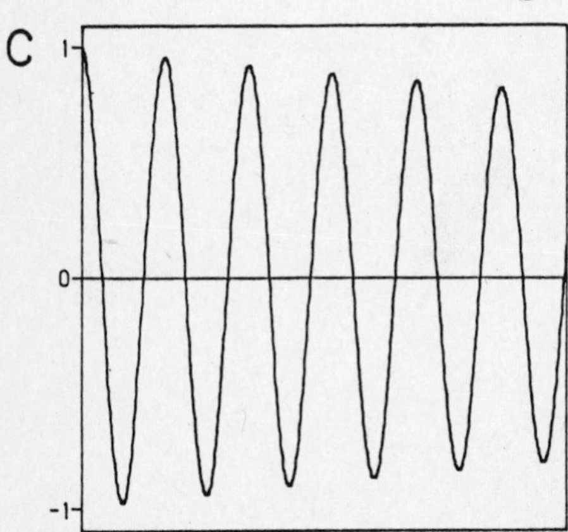

Fig.7 Eigenvalues of Λ^0 vs b. a=2, N=100. Dashed line: memory included

Fig.8 Eigenvalue plane, shift of Λ^0 with increasing b

Fig.9 Correlation function for $\varepsilon>0$

$$\frac{\mathrm{Re}\,\Lambda^0}{\mathrm{Im}\,\Lambda^0} \sim \begin{cases} |\varepsilon|\,N^0 & \varepsilon<0 \\ N^{-1/2} & \varepsilon=0 \\ 1/\varepsilon N & \varepsilon>0 \end{cases}$$

In Fig.7 the frequency and decay constant with increasing external stress is shown. Fig.8 yields the shift of the eigenvalues in the complex plane. An illustration of the correlation decay is shown in Fig.9.

The results about C(t) agree with and extend a computer calculation of $<\delta X \delta X(t)>$ by NICOLIS and MALEK-MANSOUR [22]. Some of their conjectures are confirmed analytically by the formulae (9)-(11).

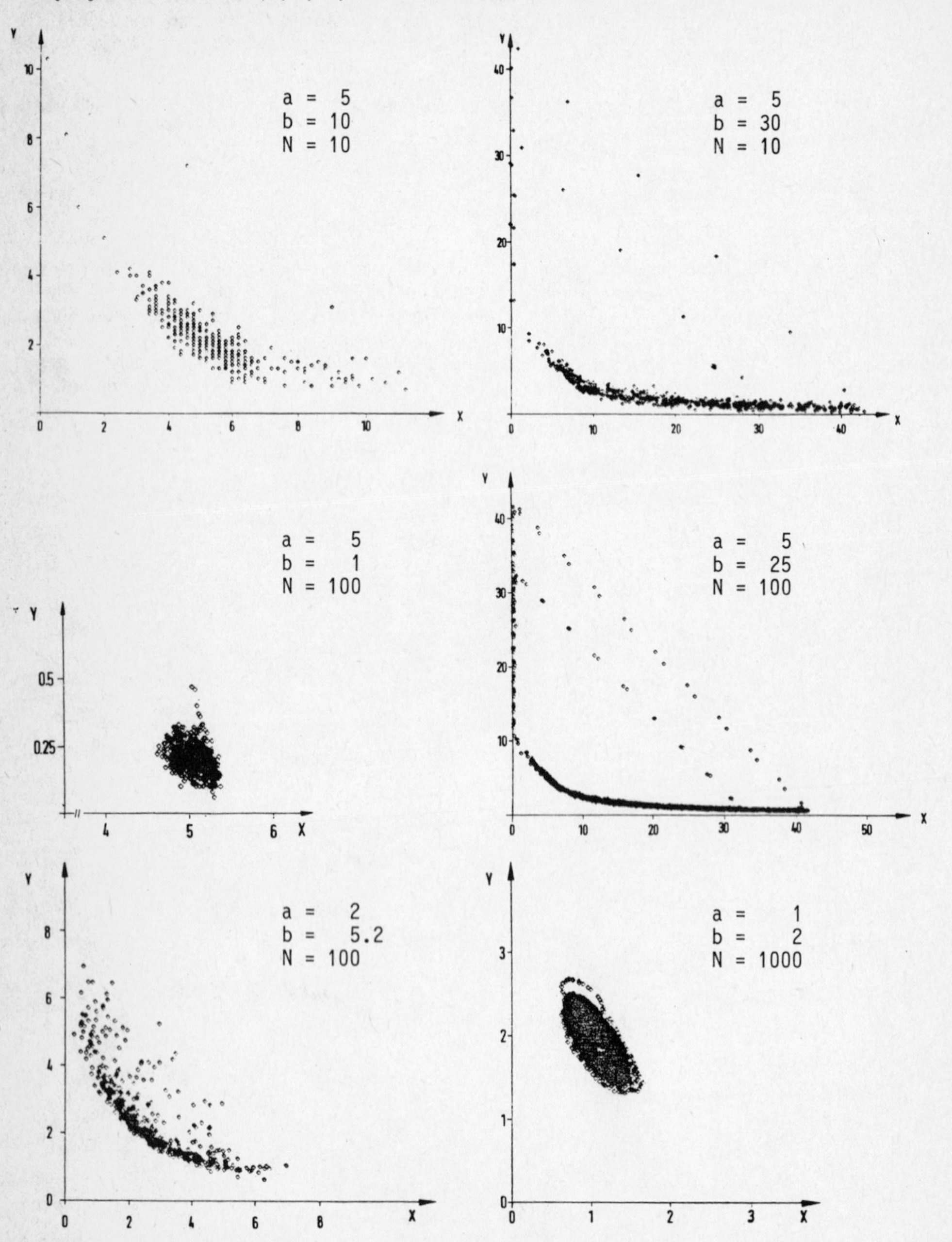

Fig.10 Phase space picture of stochastic simulation for various parameters. Note that the density of points is not directly proportional to the probability to find the system in a certain interval, but to the number of reactions taking place.

To give an impression of $P_S(x,y)$ I add Fig.10, reprinted from [2]. It has been evaluated by solving $\mathcal{D}P_S=0$ numerically for small N and by Monte Carlo calculations for medium and larger N.

5. Concluding Remarks

I have described the fluctuation dynamics avoiding unnecessary approximations like linearization, perturbation theory, or infinite volume limit. The main result is the strong relation between the stationary equal time correlations and the decay of fluctuations. Most static moments could be evaluated exactly. For more subtle details referred to [1], [2], also [23].

It is possible to extend the analysis to spatially inhomogeneous systems. - Experimental study of the correlation function, in particular of the interference between chemical fluctuations and diffusion would be very desirable.

References

1. Grossmann, S., Schranner, R.: Z.Physik B30, 325(1978)
2. Schranner, R., Grossmann, S., Richter, P.H.: Z.Physik B35, 363(1979)
3. Jähnig, F., Richter, P.H.: J.Chem.Phys. 64, 4645(1976)
4. Van Kampen, N.G.: Adv.Chem.Phys. 34, 245(1976)
5. Deker, U., Haake, F.: Phys.Rev. A12, 1629(1975);
   King, H., Deker, U., Haake, F.: Z.Physik B36, 205(1979)
6. Ziegler, K., Horner, H.: Z.Physik B37, 339(1980)
7. Zwanzig, R.: in Lect. Theor. Phys. III, Brittin, W., Dunham, L. (eds.)
   New York: Wiley 1961, p.135
8. Mori, H.: Progr.Theor.Phys. 34, 399(1965)
9. Götze, W., et al.: Excitation spectrum of HeII, Phys.Rev. B13, 3825(1976);
   Kondo problem, J.Low Temp.Phys. 6, 455(1972);
   classical liquids, Phys.Rev. A18, 1176(1978) and Phys.Rev. A20, 1603(1979);
   conductor-nonconductor transition, preprint April 1980
10. Bixon, M., Zwanzig, R.: J.Stat.Phys. 3, 245(1971)
11. Fujisaka, H., Mori, H.: Progr.Theor.Phys. 56, 754(1976)
12. Grossmann, S.: Phys.Rev. A17, 1123(1978)
13. Schlögl, F.: Z.Physik 253, 147(1972)
14. Prigogine, I., Lefever, R.: J.Chem.Phys. 48, 1695(1968)
15. Ross, J.: Berichte der Bunsenges. 80, 1112(1976)
16. Ebeling, W., Schimansky-Geier, L.: Physica 98A, 587(1979)
17. Walker, J.: Scientific American, July 1978, p.120
18. Preprint volumes I-III, Discussion Meeting Aachen, Sept. 19-22, 1979,
    Franck, U.F., Wicke, E. (eds.)
19. Lax , M.: Rev.Mod.Phys. 32, 25(1960)
20. Kuramoto, Y., Tsuzuki, T.: Progr.Theor.Phys. 52, 1399(1974); 54, 60(1975)
    Mashiyama, H., Ito, A., Ohta, T.: Progr.Theor.Phys. 54, 1050(1975)
21. Tomita, K., et al.: Progr.Theor.Phys. 51, 1731(1974); 52, 1744(1974)
22. Nicolis, G., Malek-Mansour, M.: Progr.Theor.Phys.Suppl. 64, 249(1978)
23. Schranner, R., Grossmann, S.: in [18], Vol.II, 439(1979)

Transition Phenomena in Nonlinear Optics

F.T. Arecchi
Università di Firenze and Istituto Nazionale di Ottica, Firenze

Abstract

By photon statistics methods the statistical features of optical instabilities have been measured with high accuracy over the past years. Their critical behavior can be classified with the language of equilibrium phase transitions. Furthermore, in nonequilibrium optical systems it is possible to study the transient build-up of an ordered state by a rapid passage through an instability. These transients are characterized by large fluctuations. An exact approach is described in terms of a stochastic time.

1. Introduction

Phase transitions in equilibrium systems are the result of a competition between the interparticle energy J and the thermal energy k_BT which introduces disorder. In quantum optics, even when interparticle interactions are negligible as in a very dilute gas, there may be particle correlations due to common radiation field. The transition from disorder to order consists in a passage from a regime where the atoms emit independently from one another, to a regime where the atoms emit in a strongly correlated way.

Over the past fifteen years, the introduction of the photon statistics method (for the theory see [1], for the experiments see [2]) has allowed a careful investigation of optical instabilities.

From the first experiments [2] the threshold point, where the gain due to the external excitation prevails over the internal losses, displayed the same features of a continuous phase transition in an equilibrium system (large increase in fluctuations, slowing down).

Similarly, evidence of discontinuous jumps and hysteresis effects in a laser with a saturable absorber suggested an analysis of this instability as a 1st order phase transition [3]. Recently, injecting a laser field into an interferometer filled with absorbing atoms, evidence of a hysteresis cycle suggested a region of coexistence of two stable points, hence the name of optical bistability [4]. The corresponding theory, either when the instability is due to the absorptive or the dispersive part of the atomic susceptibility, shows the characters of a 1st order phase transition [5].

Nonequilibrium systems can be driven through the instability point by a rapid passage, that is, at a rate larger than the local relaxation rate of steady state fluctuations. Such a transient situation was first observed for the laser instability [6]. A phenomenological theory [7] has shown the universality character of the anomalous transient fluctuations.

Of course, since transient phenomena are not invariant for time translation, there is no equivalent in equilibrium systems. The superfluorescence, that is the spontaneous cooperative decay of N atoms all prepared in their excited state in the absence of a classical field to drive them [8-10], is a transient collective behavior displaying a threshold.

Phase transitions in thermal systems, that is, in systems in contact with a termal reservoir, have been recently shown also to display transients by induced temperature steps. The associated instability is called spinodal decomposition [11].

Other classes of transitions in nonequilibrium systems have been recently described in hydrodynamic and chemical systems [12].

The above introductory remarks are summarized in Table 1.

Table 1. Phase transitions in quantum optics

<table>
<tr><th colspan="2" rowspan="2"></th><th rowspan="2">Thermal systems</th><th colspan="2">Nonequilibrium (pumped) systems</th></tr>
<tr><th>Quantum optics</th><th>Others</th></tr>
<tr><td rowspan="2">steady state</td><td>2nd ord.</td><td>- order/disorder
- para-ferromagnetic (H=0)</td><td>- laser threshold</td><td rowspan="3">hydrodynamic and chemical instabilities</td></tr>
<tr><td>1st ord.</td><td>- para-ferromagnetic (H=0)
- liquid gas</td><td>- laser plus saturable absorber
- optical bistability</td></tr>
<tr><td colspan="2">transient</td><td>- spinodal decomposition</td><td>- laser transient
- superfluorescence</td></tr>
</table>

2. Cooperation and Phase Transitions in Radiative Interactions

In optics, even when interparticle interactions are negligible, there may be correlations due to the common radiation field giving rise to a transition from a disordered state (or single particle emission) to an ordered state. Generally, an optical device is an open system fed by a source of energy and radiating electromagnetic energy toward a sink.

Let us consider the equations for a field coupled with N two level atoms by a resonant transition at frequency ω. Writing field a, polarization S and population inversion Δ as slowly varying variables, in the interaction volume the coupled equations reduce to (at resonance a and S are real quantities)

$$\dot{a} = gS - Ka$$
$$\dot{S} = 2ga\Delta - \gamma_{\perp}S \qquad (1)$$
$$\dot{\Delta} = -2gaS - \gamma_{\parallel}(\Delta - \Delta_0)$$

where

$$g = \left(\frac{\omega\mu^2}{2\hbar\varepsilon_0 V}\right)^{1/2}$$

is the coupling constant and K, $\gamma_{\perp}$, $\gamma_{\parallel}$ are loss rates. The normalization is such that $a^2 = n$ is the photon number and Δ the number of inverted atoms in the volume V. Δ_0 is a source term.

To give the order of magnitude, for a dilute gas of atoms with allowed transitions in the visible and for $V \sim 1\ \text{cm}^3$, it is

$$g \sim 10^4\ s^{-1} \quad , \quad \gamma_{\perp} \sim \gamma_{\parallel} \sim \gamma \sim 10^8\ s^{-1} \quad ,$$

and

$$K \sim 10^7\ s^{-1} \quad \text{or} \quad \sim 10^{10}\ s^{-1} \quad , \qquad (2)$$

depending on whether the gas is in a laser cavity or distributed over a length of some centimeters, without mirrors at the ends.

For small deviations from ground or excited state ($\Delta \sim \pm N/2$) we show by linearization of (1) that the lossless atoms-field interaction has a rate [13]

$$\gamma_c^2 \equiv g^2 N \quad . \qquad (3)$$

Losses introduce competing mechanisms with the following rates

atoms $\to(\gamma)\to$ thermal reservoirs

field $\to(K)\to$ thermal reservoirs .

The collective interaction will prevail on the separate uncorrelated dampings whenever

$$C \equiv \frac{g^2 N}{\gamma K} \equiv \frac{\gamma_c^2}{\gamma K} > 1 \quad . \qquad (4)$$

It can be shown [14] that relation (4) rules all quantum optical instabilities, namely,

i) laser threshold, optical bistability,
ii) superfluorescence,
iii) optical turbulence.

These three cases correspond to different scales of damping times [see relations (2)].

In case i), $\gamma \gg K$, hence the fast atomic variables relax to a local equilibrium which is set by the slow field amplitude which then acts as the order parameter. The evolution equation for such order parameter is, in suitable adimensional units

$$\dot{x} = (C - 1)\, x - Cx^3 \quad . \tag{5}$$

In case ii), $K \gg \gamma$, the field has a fast escape rate from the atomic medium and a detector witnesses a fast collective decay whenever atoms are prepared in an excited state in a short time.

In case iii), the damping rates are comparable, (1) must be considered simultaneously, and their solutions show chaotic behavior [12,15,16].

The interaction with termal reservoirs imposes to complete the equations for the collective variables[(1) in general, (5) for the laser case] with stochastic, or noise, sources, which can be taken in general as Gaussian processes with very short correlation times and correlation amplitudes that we call D.

The general statistical theory as well as the main experimental results are reviewed in Ref. [14]. Here we limit the discussion to the transient anomalous fluctuations, for which an exact approach is now available [17].

3. Transient Fluctuations in the Decay of an Unstable State

A nonequilibrium system, under the action of external parameters, may undergo transition in the sense that one (or a set) of its macroscopic observables have a sizable change. Usually these changes were studied by a slow setting of the external parameter, in order to measure the stationary fluctuations and their associated spectra around each equilibrium point.

More dramatic evidence, on the decay of an unstable state, can be obtained by applying sudden jumps to the driving parameter and observing the statistical transients [6]. The decay is initiated by microscopic fluctuations. In the first linear part of the decay process the fluctuations are amplified, hence during the transient, and until nonlinear saturation near the new stable point reduces them, fluctuations do not scale with the reciprocal of the system size, as it is at equilibrium.

A first experiment on the photon statistics of the laser field during its switch on [6] has opened this investigation.

Figures 1-3 give the transient photon statistics during a laser build up and the associated average photon number and variance.

Limiting to the case of one stochastic amplitude x, the most natural approach was to measure the probability density P (x,t) at a given time t after the sudden jump of the driving parameter. Under general assumptions, P (x,t) can be shown to

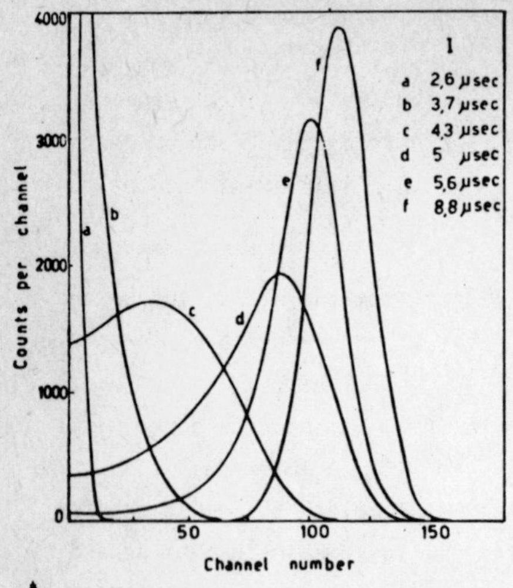

Fig.1. Experimental statistical distributions with different time delays obtained on a laser transient

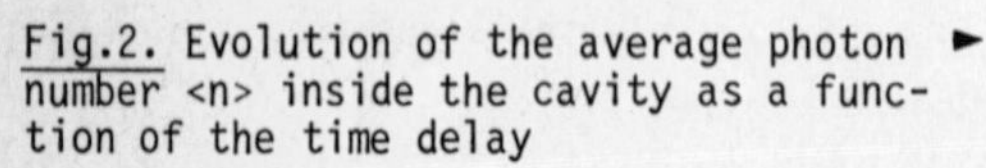

Fig.2. Evolution of the average photon number <n> inside the cavity as a function of the time delay

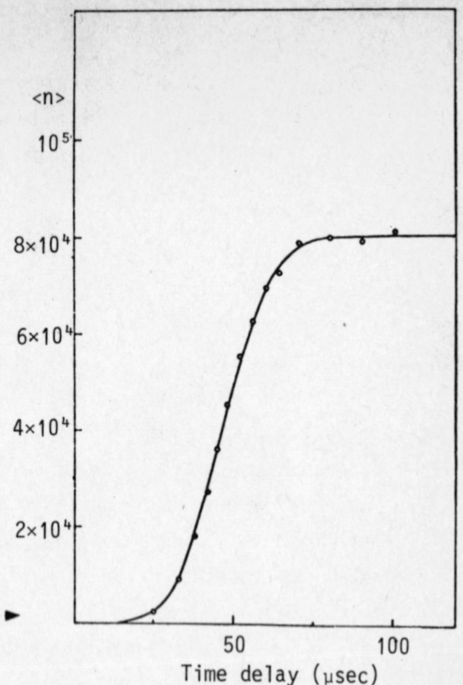

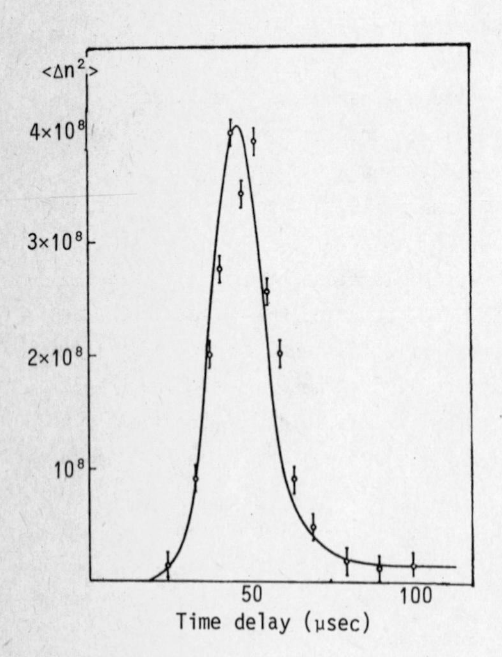

Fig.3. Evolution of the variance $\langle\Delta n^2\rangle$ of the statistical distribution of photons inside the cavity, as a function of the time delay

obey a nonlinear FOKKER-PLANCK equation (FPE). A time dependent solution in terms of an eigenfunction expansion is unsuitable for the large number of terms involved, with the exception of smalljumps near threshold [18] or the asymptotic behaviour for long times [19].

Solving for the moments $<x^k(t)>$ leads to an open hierarchy of coupled equations. A two-piece approximation first introduced for the laser [7] consists in first letting the system decay from the unstable point under the linearized part of the deterministic force, diffusing simultaneously because of the stochastic forces. This leads to a short time probability distribution of easy evaluation [20]. Then we solve for the nonlinear deterministic path and spread it over the ensemble of initial conditions previously evaluated in the linear regime.

A recent nonpiecewise treatment consisted in a 1/N expansion of the diffusion term (N being the system size) [21]. However, this approximation fails for small jumps above the threshold of instability or for nonlinear diffusion coefficients.

Another approach [22] was to trace back at any time a virtual ensemble of initial conditions, which inserted in the noise-free dynamic equations, would be responsible for the actual spread. This approach reduces the FPE to a diffusion equation, however, it fails for large deviations from the Gaussian as shown in a recent generalization [23].

Here we present an approach to transient statistics which overcomes the previous limitations. We consider the time t at which a given threshold z_F is crossed as the stochastic parameter, whose distribution $Q(t,z,z_F)$ in terms of the interval between the initial position z and z_F must be assigned. Here, the time is no longer an ordering parameter, but an interval limited by a start-stop operation. Let $P(x,t)$ be the probability density for the amplitude x which gets unstable under a force $F(x)$ and a noise delta-correlated with a correlation $D(x)$. $P(x,t)$ is solution of the FPE

$$\frac{\partial P}{\partial t} = - \frac{\partial}{\partial x} [F(x)P] + \frac{\partial^2}{\partial x^2} [D(x)P] \quad . \tag{6}$$

In order to develop an equation for the new density $Q(t,z)$ the time must be assigned as a single value parameter of z. This amounts to consider the problem of the first passage time in the Brownian motion which is ruled by the KOLMOGOROV equation [24] (KE)

$$\frac{\partial Q}{\partial t} = F(z) \frac{\partial Q}{\partial z} + D(z) \frac{\partial^2 Q}{\partial z^2} \tag{7}$$

where z is the initial value (t = 0), and the normalization is $\int_0^\infty Q \, dt = 1$.

Since we are studying the space evolution of the time distribution, (7) must be considered as a second order differential equation and we need two boundary conditions, that is the final value (the threshold) z_F, and the value α above which the process has to remain limited during the evolution. Like for the usual

FPE, the evaluation of the moments $T_m = \langle t^m \rangle$ is formally equivalent to the solution of the equation, but in this case we have a simple recurrence formula as

$$F(z)T_m' + D(z)T_m'' = -m\, T_{m-1} \tag{8}$$

(the apex denoting differentiation with respect to z).

In particular, we have for the mean time T_1

$$T_1(z) = \int_z^{z_F} \frac{dy}{W(y)} \int_{-\infty}^{y} dx\, W(x)/D(x) \tag{9}$$

where

$$W(x) \equiv \exp \int_{x_0}^{x} d\xi F(\xi)/D(\xi) \quad .$$

Equation (9) holds for $z_F > z$ and $\alpha = -\infty$. For a spread in the initial position z, $T_1(z)$ should be still averaged over the set of z. In a similar way we obtain $T_2(z)$, etc.

When we apply this formalism to the decay of unstable states, since D scales with the inverse system size, we can expand the above results in D series and display the first relevant correction to the deterministic solution. We find immediately that

$$T_1(z) = \int_z^{z_F} \frac{dy}{F(y)} + \int_z^{z_F} dy\, D \frac{dF}{dy} \frac{1}{F^3} \tag{10}$$

where the first term on the righthand side is the deterministic part. Similarly, performing the approximation for T_2 we obtain for the variance $\Delta T \equiv T_2 - T_1^2$ the following relation

$$\Delta T = 2 \int_z^{z_F} dy \frac{D(y)}{F^3(y)} \tag{11}$$

In order to show the power of this approach, we have measured the crossing time probability distributions for an electronic oscillator driven from below to above threshold [25].

Figure 4 gives the mean oscillator amplitude and its variance versus time as in the usual stochastic treatment of transients, Fig.5 gives the variance of crossing times for increasing thresholds as defined here.

The following comments convey some of the relevant physics:
i) The first term of (10) yields an average decay time which scales as $T_1 \sim \ln(N)$, that is, a logarithmic divergence with the system size N; ii) a constant variance for increasing threshold means that the various trajectories are shifted versions of the same deterministic curve, and the noise scaling as 1/N plays a role only in spreading the initial condition; iii) introduction of an external noise D_0 adds a fluctuation peculiar for each path, giving an ΔT dependent on z_F.

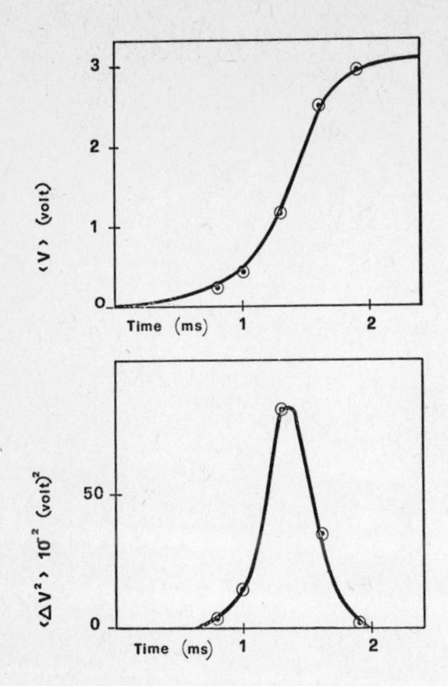

Fig.4. Transient statistical evolution of an electronic oscillator driven from below to above threshold by a sudden jump. No external noise added. Average amplitude <V> and variance $<\Delta V^2>$. Dots: experiment; line: theory

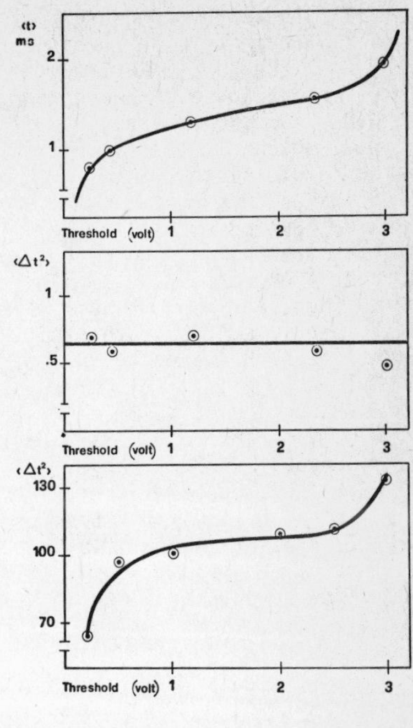

Fig.5

Fig.5. Transient oscillator as in Fig.1. Statistical distribution of the time intervals between the initial condition and the crossing of the threshold V. Average crossing time (a), variance under the action of the internal noise (b), variance for an added external noise (c). In (b) and (c) the scale is in units of 10^{-9} s^2. Dots: experiment; line: theory

In conclusion, we have shown a clear separation between the role of the initial spread and the noise along each path, and have introduced a new experimental characterization of a statistical transient which can be dealt with in an exact way.

As here presented, the method seems limited to discrete variables. However, by a suitable mode expansion and selection of the lowest threshold modes one can reduce field problems, (diffusive instabilities) to a set of few discrete coupled variables which can be dealt with by our method.

References

1. R.J. Glauber: in *Quantum Optics and Electronics*, Proc. of 1964 Les Houches School (McGraw-Hill, New York 1965) pp.65-185
2. F.T. Arecchi: in *Quantum Optics*, Proc. of 1967 Varenna School (Academic, New York 1969) pp.57-110
3. L.A. Lugiato, P. Mandel, S.T. Demlinski, A. Kossakowski: Phys. Rev. A*18*, 238 (1978)
4. H.M. Gibbs, S.L. McCall, T. Venkatesan: Phys. Rev. Letters *36*, 1135 (1976)
5. R. Bonifacio, L.A. Lugiato: Optc. Comm. *19*, 172 (1976)

6. F.T. Arecchi, V. Degiorgio, B. Querzola: Phys. Rev. Lett. *19*, 1168 (1967)
7. F.T. Arecchi, V. Degiorgio: Phys. Rev. A*3*, 1108 (1971)
8. R.H. Dicke: Phys. Rev. *93*, 99 (1954)
9. R. Bonifacio, P. Schwendimann, F. Haake: Phys. Rev. A*4*, 302, 804 (1971)
10. H.M. Gibbs, Q. Vrehen, H. Hikspoors: Phys. Rev. Lett. *39*, 547 (1978)
11. W.G. Goldburg, J.S. Huang: in *Fluctuations, Instabilities and Phase Transitions*, ed. by T. Riste (Plenum Press, New York 1975)
12. H. Haken: Synergetics. An Introduction, Springer Series in Synergetics, Vol.1, 2ed. (Springer, Berlin, Heidelberg, New York 1978)
13. F.T. Arecchi, E. Courtens: Phys. Rev. A*2*, 1730 (1970)
14. F.T. Arecchi: in *Dynamical Critical Phenomena*, Lecture Notes in Physics, Vol.104, ed. by C.P. Enz (Springer, Berlin, Heidelberg, New York 1979)
15. V. Benza, L.A. Lugiato: Zeit. Phys. B*35*, 383 (1979)
16. K. Ikeda, H. Daido, D. Akimoto: Phys. Rev. Lett. *45*, 709 (1980)
17. F.T. Arecchi, A. Politi: Phys. Rev. Lett., to be published
18. H. Risken, H.D. Vollmer: Z. Phys. *204*, 240 (1967)
19. N.G. Van Kampen: J. Stat. Phys. *17*, 71 (1977)
20. M. Suzuki: J. Stat. Phys. *16*, 447 (1977)
21. F. Haake: Phys. Rev. Lett. *41*, 1685 (1978)
22. F. De Pasquale, P. Tombesi: Phys. Lett. *72*A, 7 (1979)
23. F.T. Arecchi, A. Politi: Lett. N. Cimento *27*, 486 (1980)
24. R.L. Stratonovich: *Topics in the Theory of Random Noise*, Vol.1 (Gordon and Breach Science Publ. Inc., New York 1963)
25. F.T. Arecchi, M. Cetica, F. Francini, L. Ulivi: to be published

Part IX

Markov Processes and Time Reversibility

Time-Reversibility in Dynamical Systems

J.C. Willems

Mathematics Institute, University of Groningen,
B.O. Box 800, 9700 AV Groningen, The Netherlands

In this talk a system theoretic setting is outlined for treating time-reversibility and other symmetry relations in dynamical systems. A detailed exposition of this approach containing a number of examples may be found in Ref.[1,2,3]. The present paper is limited to brief exposition of the main ideas.

Let W be a set called the (external) *signal alphabet* and $T \subset \mathbb{R}$ be the *time axis*. Assume for simplicity that $T = \mathbb{R}$. A dynamical system Σ (on W) is then simply a subset of W^T. H is said to be *linear* if W is a vector space and Σ is a linear subspace of W^T. Let S_Δ be the shift, i.e. $(S_\Delta f)(t) := f(t-\Delta)$. The system Σ is said to be *time-invariant* if $S_\Delta \Sigma = \Sigma$ for all $\Delta \in T$.

Let X be a set called the *state space* and $\Sigma_i \subset (X \times W)^T$ be a dynamical system on (XxW). Then it is called a *state space system* if $(x_i, w_i) \in \Sigma$, i=1,2, with $x_1(t) = x_2(t)$ for some $t \in T$ yields $(x,w) \in \Sigma$ where (x,w) is defined by $(x,w)(\tau) := (x_1, w_1)(\tau)$ for $\tau < t$ and $(x,w)(\tau) := (x_2, w_2)(\tau)$ for $\tau \geq t$.

Let Σ_i be a state space system. Then $\Sigma_e := \{w \in W^T \mid \exists x \in X^T$ such that $(x,w) \in \Sigma_i\}$ is called the *external behavior* of Σ_i, denoted by $\Sigma_i \Rightarrow \Sigma_e$. Conversely, one calls Σ_i a *state space representation* or *realization* of Σ_e.

The problem of state space realization is one of the basic and much studied problems in system theory. There is a notion of *minimal* realization which intuitively means that the state space X which is appended to the signal letter W contains as few as possible elements. One may then prove that under reasonable conditions *all minimal realizations are equivalent*. That is to say that if $\Sigma_I' \subset (X' \times W)^T$ and $\Sigma_i'' \subset (X'' \times W)^T$ are two minimal realizations of the same external behavior Σ_e the there exists a bijection $S: X' \rightarrow X''$ such that $(x', w) \in \Sigma_i \Leftrightarrow (Sx', w) \in \Sigma_i''$.

We now come to the concept of time-reversibility. Let R be the time-reversal operator defined by $(Rf)(t) := f(-t)$. Let Σ be given. Then $R\Sigma$ is called its time-reverse and Σ is said to be *time-reversible* if $R\Sigma = \Sigma$. The question now arises whether a time-reversible system admits a (minimal) time-reversible realization. This is in general not the case but by extending the notion of time-reversibility somewhat we may obtain such a result: A system is said to be *quasi-time- reversible* (sometimes called dynamically time-reversible) if there exist an involution S $(S = S^{-1})$ on W such that $RS\Sigma = \Sigma$. Now it is possible to prove, under the conditions which ensured all minimal realizations to be equivalent, that *a dynamical is quasi-time-reversible if it admits a minimal quasi-time-reversible state space realizations*. Details are given in [1,2].

An analogous result may be derived for a class of stochastic systems but there the result is a much more difficult one te prove because it is not true that for stochastic systems all minimal state space realizations are equivalent. We limit ourselves to explaning this result in the case of Gaussian processes.

Let $(y_t, t \in \mathbb{R})$ be a process on a probability space $\{\Omega, A, P\}$ with values in the measurable space Y. Similarly, let $(y_t', t \in \mathbb{R})$ be a process on $\{\Omega', A', P'\}$ also with values in Y. Then y is said to be *stochastically equivalent* to y' (denoted $y\sim y'$) if $\forall t_1,\ldots,t_n$ and measurable sets $B_1,\ldots,B_n$ there holds $P(y_{t_1}\in B_1,\ldots,y_{t_n}\in B_n) = P(y'_{t_1}\in B_1,\ldots,y'_{t_n}\in B_n)$. A process is said to be *stationary* if $y\sim S_\Delta y$ for all $\Delta\in\mathbb{R}$, and *time-reversible* if it is stationary and if $y\sim Ry$. This definition is easily extended to quasi-time-reversibility: a process is said to be *quasi-time-reversible* if it is stationary and if there exists an involution S on Y such $y\sim RSy$.

The notion of state space realization for smooth processes reduces to the following: let $(y_t,\ t\in\mathbb{R})$ be a given process with values in Y. A *state space realization* is then a pair (x,h) with $x = (x_t, t\in\mathbb{R})$ a Markov process on X and h a map, $h:X\to Y$, such that $h(x)\sim y$.

For Gaussian processes it is possible to prove the following: let $(y_t,\ t\in\mathbb{R})$ be a zero mean Gaussian stochastic process with rational spectral density matrix and taking values in $\mathbb{R}^m$. Assume that y is quasi-time-reversible with linear involution S_e. Then there exists a minimal $\mathbb{R}^n$-valued (minimality may be interpreted here as meaning that n is as small as possible) Gauss-Markov process x and a matrix $C:\mathbb{R}^n\to\mathbb{R}^m$ such that $Cx\sim y$ and such that x is quasi-time-reversible with respect to a linear involution S_i, i.e., $(x,y)\sim R(S_i x, S_e y)$. This result is not easy to prove and leads to an interesting decomposition of the state of a dynamical system in *even* and *odd* variables. Details may be found in [3]. Related ideas and results are developed in [4-6].

REFERENCES

[1] J.C. Willems: "System Theoretic Models for the Analysis of Physical Systems", *Ricerche di Automatica*, Vol. 10, to appear.

[2] J.C. Willems: "Systems Theoretic Foundations for Modelling Physical Systems", in *Dynamical Systems and Microphysics*, ed. by A. Blaquière, F. Fer, A. Marzollo, CISM Courses and Lectures No. 261, Springer, Berlin, Heidelberg, New York 1980, pp. 279-289.

[3] J.C. Willems:"Time Reversibility in Deterministic and Stochastic Dynamical Systems" in *Recent Develoments in Variable Structure Systems, Economics and Biology*, ed. by R.R. Mohler, A. Ruberti, Lecture Notes in Economics and Mathematical Systems, Vol. 162, Springer, Berlin, Heidelberg, New York 1978, pp. 318-326.

[4] P. Whittle: "Reversibility and Acyclicity" in *Perspectives in Probability and Statistics* (Ed.: J. Gani), Applied Probability Trust, 1975, pp. 217-224.

[5] F.P. Kelly: *Reversibility and Stochastic Networks*, Wiley, 1979.

[6] B.D.O. Anderson and T. Kailath: "Forwards, Backwardsand Dynamically Reversible Markovian Models of Second Order Processes", *IEEE Trans. on Circuits and Systems*, 1980.

Index of Contributors

List of Participants

Arecchi, F.T. (Firenze)
Arnold, L. (Bremen)
Bhattacharya, R.N. (Tucson)
Binder, K. (Jülich)
Brand, H. (Essen)
Brenig, L. (Brussels)
Chaturvedi, S. (Stuttgart)
Crauel, H. (Bremen)
Curtain, R. (Groningen)
Dawson, D.A. (Ottawa)
Dewel, G. (Brussels)
Dress, A. (Bielefeld)
Eckmann, J.-P. (Genève)
Ehrhardt, M. (Bremen)
Föllmer, H. (Zürich)
Gardiner, C.W. (Waikato)
Graham, R. (Essen)
Grossmann, S. (Marburg)
Haken, H. (Stuttgart)
Hongler, M.O. (Genève)
Horsthemke, W. (Brussels)
Kampen, N.G. van (Utrecht)
Kliemann, W. (Bremen)
Koch, G. (Roma)
Kotelenez, P. (Bremen)
Kurtz, T.G. (Madison)
Lang, R. (Heidelberg)
Lefever, R. (Brussels)
Malek-Mansour, M. (Brussels)
Maruyama, T. (Houston)
Miller, D.E. (Bielefeld)
Misra, B. (Brussels)
Nicolis, G. (Brussels)
Oelschläger, K. (Heidelberg)
Papanicolaou, G. (New York)
Pardoux, E. (Marseille)
Pomeau, Y. (Gif-sur-Yvette)
Prigogine, I. (Brussels)
Rost, H. (Heidelberg)
Rümelin, W. (Bremen)
Sancho, J.M. (Barcelona)
San Miguel, M. (Philadelphia)
Sawitzki, G. (Bochum)
Schenzle, A. (Essen)
Sommer, U. (Bremen)
Spohn, H. (München)
Tuckwell, H.C. (Vancouver)
Turner, J.W. (Brussels)
Velarde, M.G. (Madrid)
Walgraef, D. (Brussels)
Weizsäcker, H. von (Kaiserslautern)
Wihstutz, V. (Bremen)
Willems, J.C. (Groningen)

C. P. Slichter

Principles of Magnetic Resonance

Corrected 2nd Printing of the 2nd Revised and Expanded Edition. 1980. 115 figures, 9 tables. XII, 397 pages
(Springer Series in Solid-State Sciences, Volume 1)
ISBN 3-540-08476-2

Contents:
Elements of Resonance. – Basic Theory. – Magnetic Dipolar Broadening of Rigid Lattices. – Magnetic Interactions of Nuclei with Electrons. – Spin-Lattice Relaxation and Motional Narrowing of Resonance Lines. – Spin Temperature in Magnetism and in Magnetic Resonance. – Double Resonance. – Advanced Concepts in Pulsed Magnetic Resonance. – Electric Quadrupole Effects. – Electron Spin Resonance. – Summary. – Problems. – Appendices. – Selected Bibliography. – References. – Author Index. – Subject Index.

Synchrotron Radiation

Techniques and Applications

Editor: C. Kunz
1979. 162 figures, 28 tables. XVI, 442 pages
(Topics in Current Physics, Volume 10)
ISBN 3-540-09149-1

Contents:
C. Kunz: Introduction – Properties of Synchrotron Radiation. – *E. M. Rowe:* The Synchrotron Radiation Source. – *W. Gudat, C. Kunz:* Instrumentation for Spectroscopy and Other Applications. – *A. Kotani, Y. Toyozawa:* Theoretical Aspects of Inner Level Spectroscopy. – *K. Codling:* Atomic Spectroscopy. – *E. E. Koch, B. F. Sonntag:* Molecular Spectroscopy. – *D. W. Lynch:* Solid-State Spectroscopy.

Springer-Verlag
Berlin
Heidelberg
New York

Mössbauer Spectroscopy

Editor: U. Gonser
1975. 96 figures. XVIII, 241 pages
(Topics in Applied Physics, Volume 5)
ISBN 3-540-07120-2

Contents:
U. Gonser: From a Strange Effect to Mössbauer Spectroscopy. – *P. Gütlich:* Mössbauer Spectroscopy in Chemistry. – *R. W. Grant:* Mössbauer Spectroscopy in Magnetism: Characterization of Magnetically-Ordered Compounds. – *C. E. Johnson:* Mössbauer Spectroscopy in Biology. – *S. S. Hafner:* Mössbauer Spectroscopy in Lunar Geology and Mineralogy. – *F. E. Fujita:* Mössbauer Spectroscopy in Physical Metallurgy.